Edexcel A2 Chemistry

Ann Fullick Bob McDuell

STUDENTS' BOOK

A PEARSON COMPANY

How to use this book

This book contains a number of great features that will help you find your way around your A2 Chemistry course and support your learning.

Introductory pages

Each unit has two introductory pages, giving you an idea of the chemistry to come and linking the content to the chemical ideas, how chemists work and chemistry in action that you will encounter in the unit.

HSW boxes

How Science Works is a key feature of your course. The many HSW boxes within the text will help you cover all the new aspects of How Science Works that you need. These include how scientists investigate ideas and develop theories, how to evaluate data and the design of studies to test their validity and reliability, and how science affects the real world including informing decisions that need to be taken by individuals and society.

Main text

The main part of the book covers all you need to learn for your course. The text is supported by many diagrams and photographs that will help you understand the concepts you need to learn.

Key terms in the text are shown in bold type. These terms are defined in the interactive glossary that can be found on the software using the 'search glossary' feature.

UNIT 4 General principles of chemistry I – Rates, equilibria and further organic chemistry

Unit 4 General principles of chemistry I – Rates, equilibria and further organic chemistry

In Unit 4 you will consider the factors which determine both the rates of chemical reactions and how far these reactions go. You will explore how understanding rates and equilibria enables chemists to solve many practical problems in industrial chemistry.

You will also study some more groups of organic chemicals, and use that knowledge to understand how chemists develop the polymers that are so useful in everyday life. And finally you will learn more about spectroscopy and chromatography and how they are used.

Chemical ideas

Studying the rates at which chemical reactions take place is known as kinetics. You will use kinetics to understand both how to control the rates of chemical reactions and how develop models for the mechanism of those reactions. You will develop an understandir of the concept of entropy – the level of order or disorder in a system – and see its importance in predicting the likelihood of a reaction taking place. You will also revisit t ideas of chemical equilibrium met during your AS course and discover how chemists ca manipulate equilibrium positions both in the laboratory and in industry. You will explor several theories of acids and bases, and find out more about their uses in chemistry.

You will study the importance of optical isomerism in organic chemistry and drug safet and learn more about organic groups such as the carbonyl compounds and carboxylic acids. Finally in this unit you will learn more about spectroscopy and chromatography – vital tools for the modern chemist in building up an understanding of many large, comp molecules.

How chemists work

You will develop your practical skills to increase your understanding of the factors affecting both rate and equilibria in laboratory chemistry. You will use colorimeters and pH meters, as well as carry out a wide range of titrations and other experimental techniques to investigate acid/base reactions among others. You will study how chemist develop models of reaction mechanisms and how they seek practical evidence to suppor these models. You will find out how chromatography can be used to understand compounds such as amino acids, and through interpreting spectra you will begin to understand just how complex this area of chemistry can be.

Chemistry in action

In the chemical industry it is important to be able to control the rate of reactions and manipulate their equilibrium position so as to maximise yield. The first four chapters of this unit will help you to understand and predict the conditions used in a variety of different industrial processes. You will develop an understanding of the reasons for choosing particular acids and bases – and indicators – in titrations. You will explore the role of chemists in developing new polymers and fuels which perform effectively, yet are environmentally friendly. Finally, chemists need to be able to identify and analyse chemicals; nowadays this is often done using techniques such as spectroscopy and chromatography. You will discover just how these spectra are produced and have an opportunity to interpret some of the data, giving you an insight into chemistry in action the vital areas of chemical and medicines research.

8

Introductory pages

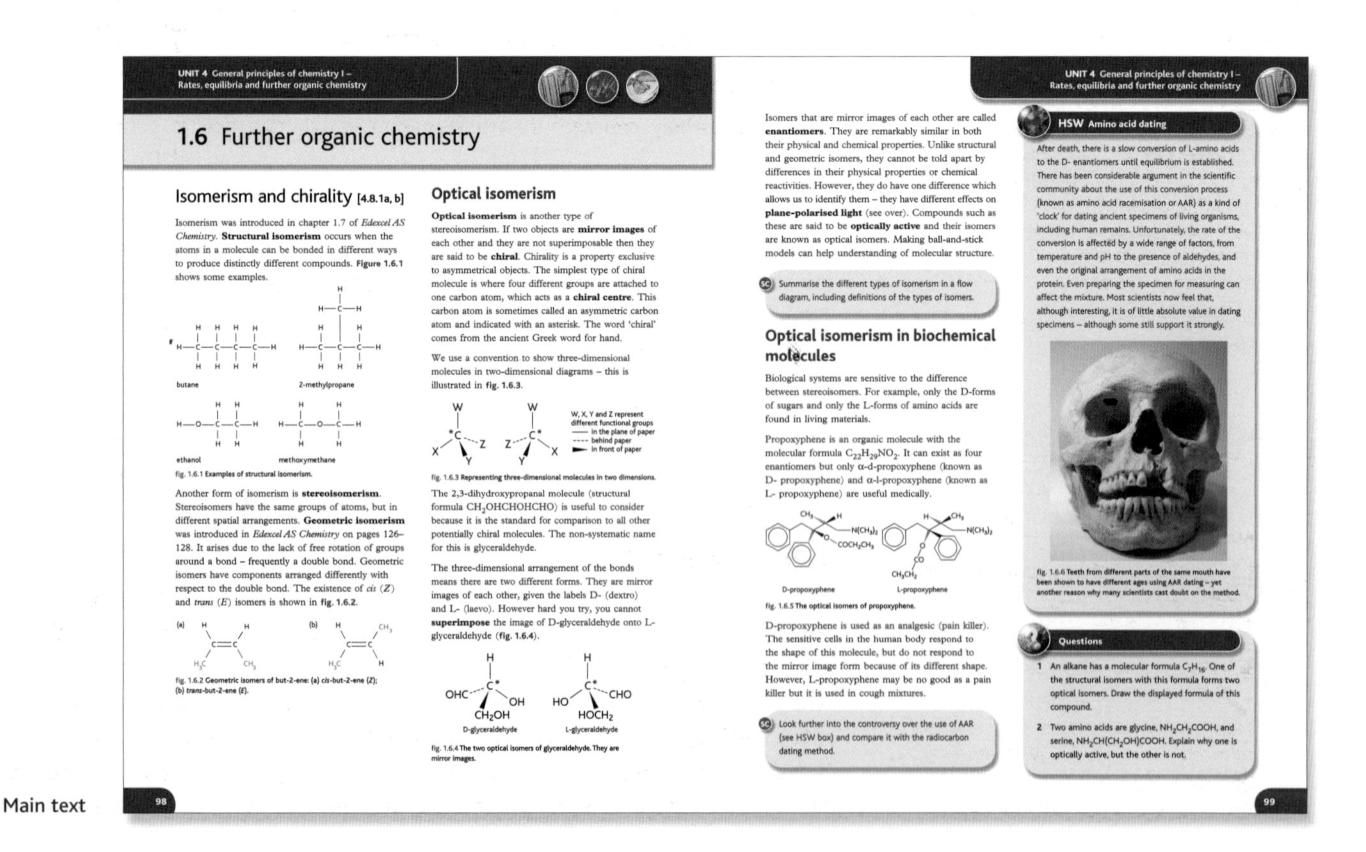

UNIT 4 General principles of chemistry I – Rates, equilibria and further organic chemistry

1.6 Further organic chemistry

Isomerism and chirality [4.8.1a, b]

Isomerism was introduced in chapter 1.7 of *Edexcel AS Chemistry*. **Structural isomerism** occurs when the atoms in a molecule can be bonded in different ways to produce distinctly different compounds. **Figure 1.6.1** shows some examples.

fig. 1.6.1 Examples of structural isomerism.

Another form of isomerism is **stereoisomerism**. Stereoisomers have the same groups of atoms, but in different spatial arrangements. **Geometric isomerism** was introduced in *Edexcel AS Chemistry* on pages 126–128. It arises due to the lack of free rotation of groups around a bond – frequently a double bond. Geometric isomers have components arranged differently with respect to the double bond. The existence of *cis* (*Z*) and *trans* (*E*) isomers is shown in **fig. 1.6.2**.

fig. 1.6.2 Geometric isomers of but-2-ene: (a) *cis*-but-2-ene (*Z*); (b) *trans*-but-2-ene (*E*).

Optical isomerism

Optical isomerism is another type of stereoisomerism. If two objects are **mirror images** of each other and they are not superimposable then they are said to be **chiral**. Chirality is a property exclusive to asymmetrical objects. The simplest type of chiral molecule is where four different groups are attached to one carbon atom, which acts as a **chiral centre**. This carbon atom is sometimes called an asymmetric carbon atom and indicated with an asterisk. The word 'chiral' comes from the ancient Greek word for hand.

We use a convention to show three-dimensional molecules in two-dimensional diagrams – this is illustrated in **fig. 1.6.3**.

fig. 1.6.3 Representing three-dimensional molecules in two dimensions.

The 2,3-dihydroxypropanal molecule (structural formula $CH_2OHCHOHCHO$) is useful to consider because it is the standard for comparison to all other potentially chiral molecules. The non-systematic name for this is glyceraldehyde.

The three-dimensional arrangement of the bonds means there are two different forms. They are mirror images of each other, given the labels D- (dextro) and L- (laevo). However hard you try, you cannot **superimpose** the image of D-glyceraldehyde onto L-glyceraldehyde (**fig. 1.6.4**).

fig. 1.6.4 The two optical isomers of glyceraldehyde. They are mirror images.

Isomers that are mirror images of each other are called **enantiomers**. They are remarkably similar in both their physical and chemical properties. Unlike structural and geometric isomers, they cannot be told apart by differences in their physical properties or chemical reactivities. However, they do have one difference which allows us to identify them – they have different effects on **plane-polarised light** (see over). Compounds such as these are said to be **optically active** and their isomers are known as optical isomers. Making ball-and-stick models can help understanding of molecular structure.

SC Summarise the different types of isomerism in a flow diagram, including definitions of the types of isomers.

Optical isomerism in biochemical molecules

Biological systems are sensitive to the difference between stereoisomers. For example, only the D-forms of sugars and only the L-forms of amino acids are found in living materials.

Propoxyphene is an organic molecule with the molecular formula $C_{22}H_{29}NO_2$. It can exist as four enantiomers but only α-d-propoxyphene (known as D- propoxyphene) and α-l-propoxyphene (known as L- propoxyphene) are useful medically.

fig. 1.6.5 The optical isomers of propoxyphene.

D-propoxyphene is used as an analgesic (pain killer). The sensitive cells in the human body respond to the shape of this molecule, but do not respond to the mirror image form because of its different shape. However, L-propoxyphene may be no good as a pain killer but it is used in cough mixtures.

SC Look further into the controversy over the use of AAR (see HSW box) and compare it with the radiocarbon dating method.

UNIT 4 General principles of chemistry I – Rates, equilibria and further organic chemistry

HSW Amino acid dating

After death, there is a slow conversion of L-amino acids to the D- enantiomers until equilibrium is established. There has been considerable argument in the scientific community about the use of this conversion process (known as amino acid racemisation or AAR) as a kind of 'clock' for dating ancient specimens of living organisms, including human remains. Unfortunately, the rate of the conversion is affected by a wide range of factors, from temperature and pH to the presence of aldehydes, and even the original arrangement of amino acids in the protein. Even preparing the specimen for measuring can affect the mixture. Most scientists now feel that, although interesting, it is of little absolute value in dating specimens – although some still support it strongly.

fig. 1.6.6 Teeth from different parts of the same mouth have been shown to have different ages using AAR dating – yet another reason why many scientists cast doubt on the method.

Questions

1 An alkane has a molecular formula C_7H_{16}. One of the structural isomers with this formula forms two optical isomers. Draw the displayed formula of this compound.

2 Two amino acids are glycine, NH_2CH_2COOH, and serine, $NH_2CH(CH_2OH)COOH$. Explain why one is optically active, but the other is not.

98 99

Main text

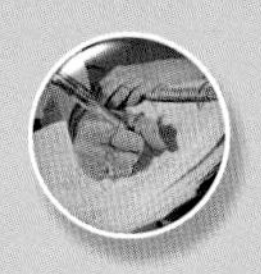

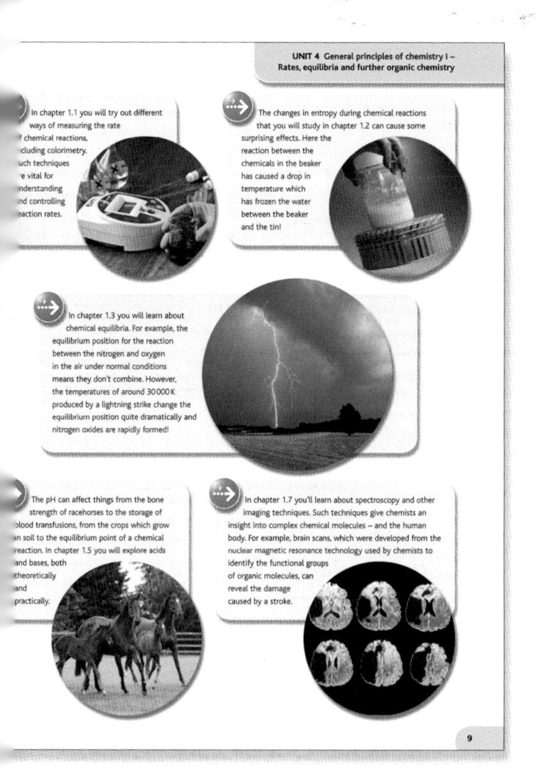
UNIT 4 General principles of chemistry I –
Rates, equilibria and further organic chemistry

In chapter 1.1 you will try out different ways of measuring the rate of chemical reactions, including colorimetry. Such techniques are vital for understanding and controlling reaction rates.

The changes in entropy during chemical reactions that you will study in chapter 1.2 can cause some surprising effects. Here the reaction between the chemicals in the beaker has caused a drop in temperature which has frozen the water between the beaker and the tin!

In chapter 1.3 you will learn about chemical equilibria. For example, the equilibrium position for the reaction between the nitrogen and oxygen in the air under normal conditions means they don't combine. However, the temperatures of around 30 000 K produced by a lightning strike change the equilibrium position quite dramatically and nitrogen oxides are rapidly formed!

The pH can affect things from the bone strength of racehorses to the storage of blood transfusions, from the crops which grow in soil to the equilibrium point of a chemical reaction. In chapter 1.5 you will explore acids and bases, both theoretically and practically.

In chapter 1.7 you'll learn about spectroscopy and other imaging techniques. Such techniques give chemists an insight into complex chemical molecules – and the human body. For example, brain scans, which were developed from the nuclear magnetic resonance technology used by chemists to identify the functional groups of organic molecules, can reveal the damage caused by a stroke.

9

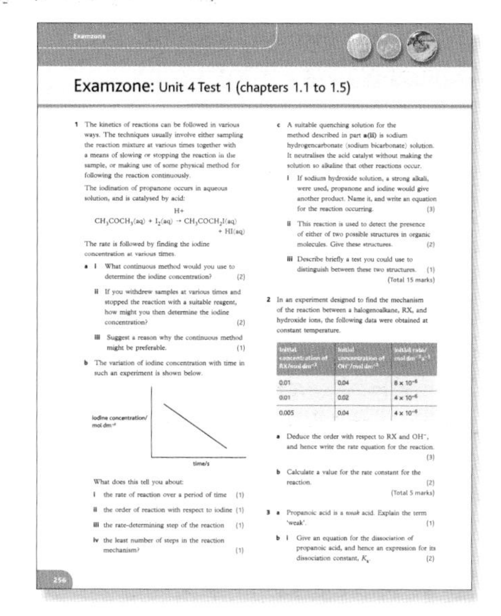
Examzone

Examzone: Unit 4 Test 1 (chapters 1.1 to 1.5)

1 The kinetics of reactions can be followed in various ways. The techniques usually involve either sampling the reaction mixture at various times together with a means of slowing or stopping the reaction in the sample, or making use of some physical method for following the reaction continuously.

The iodination of propanone occurs in aqueous solution, and is catalysed by acid:

$$CH_3COCH_3(aq) + I_2(aq) \xrightarrow{H+} CH_3COCH_2I(aq) + HI(aq)$$

The rate is followed by finding the iodine concentration at various times.

a i What continuous method would you use to determine the iodine concentration? (2)

ii If you withdrew samples at various times and stopped the reaction with a suitable reagent, how might you then determine the iodine concentration? (2)

iii Suggest a reason why the continuous method might be preferable. (1)

b The variation of iodine concentration with time in such an experiment is shown below.

What does this tell you about:

i the rate of reaction over a period of time (1)

ii the order of reaction with respect to iodine (1)

iii the rate-determining step of the reaction (1)

iv the least number of steps in the reaction mechanism? (1)

c A suitable quenching solution for the method described in part a(ii) is sodium hydrogencarbonate (sodium bicarbonate) solution. It neutralises the acid catalyst without making the solution so alkaline that other reactions occur.

i If sodium hydroxide solution, a strong alkali, were used, propanone and iodine would give another product. Name it, and write an equation for the reaction occurring. (3)

ii This reaction is used to detect the presence of either of two possible structures in organic molecules. Give these structures. (2)

iii Describe briefly a test you could use to distinguish between these two structures. (1)

(Total 15 marks)

2 In an experiment designed to find the mechanism of the reaction between a halogenoalkane, RX, and hydroxide ions, the following data were obtained at constant temperature.

Initial concentration of RX/mol dm⁻³	Initial concentration of OH⁻/mol dm⁻³	Initial rate/ mol dm⁻³ s⁻¹
0.01	0.04	8×10^{-6}
0.01	0.02	4×10^{-6}
0.005	0.04	4×10^{-6}

a Deduce the order with respect to RX and OH^-, and hence write the rate equation for the reaction. (3)

b Calculate a value for the rate constant for the reaction. (2)

(Total 5 marks)

3 a Propanoic acid is a *weak* acid. Explain the term 'weak'. (1)

b i Give an equation for the dissociation of propanoic acid, and hence an expression for its dissociation constant, K_a. (2)

256

Examzone page

HSW Supporting a theory

A reaction mechanism worked out for a reaction is only a theory, because in most cases chemists cannot detect or isolate the proposed intermediates. It has to be 'guessed' from the raw data. So to support a model of the mechanism of a particular reaction, the predicted rate law from the mechanism must match the rate law ach[illegible]ed from experimental data. If they don't match,

Worked example

Calculate the entropy change in making 1 mole of sodium chloride from its elements:

$$Na(s) + \tfrac{1}{2}Cl_2(g) \rightarrow NaCl(s)$$

The entropy change of the system is given by:

$$\Delta S^{\ominus}_{sys} = \Sigma S^{\ominus}_{products} - \Sigma S^{\ominus}_{reactants}$$
$$= 72.4\,J\,mol^{-1}\,K^{-1} - (51.0 + \tfrac{1}{2} \times 223.0)\,J\,mol^{-1}\,K^{-1}$$

Measuring K_c

You may carry out a similar experiment with ethanol and ethanoic acid. You can use an Excel spreadsheet to calculate your values for K_c.

Questions

1 An alkane has a molecular formula C_7H_{16}. One of the structural isomers with this formula forms two optical isomers. Draw the displayed formula of this compound.

2 Two amino acids are glycine, NH_2CH_2COOH, and serine, $NH_2CH(CH_2OH)COOH$. Explain why one is optically active, but the other is not.

Practical boxes

Your course contains a number of core practicals that you may be tested on. These boxes indicate links to practical work. Your teacher will give you opportunities to cover these investigations.

Worked example boxes

These boxes give example questions and answers, taking you through the calculations step-by-step and showing how you could set out your answers.

Maths boxes

These boxes help you with the mathematics you'll encounter in your chemistry studies.

Question boxes

At the end of each section of text you will find a box containing questions that cover what you have just learnt. You can use these questions to help you check whether you have understood what you have just read, and whether there is anything that you need to look at again.

Stretch and challenge SC

Throughout the text there are opportunities to stretch and challenge yourself. This includes extra questions and activities to help you practise different types of assessment and to link together topics.

Examzone pages

At the end of the book you will find 8 pages of exam questions from past papers. You can use these questions to test how fully you have understood the units, as well as to help you practise for your exams. There are 2 tests for each unit.

The contents list shows you that there are two units in the book, matching the Edexcel AS specification for Chemistry. There are 7 chapters in unit 4 (chapters 1.1 to 1.7) and 5 chapters in unit 5 (chapters 2.1 to 2.5). Page numbering in the contents list, and in the index at the back of the book, will help you find what you are looking for.

How to use your ActiveBook

Click on this tab to see menus which list all the electronic files on the ActiveBook.

The ActiveBook is an electronic copy of the book, which you can use on a compatible computer. The CD-ROM will only play while the disc is in the computer. Investigate your ActiveBook's features:

Student Book tab

Click this tab at the top of the screen to access the electronic version of the book.

Key words

Click on any of the words in **bold** to see a box with the word and what it means. Click 'play' to listen to someone read it out for you to help you pronounce it.

Interactive view

Click this button to see all the icons on the page that link to electronic files, such as documents and spreadsheets. You have access to all of the features that are useful for you to use at home on your own. If you don't want to see these links you can return to **Book view**.

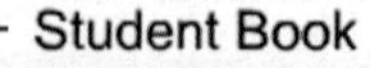

Find Resources | Glossary | Help

UNIT 4 General principles of chemistry I – Rates, equilibria and further organic chemistry

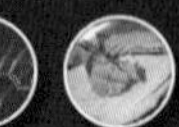

1.6 Further organic chemistry

Isomerism and chirality [4.8.1a, b]

Isomerism was introduced in chapter 1.7 of *Edexcel AS Chemistry*. **Structural isomerism** occurs when the atoms in a molecule can be bonded in different ways to produce distinctly different compounds. **Figure 1.6.1** shows some examples.

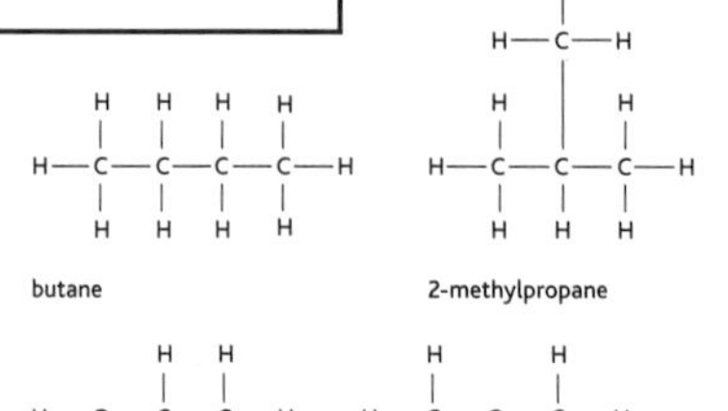

ethanol methoxymethane

fig. 1.6.1 Examples of structural isomerism.

Another form of isomerism is **stereoisomerism**. Stereoisomers have the same groups of atoms, but in different spatial arrangements. **Geometric isomerism** was introduced in *Edexcel AS Chemistry* on pages 126–128. It arises due to the lack of free rotation of groups around a bond – frequently a double bond. Geometric isomers have components arranged differently with respect to the double bond. The existence of *cis* (*Z*) and *trans* (*E*) isomers is shown in **fig. 1.6.2**.

fig. 1.6.2 Geometric isomers of but-2-ene: (a) *cis*-but-2-ene (*Z*); (b) *trans*-but-2-ene (*E*).

Optical isomerism

Optical isomerism is another type of stereoisomerism. If two objects are **mirror images** of each other and they are not superimposable then they are said to be **chiral**. Chirality is a property exclusive to asymmetrical objects. The simplest type of chiral molecule is where four different groups are attached to one carbon atom, which acts as a **chiral centre**. This carbon atom is sometimes called an asymmetric carbon atom and indicated with an asterisk. The word 'chiral' comes from the ancient Greek word for hand.

We use a convention to show three-dimensional molecules in two-dimensional diagrams – this is illustrated in **fig. 1.6.3**.

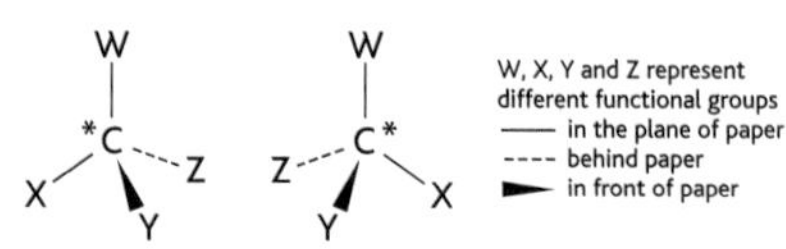

fig. 1.6.3 Representing three-dimensional molecules in two dimensions.

The 2,3-dihydroxypropanal molecule (structural formula $CH_2OHCHOHCHO$) is useful to consider because it is the standard for comparison to all other potentially chiral molecules. The non-systematic name for this is glyceraldehyde.

The three-dimensional arrangement of the bonds means there are two different forms. They are mirror images of each other, given the labels D- (dextro) and L- (laevo). However hard you try, you cannot **superimpose** the image of D-glyceraldehyde onto L-glyceraldehyde (**fig. 1.6.4**).

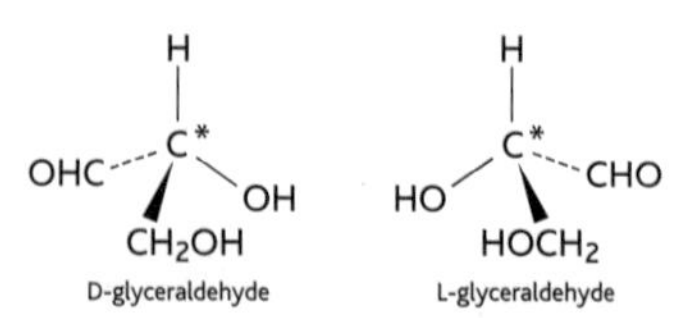

fig. 1.6.4 The two optical isomers of glyceraldehyde. They are mirror images.

98

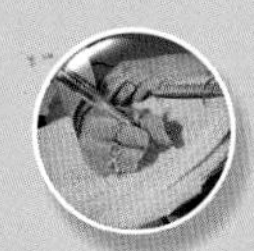

Glossary

Click this tab to see all of the key words and what they mean. Click 'play' to listen to someone read them out to help you pronounce them.

Help

Click on this tab at any time to search for help on how to use the ActiveBook.

Isomers that are mirror images of each other are called **enantiomers**. They are remarkably similar in both their physical and chemical properties. Unlike structural and geometric isomers, they cannot be told apart by differences in their physical properties or chemical reactivities. However, they do have one difference which allows us to identify them – they have different effects on **plane-polarised light** (see over). Compounds such as these are said to be **optically active** and their isomers are known as optical isomers. Making ball-and-stick models can help understanding of molecular structure.

SC Summarise the different types of isomerism in a flow diagram, including definitions of the types of isomers.

Optical isomerism in biochemical molecules

Biological systems are sensitive to the difference between stereoisomers. For example, only the D-forms of sugars and only the L-forms of amino acids are found in living materials.

Propoxyphene is an organic molecule with the molecular formula $C_{22}H_{29}NO_2$. It can exist as four enantiomers but only α-d-propoxyphene (known as D- propoxyphene) and α-l-propoxyphene (known as L- propoxyphene) are useful medically.

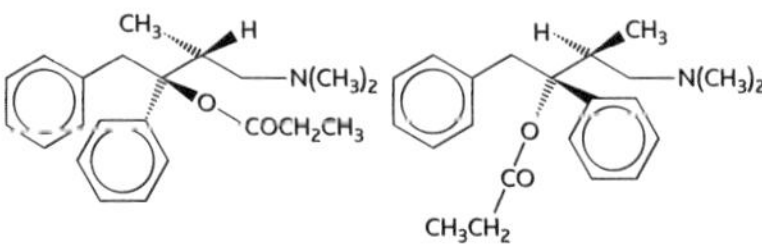

fig. 1.6.5 The optical isomers of propoxyphene.

D-propoxyphene is used as an analgesic (pain killer). The sensitive cells in the human body respond to the shape of this molecule, but do not respond to the mirror image form because of its different shape. However, L-propoxyphene may be no good as a pain killer but it is used in cough mixtures.

SC Look further into the controversy over the use of AAR (see HSW box) and compare it with the radiocarbon dating method.

HSW Amino acid dating

After death, there is a slow conversion of L-amino acids to the D- enantiomers until equilibrium is established. There has been considerable argument in the scientific community about the use of this conversion process (known as amino acid racemisation or AAR) as a kind of 'clock' for dating ancient specimens of living organisms, including human remains. Unfortunately, the rate of the conversion is affected by a wide range of factors, from temperature and pH to the presence of aldehydes, and even the original arrangement of amino acids in the protein. Even preparing the specimen for measuring can affect the mixture. Most scientists now feel that, although interesting, it is of little absolute value in dating specimens – although some still support it strongly.

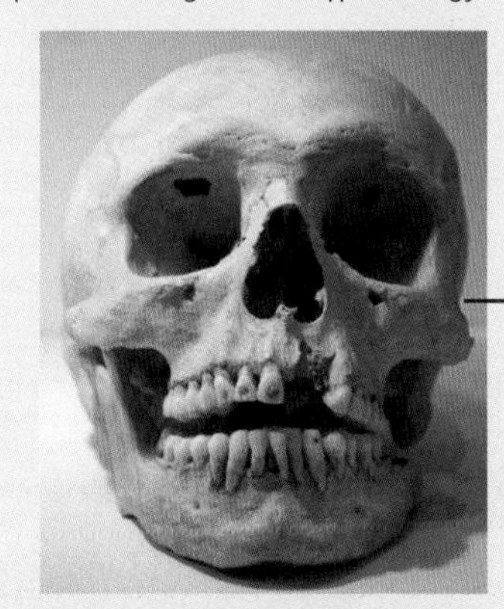

fig. 1.6.6 Teeth from different parts of the same mouth have been shown to have different ages using AAR dating – yet another reason why many scientists cast doubt on the method.

Questions

1 An alkane has a molecular formula C_7H_{16}. One of the structural isomers with this formula forms two optical isomers. Draw the displayed formula of this compound.

2 Two amino acids are glycine, NH_2CH_2COOH, and serine, $NH_2CH(CH_2OH)COOH$. Explain why one is optically active, but the other is not.

Zoom feature

Just click on a section of the page and it will magnify so that you can read it easily on screen. This also means that you can look closely at photos and diagrams.

CONTENTS

Unit 4 General principles of chemistry I – Rates, equilibria and further organic chemistry – p8

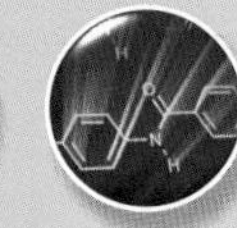
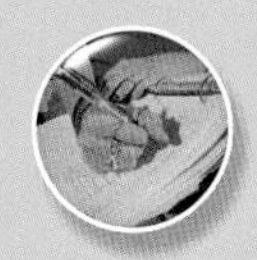
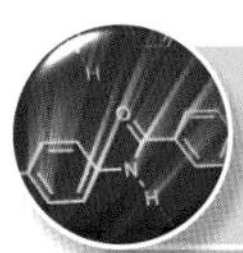

Unit 5 General principles of chemistry II – Transition metals and organic nitrogen chemistry – p144

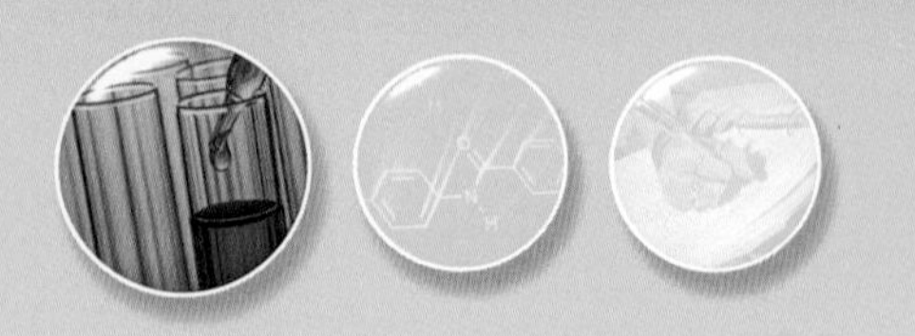

Unit 4 General principles of chemistry I – Rates, equilibria and further organic chemistry

In Unit 4 you will consider the factors which determine both the rates of chemical reactions and how far these reactions go. You will explore how understanding rates and equilibria enables chemists to solve many practical problems in industrial chemistry.

You will also study some more groups of organic chemicals, and use that knowledge to understand how chemists develop the polymers that are so useful in everyday life. And finally you will learn more about spectroscopy and chromatography and how they are used.

Chemical ideas

Studying the rates at which chemical reactions take place is known as kinetics. You will use kinetics to understand both how to control the rates of chemical reactions and how to develop models for the mechanism of those reactions. You will develop an understanding of the concept of entropy – the level of order or disorder in a system – and see its importance in predicting the likelihood of a reaction taking place. You will also revisit the ideas of chemical equilibrium met during your AS course and discover how chemists can manipulate equilibrium positions both in the laboratory and in industry. You will explore several theories of acids and bases, and find out more about their uses in chemistry.

You will study the importance of optical isomerism in organic chemistry and drug safety, and learn more about organic groups such as the carbonyl compounds and carboxylic acids. Finally in this unit you will learn more about spectroscopy and chromatography – vital tools for the modern chemist in building up an understanding of many large, complex molecules.

How chemists work

You will develop your practical skills to increase your understanding of the factors affecting both rate and equilibria in laboratory chemistry. You will use colorimeters and pH meters, as well as carry out a wide range of titrations and other experimental techniques to investigate acid/base reactions among others. You will study how chemists develop models of reaction mechanisms and how they seek practical evidence to support these models. You will find out how chromatography can be used to understand compounds such as amino acids, and through interpreting spectra you will begin to understand just how complex this area of chemistry can be.

Chemistry in action

In the chemical industry it is important to be able to control the rate of reactions and manipulate their equilibrium position so as to maximise yield. The first four chapters of this unit will help you to understand and predict the conditions used in a variety of different industrial processes. You will develop an understanding of the reasons for choosing particular acids and bases – and indicators – in titrations. You will explore the role of chemists in developing new polymers and fuels which perform effectively, yet are environmentally friendly. Finally, chemists need to be able to identify and analyse chemicals; nowadays this is often done using techniques such as spectroscopy and chromatography. You will discover just how these spectra are produced and have an opportunity to interpret some of the data, giving you an insight into chemistry in action in the vital areas of chemical and medicines research.

In chapter 1.1 you will try out different ways of measuring the rate of chemical reactions, including colorimetry. Such techniques are vital for understanding and controlling reaction rates.

The changes in entropy during chemical reactions that you will study in chapter 1.2 can cause some surprising effects. Here the reaction between the chemicals in the beaker has caused a drop in temperature which has frozen the water between the beaker and the tin!

In chapter 1.3 you will learn about chemical equilibria. For example, the equilibrium position for the reaction between the nitrogen and oxygen in the air under normal conditions means they don't combine. However, the temperatures of around 30 000 K produced by a lightning strike change the equilibrium position quite dramatically and nitrogen oxides are rapidly formed!

The pH can affect things from the bone strength of racehorses to the storage of blood transfusions, from the crops which grow in soil to the equilibrium point of a chemical reaction. In chapter 1.5 you will explore acids and bases, both theoretically and practically.

In chapter 1.7 you'll learn about spectroscopy and other imaging techniques. Such techniques give chemists an insight into complex chemical molecules – and the human body. For example, brain scans, which were developed from the nuclear magnetic resonance technology used by chemists to identify the functional groups of organic molecules, can reveal the damage caused by a stroke.

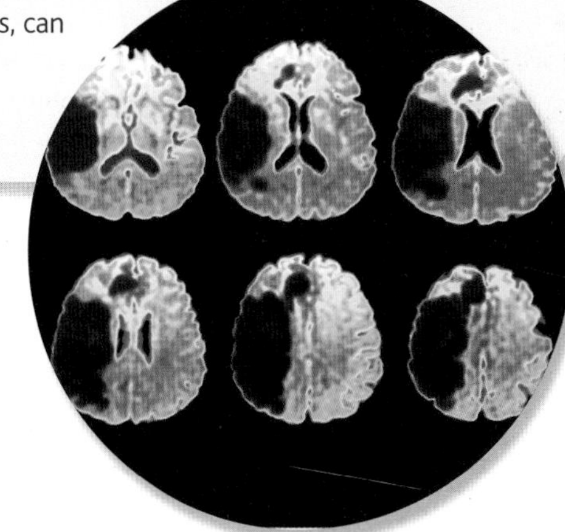

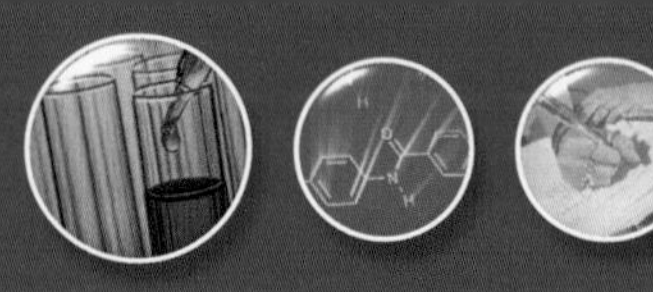

1.1 How fast? Rates of chemical change

Techniques to measure rate of reaction

[4.3b]

The **rate of chemical reactions** – and the factors affecting the rate of chemical change – have been studied closely for many years. Why do chemists study reaction rates (kinetics)?

- The fact that reactions can occur at very different rates is intriguing and chemists want to understand what is going on.
- Chemists need to understand how to change the rate of a reaction. In industry, it is essential to understand rates of reaction to help work out the economics of a manufacturing process.
- Reaction rates can provide evidence of the mechanism of a chemical reaction – the individual steps by which a reaction takes place.

The ability to manipulate the rate of a chemical reaction is as vital in the school lab as it is in the chemical industry. As you saw in your AS level Chemistry book (pages 196–7), concentration, pressure, temperature, the surface area of the reacting substances and catalysts all affect the rate at which chemical reactions happen. Using the chemistry of fireworks as an example, it is important that a firework fuse burns for long enough to allow the person who lit the firework to get out of the way. However, it mustn't burn too long or people are tempted to go back and make sure it hasn't gone out. The Health and Safety Executive has fixed the minimum burn time at 5 seconds, and the maximum at 15 seconds. The 'blue touch paper' which acts as a fuse is paper impregnated with potassium nitrate – it will burn even in wet, windy conditions. The speed at which it burns is controlled by the potassium nitrate content of the paper – the higher the concentration, the faster it burns.

fig. 1.1.1 Controlling the rate at which the fuse on a firework burns is crucial for everyone's safety.

To understand and control the rate of chemical reactions, chemists need to be able to measure how fast reactions occur. You can calculate the rate of a reaction by measuring the change in concentration of one of the reactants or products and the time taken:

$$\text{rate of reaction} = \frac{\text{change in concentration}}{\text{change in time}}$$

$$\text{or rate} = \frac{\Delta c}{\Delta t}$$

A number of different methods can be used to collect the data we need to measure the rate of a reaction. They all investigate changes in concentration, either directly or indirectly. As you saw in your AS course, the techniques largely depend on measuring either how quickly the reactants are used up or how quickly the products are formed. You will be considering and using many of these methods.

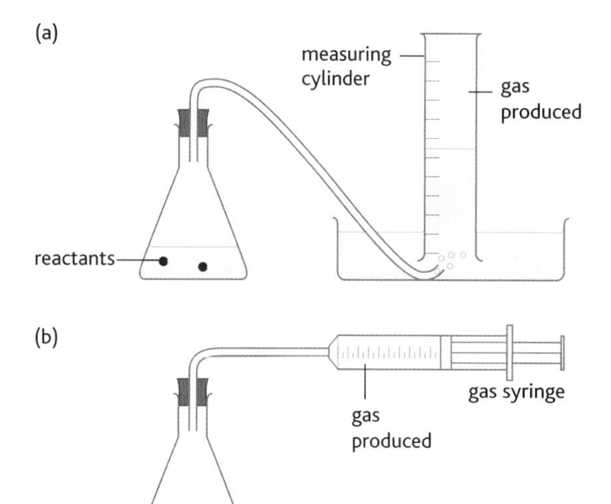

fig. 1.1.2 Using a gas syringe is a method of measuring the volume of gas produced during a reaction quite accurately.

fig. 1.1.3 Measuring changes in mass is a very useful technique for investigating the effect of surface area on reaction rate.

fig. 1.1.4 In reactions with a colour change, colorimetry is an excellent way of measuring the rate of the reaction.

Measuring the volume of a gas produced

In a reaction in which a gas is given off, the progress of the reaction can be measured by monitoring the rate at which the gaseous product is given off. The technique you use will depend on the level of accuracy demanded – the simplest is simply to count the number of bubbles produced in a certain time.

For more accurate results you can measure the volume of gas given off at regular time intervals, collecting the gas in a measuring cylinder by water displacement – particularly useful for larger volumes of gas – or in a gas syringe (**fig. 1.1.2**). You plot a graph of the total volume of gas produced against time to show how the rate of reaction changes. By changing the reaction conditions – for example the concentrations of the reactants or the temperature – you can investigate how these factors affect reaction rates.

Measuring the change in mass of a reaction mixture

This technique also depends on having gaseous substances as products of the reaction. You measure the decrease in mass of the total reacting mixture as the reaction proceeds (**fig. 1.1.3**). It is relatively easy to carry out in the laboratory, needing only a sensitive balance, but there are many opportunities for experimental error to creep in unless great care is taken to avoid splashes etc.

Monitoring a colour change (colorimetry)

The techniques involved in using colour changes to measure reaction rate range from simple observation to the use of a colorimeter (**fig. 1.1.4**). The appearance of a coloured product, or the loss of a coloured reactant, is used. By changing the reaction conditions and measuring the effect on the appearance or loss of colour, the impact of the change on the rate of the reaction can be measured. The iodine clock, an experimental technique you will look at later, is an example where you use your eyes – iodine is formed in a reaction and this reacts with starch indicator to form a noticeable blue-black colour.

In a similar investigation, the reaction of potassium iodide with potassium peroxodisulfate is carried out in a colorimeter. As the oxidation takes place and iodine is produced, the brown colour of the iodine is monitored extremely precisely in the colorimeter. The change detected by the colorimeter is far more subtle than the human eye could observe.

fig. 1.1.5 This technique gives a detailed picture of changes in rate during a reaction – also enabling detailed comparisons when investigating the effect of different factors on the rate of the reaction.

Titrimetric analysis

This involves removing small portions (aliquots) of the reacting mixture at regular intervals. These aliquots are usually added to another reagent, which immediately stops or quenches the reaction.

Alternatively, the reaction is slowed by immersing the portion in an ice bath, so that there are no further changes to the concentrations in the reacting mixture until further analysis can be carried out. The quenched aliquots are then titrated to find the concentrations of known compounds in them.

In a reaction in which acid is being used up, titration could be carried out with a standard solution of sodium hydroxide. A similar technique of titrimetric analysis is useful for investigating the reaction between iodine and propanone (**fig. 1.1.5**), which is catalysed by acid. The samples are run into excess sodium hydrogencarbonate to stop the reaction. By titrating with sodium thiosulfate you can measure the concentration of iodine in the reacting mixture (see page 14):

$$CH_3COCH_3(aq) + I_2(aq) \rightarrow CH_3COCH_2I(aq) + H^+(aq) + I^-(aq)$$

$$2Na_2S_2O_3(aq) + I_2(aq) \rightarrow Na_2S_4O_6(aq) + 2NaI(aq)$$

Conductimetric analysis

The number and type of ions in a solution affect its electrical conductivity. As some chemical reactions take place in solution, the ionic balance changes and the resulting change in conductivity can be used to measure the rate of the reaction (**fig. 1.1.6**). These reflect the changes in the ions present in the solution and so can be used to measure the changes in concentration of the various components of the mixture. For example, when bromoethane reacts with hydroxide ions in alkaline solution, the small, mobile hydroxide ions in the starting solution are replaced by larger, slower moving bromide ions as the reaction progresses. This is reflected in a change in the conductivity of the solution as the ionic concentrations change – which in turn gives us a measure of the rate at which the reaction is taking place.

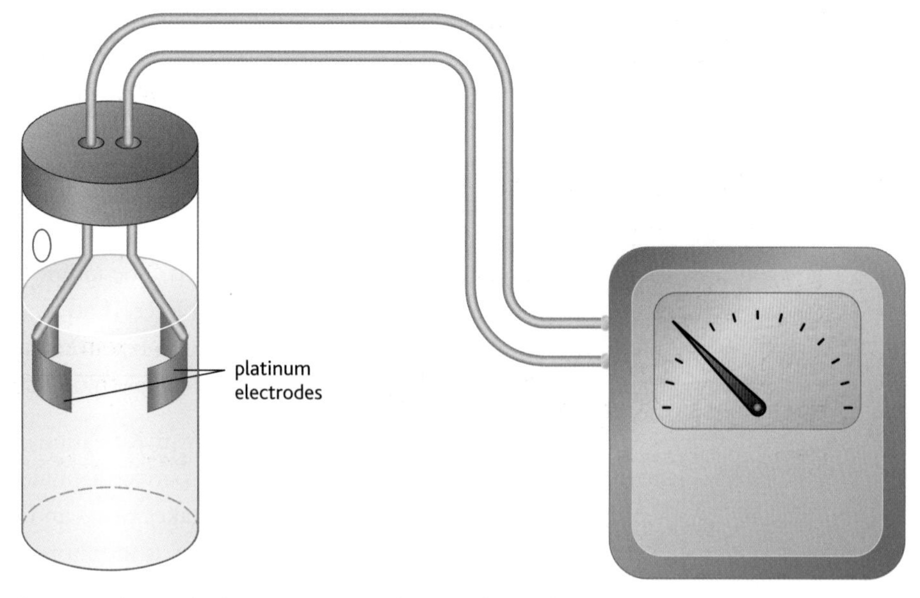

fig. 1.1.6 Using conductimetry to measure the rate of a reaction.

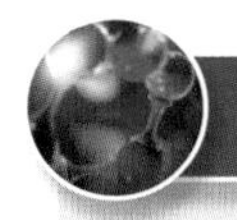

HSW Reaction rates and brewing beer

Understanding reaction rates and how to change them is very important in the brewing industry. There are many different types of beers, from fruit beers to dark stouts, light lagers and heavy ales. One important factor affecting their alcohol content and taste is the rate at which the fermentation takes place. Fermentation – the conversion of sugar into ethanol – is the key reaction in making beer. It is brought about by the zymase enzyme found in yeast; controlling the rate of this reaction is crucial to the beer-making process. Brewers want fermentation to be as economical as possible while producing a good, drinkable beer. Within limits, increasing the temperature of the fermenting mixture would speed up the process so the beer could be made and sent for sale more quickly – but it isn't that easy!

Scientists have identified two main factors affecting the fermentation reaction in a commercial vat. The start temperature – the temperature of the reaction mixture (or wort) when the yeast (catalyst) is added – is very important. Too low and the fermentation rate will be too slow; too high and the yeast will be 'shocked', producing beer with poor flavour.

The maximum temperature of the process is also crucial to the taste and flavour of the beer. Too low and not enough fermentation takes place – the beer tastes sweet with low alcohol levels because lots of sugar is left. Too high and too much fermentation occurs – the beer tastes 'thin' with high alcohol levels. And if it is really hot then the yeast is 'killed' and fermentation halts completely. Most fermentations are carried out at 14–22 °C, depending on the type of beer wanted. The alcohol concentration in the beer is also very important – if excessive the yeast will be inhibited and fermentation will stop. By using different strains of yeast, brewers can vary the alcohol levels of the beers they produce.

Scientists have developed several different ways of monitoring the rate of a fermentation in commercial breweries. The temperature is monitored constantly, the rate of carbon dioxide production is measured using chemical sensors, the amount of yeast in the brew can be measured by its turbidity (cloudiness) using a colorimeter and the pH, oxygen levels and more can all be recorded to make sure that the rate of the reaction is always as close as possible to ideal.

Questions

1 How would you change the conditions to control the following reaction rates:

 a slow down the souring of milk

 b speed up the fermentation of sugar to carbon dioxide and ethanol

 c slow down the reaction of iron with air and water

 d slow down the rate of carbon dioxide formation in the reaction between calcium carbonate and hydrochloric acid?

2 Suggest suitable experimental techniques to obtain rate data for the following reactions and explain your choices:

 a magnesium with dilute hydrochloric acid

 b ethyl ethanoate with sodium hydroxide:
 $CH_3COOC_2H_5(l) + NaOH(aq) \rightarrow CH_3COONa(aq) + C_2H_5OH(aq) + H_2O(l)$

 c copper(II) ions with ammonia molecules:
 $Cu^{2+}(aq) + 4NH_3(aq) \rightarrow Cu(NH_3)_4{}^{2+}(aq)$

3 Why would the method involving measuring the mass of the reaction vessel not work well in the reaction of magnesium and dilute hydrochloric acid? Think about the density of hydrogen.

4 a Which method in **fig. 1.1.2** do you think is likely to give the more accurate results?

 b If the gas produced were soluble, which method would be better and why?

Rate equations, rate constants and the order of a reaction [4.3a]

When measuring the rate of reaction, a chemist will investigate the increase in the concentration of one of the products or a decrease in the concentration of one of the reactants.

So in the reaction

$$A + B \rightarrow C + D$$

the results might show that the rate is related to the concentrations of A and B as follows:

$$\text{rate} = k[A]^m[B]^n$$

This is known as a **rate equation**. [A] and [B] represent the concentrations of A and B in mol dm^{-3}, and k is called the **rate constant**.

The indices m and n are usually whole numbers (1, 2 ...) but they can be fractional or zero. The index m is called the **order of reaction** with respect to A. The overall order of a reaction is the sum of the indices $(m + n)$.

Many reactions involve several steps. The slowest step controls the overall rate of a reaction and this is called the **rate-determining step**. A useful way of visualising the idea of a rate-determining step is to imagine that a teacher has prepared some pages of notes and wants to collect them into sets with the help of three students. The pages of the notes are arranged in 10 piles. The first student collects a page from each of the piles (step 1). The second takes this set of 10 pages from the first and tidies them ready for stapling (step 2). The third student staples this sets of notes together (step 3). It is not hard to see that the overall rate of the process (the rate at which the sets of notes are prepared) depends on step 1, the collecting of the sheets of notes, because this is the slowest step. It does not matter, within reason, how quickly the tidying or the stapling is done.

The overall order of the reaction is related to the **molecularity** of the rate-determining step – that is the number of particles involved. In a reaction between A and B, if the rate-determining step involves the collision of one particle of A with one particle of B, the molecularity is 2.

Clock reactions

By far the best way to study the effect of changing the concentration of the reactants on the reaction rate is to carry out some experiments. A reaction that is easy to follow experimentally is the reaction between iodide ions and hydrogen peroxide in the presence of an acid, starch and sodium thiosulfate – often referred to as an iodine 'clock' reaction.

Iodine clock reaction

This is an experiment first devised by Hans Heinrich Landolt in 1886. Two colourless solutions are mixed and at first there is no visible reaction, but after a short time delay the liquid suddenly turns dark blue (sometimes described as blue-black) (**fig. 1.1.7**). The time, t, is measured from mixing until the first formation of the blue colour. The rate of reaction is proportional to $\frac{1}{t}$.

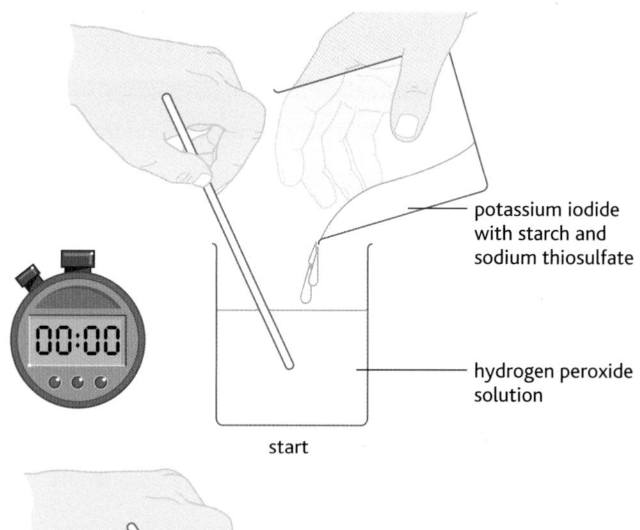

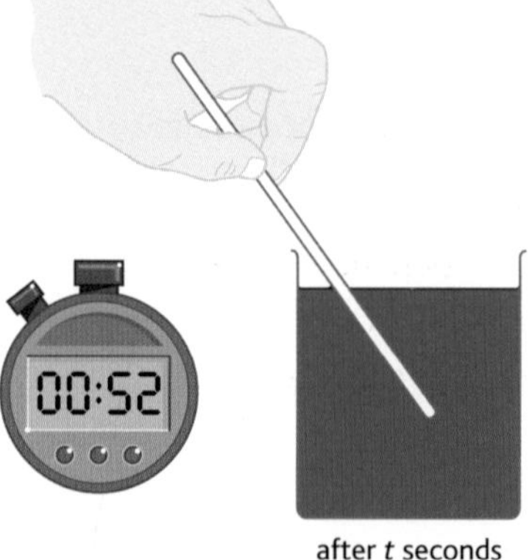

fig. 1.1.7 An iodine clock reaction.

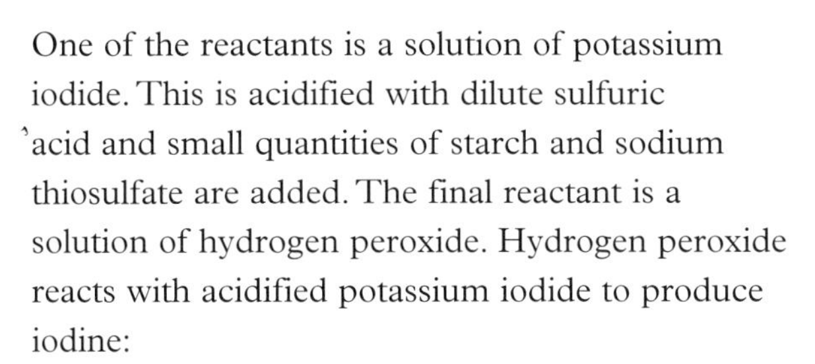

One of the reactants is a solution of potassium iodide. This is acidified with dilute sulfuric acid and small quantities of starch and sodium thiosulfate are added. The final reactant is a solution of hydrogen peroxide. Hydrogen peroxide reacts with acidified potassium iodide to produce iodine:

Step 1

$$H_2O_2(aq) + 2I^-(aq) + 2H^+(aq) \rightarrow 2H_2O(l) + I_2(aq)$$

As soon as any iodine is produced, it reacts with sodium thiosulfate, forming iodide ions again:

Step 2

$$I_2(aq) + 2S_2O_3^{2-}(aq) \rightarrow 2I^-(aq) + S_4O_6^{2-}(aq)$$

When all the sodium thiosulfate is used up, a blue colour is suddenly formed because the iodine forms a complex with the starch that has been added.

In this reaction, step 2 has no effect on the overall rate – it is much faster than step 1. Step 1 is the rate-determining step.

By repeating the experiment using several different concentrations of hydrogen peroxide, the effect on the rate of changing the concentration of hydrogen peroxide can be investigated. The rate equation can be represented as:

$$\text{rate} = k[H_2O_2]^x\ [I^-]^y\ [H^+]^z$$

The order with respect to one reactant (for example H_2O_2) can be determined by using a large excess of the other reactants (I^- and H^+) so that their concentrations remain effectively constant throughout the reaction.

Under these conditions the rate of reaction can be expressed as:

$$\text{rate} = k'[H_2O_2]^x$$

where k' is a modified rate constant that includes the constant concentration of the other reactants.

The units of a rate constant depend on the form of its rate equation. For example, if rate = k[A] then the units of the rate constant are s^{-1} because:

$$k = \frac{\text{rate}}{[A]}$$

$$= \frac{\text{mol dm}^{-3}\,\text{s}^{-1}}{\text{mol dm}^{-3}} = \text{s}^{-1}$$

Worked examples

1 The reaction of peroxodisulfate ions and iodide ions can be represented as:

$$S_2O_8^{2-}(aq) + 2I^-(aq) \rightarrow 2SO_4^{2-}(aq) + I_2(aq)$$

The rate equation determined by experiment is:

$$\text{rate} = k[S_2O_8^{2-}][I^-]$$

The values of the order indices are both 1 in this reaction – so the order of reaction with respect to peroxodisulfate ions is 1, and with respect to iodide ions is also 1. The overall order of the reaction is 2. You can see the order is not related to the equation.

2 The reaction of hydrogen and bromine can be represented by:

$$H_2(g) + Br_2(g) \rightarrow 2HBr(g)$$

The rate equation determined by experiment is:

$$\text{rate} = k[H_2][Br_2]^{0.5}$$

The order with respect to hydrogen is 1, and with respect to bromine is 0.5 (or ½). The overall order of reaction is 1½. Notice that the order of reaction can be a whole number or a fraction – whatever fits the mathematical relationship.

Questions

1 Why is it important in the iodine clock reaction that there is only a small quantity of sodium thiosulfate?

2 The reaction between NO(g) and Cl_2(g) has been studied at 50 °C.

$$NO(g) + ½Cl_2(g) \rightarrow NOCl(g)$$

The table shows the initial concentrations of reactants and the initial rate of formation of NOCl(g).

[NO(g)]/ mol dm^{-3}	[Cl_2(g)]/ mol dm^{-3}	Initial rate/ mol dm^{-3} s^{-1}
0.250	0.250	1.43 × 10^{-6}
0.250	0.500	2.86 × 10^{-6}
0.500	0.500	11.44 × 10^{-6}

a What is the order of reaction with respect to Cl_2(g)?

b What is the order of reaction with respect to NO(g)?

c What is the overall order of reaction?

d Write the rate equation for this reaction.

e Calculate the value of the rate constant k. Give its units.

Determining the order of a reaction and the rate equation from experimental data [4.3f (ii), (iii)]

We have seen that rate constants appear in rate equations, and that the units of rate constants can vary. In this section we are going to find out how to calculate a rate of reaction graphically by drawing tangents to the curve of its reaction rate graph. Then we can use the data to calculate the order of the reaction and the rate constant.

It is not possible to predict the order of reaction from a balanced equation. It *has* to be found out by experiment.

Interpreting experimental data

Once the changes in concentration have been measured in an experiment, the reaction rate can be obtained by plotting concentration against time. **Figure 1.1.8** shows a graph of the concentration of A against time for a slow reaction, A(aq) + B(aq) → C(aq) + D(aq).

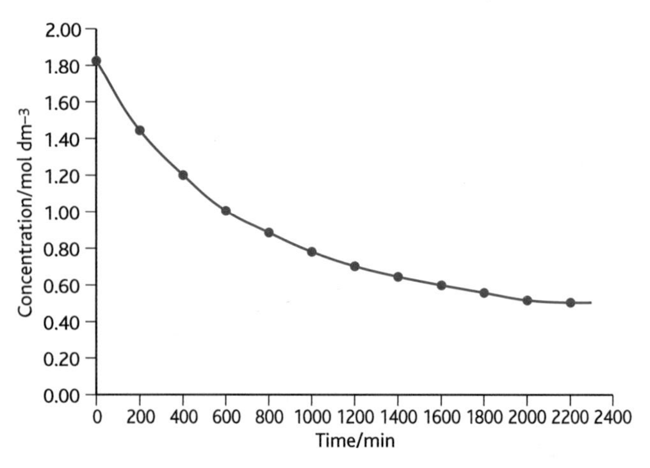

fig. 1.1.8 Graph to show the change in concentration of A with time.

You will notice that the graph is steepest at the start of the reaction, and it gradually gets less steep as the reaction progresses. If such a graph becomes horizontal then the reaction has finished. The rate of reaction decreases with time because the number of particles of A and B decrease and so there are fewer collisions.

To find the rate of reaction at a particular time, a tangent has to be drawn to the curve at that time and its gradient calculated, as shown in **fig. 1.1.9**.

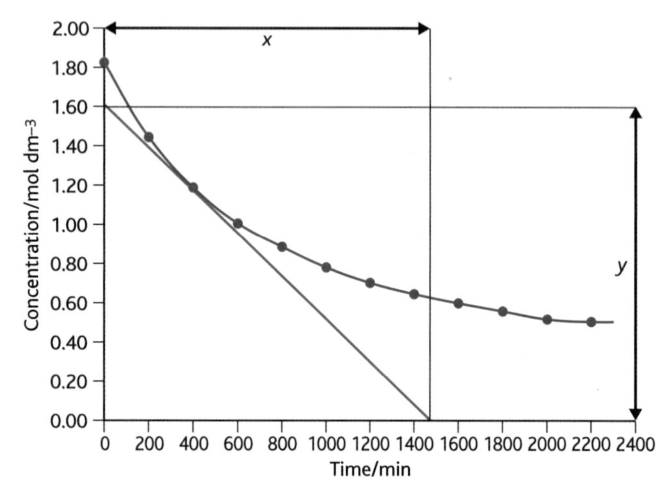

fig. 1.1.9 Calculating the gradient of the graph.

To draw a tangent, hold a ruler to the curve so that it just touches the curve at the time required. Draw a long straight line – a large triangle is required to get accurate results. In the example in **fig. 1.1.9**, the rate at 400 minutes is $-1.09 \times 10^{-3}\,\text{mol dm}^{-3}\,\text{min}^{-1}$:

$$\text{Gradient} = \frac{y}{x}$$

$$= \frac{-1.60\,\text{mol dm}^{-3}}{1470\,\text{min}}$$

$$= -1.09 \times 10^{-3}\,\text{mol dm}^{-3}\,\text{min}^{-1}$$

The value of the rate of reaction is negative and this indicates that the concentration of A is decreasing as time goes on.

Worked example

The equation for the decomposition of sulfur dichloride oxide, SO_2Cl_2, is:

$$SO_2Cl_2(g) \rightarrow SO_2(g) + Cl_2(g)$$

The rate of this reaction can be followed by monitoring the pressure of the gases in the reaction vessel. This will increase because the number of moles of gas doubles on going from left to right.

The pressure results were used to calculate the concentration of sulfur dichloride oxide at certain times – the results are summarised in **table 1.1.1**. **Figure 1.1.10** shows the graph of concentration against time.

The graph shows how $[SO_2Cl_2(g)]$ falls during the reaction. We can therefore write an expression for the rate of this reaction with respect to $[SO_2Cl_2]$:

$$\text{rate} = \frac{-\Delta[SO_2Cl_2]}{\Delta t}$$

The negative sign in this expression is necessary because $[SO_2Cl_2]$ *decreases* with time, so the change in $[SO_2Cl_2]$ divided by the change in time is *negative*. By definition, a rate of reaction is *positive* and the negative sign ensures that this is so. Now look at the graph in **fig. 1.1.11** and **table 1.1.2**. The values in the right-hand column were calculated by drawing tangents to the curve at each point and calculating the gradient.

Time/s	Concentration of SO_2Cl_2/mol dm^{-3}
0	0.50
500	0.43
1000	0.37
2000	0.27
3000	0.20
4000	0.15

table 1.1.1 Data for fig. 1.1.10.

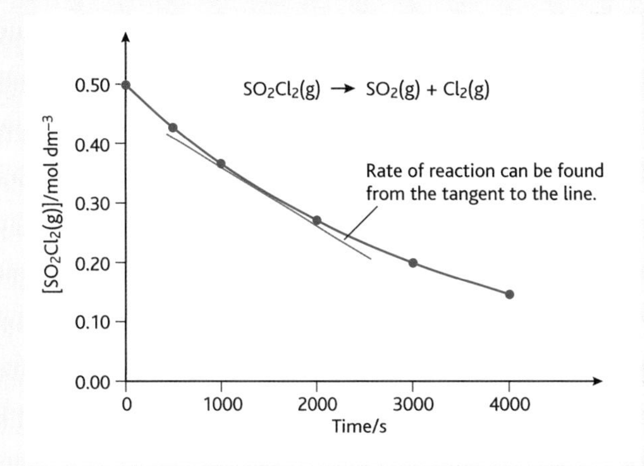

fig. 1.1.10 The rate of dissociation of SO_2Cl_2 decreases as the concentration of SO_2Cl_2 falls.

Concentration of SO_2Cl_2/mol dm^{-3}	$\frac{-\Delta[SO_2Cl_2]}{\Delta t}$/mol dm^{-3} s^{-1}
0.45	1.35×10^{-4}
0.39	1.17×10^{-4}
0.34	1.02×10^{-4}
0.28	8.40×10^{-5}
0.23	6.90×10^{-5}
0.18	5.40×10^{-5}

table 1.1.2 Data for fig. 1.1.11.

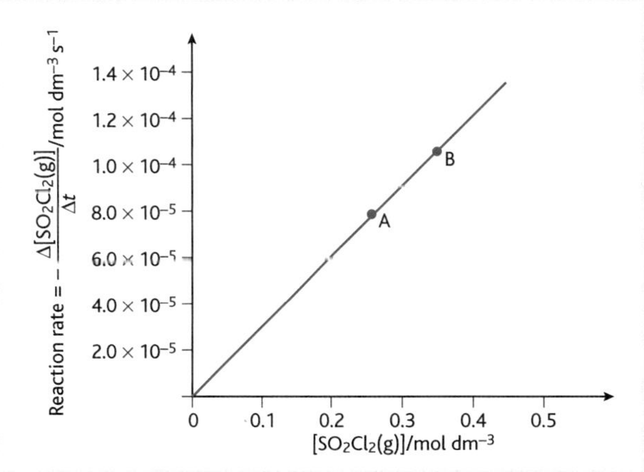

fig. 1.1.11 The variation of reaction rate with concentration of SO_2Cl_2 for the dissociation of SO_2Cl_2.

The graph of $\frac{-\Delta[SO_2Cl_2]}{\Delta t}$ against $[SO_2Cl_2]$ in **fig. 1.1.11** on the previous page is a straight line. Since reaction rate is plotted on the *y*-axis of the graph and $[SO_2Cl_2(g)]$ is plotted on the *x*-axis, it follows that

$$\text{rate} = k[SO_2Cl_2(g)]$$

Straight line graphs

Graphs are extremely useful for finding and confirming relationships between different variables – for example, the variation of the volume of a gas with its temperature. The simplest type of relationship is one which is linear, in which a graph of one variable against another is a straight line.

The graph in **fig. 1.1.11** is an example of a linear graph. The reaction rate is plotted on the vertical axis (referred to as the *y*-axis or the ordinate) and the concentration of $SO_2Cl_2(g)$ is plotted on the horizontal axis (referred to as the *x*-axis or the abscissa).

The general form of the equation for a straight line is:

$$y = mx + c$$

where *m* is the slope or gradient of the line and *c* is the intercept on the *y*-axis (where the line crosses the *y*-axis).

In this case $c = 0$, so the equation has the form:

$$y = mx$$

Rate expressions

The rate of dissociation of SO_2Cl_2 follows a law which can be written as:

$$\text{rate} = k[SO_2Cl_2(g)]^1$$

or

$$\text{rate} = k[SO_2Cl_2(g)]$$

The reaction is said to be first order with respect to SO_2Cl_2 because the concentration of SO_2Cl_2 appears in the rate expression raised to the power of 1.

We can calculate the rate constant for the reaction from gradient of the graph in **fig. 1.1.11**.

The coordinates for points A and B are ($0.26\,\text{mol dm}^{-3}$, $7.8 \times 10^{-5}\,\text{mol dm}^{-3}\,\text{s}^{-1}$) and ($0.35\,\text{mol dm}^{-3}$, $1.05 \times 10^{-4}\,\text{mol dm}^{-3}\,\text{s}^{-1}$) respectively, so

$$k = \frac{1.05 \times 10^{-4}\,\text{mol dm}^{-3}\,\text{s}^{-1} - 7.8 \times 10^{-5}\,\text{mol dm}^{-3}\,\text{s}^{-1}}{0.35\,\text{mol dm}^{-3} - 0.26\,\text{mol dm}^{-3}}$$

$$= 3.0 \times 10^{-4}\,\text{s}^{-1}$$

Remember that the rate constant for a reaction varies with temperature, so the value of *k* calculated from a particular investigation applies only to the temperature at which the investigation was carried out.

For reactions with more than two reactants, the rate expression can be extended. For example, the reaction:

$$2HCrO_4^-(aq) + 3HSO_3^-(aq) + 5H^+(aq) \rightarrow 2Cr^{3+}(aq) + 3SO_4^{2-}(aq) + 5H_2O(l)$$

has the rate expression:

$$\text{rate} = k[HCrO_4^-(aq)][HSO_3^-(aq)]^2[H^+(aq)]$$

Notice that this reaction is fourth order overall – first order with respect to $HCrO_4^-$ and to H^+, and second order with respect to HSO_3^-.

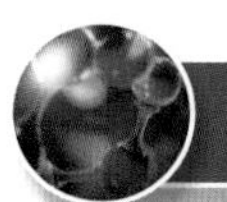

HSW Predicting a rate expression

As a general rule, it is not possible to obtain the rate expression for a reaction from its balanced equation. This is because a great many reactions happen in several steps – for example, the decomposition of ozone (O_3) to oxygen (O_2) occurs in two steps – the first of which is rapid, the second of which is much slower:

$$O_3 \rightleftharpoons O_2 + O$$

$$O + O_3 \rightleftharpoons 2O_2$$

The second step is slow, and therefore determines the rate of the reaction – it is the rate-determining step. We can now write a rate expression for this reaction, since the rate of formation of O_2 will depend on [O] and $[O_3]$:

$$\text{rate} = k[O][O_3]$$

From this we can deduce that the rate of decomposition of ozone in the upper atmosphere increases as the concentration of ozone increases. Since ozone is used up in the second step, this is a means of ensuring that the level of ozone does not become too high.

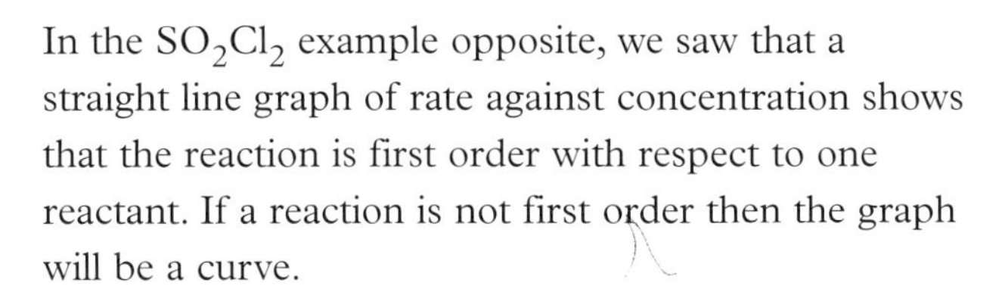

In the SO_2Cl_2 example opposite, we saw that a straight line graph of rate against concentration shows that the reaction is first order with respect to one reactant. If a reaction is not first order then the graph will be a curve.

To find out if it is second order, you have to plot a graph of rate of reaction against $(\text{concentration})^2$. If this is a straight line the reaction is second order with respect to that reactant. Obviously this is a very long method involving the use of trial and error to find the order of reaction. **Figure 1.1.12** shows the typical shapes of concentration against time curves for zero order, first order and second order reactions. It also shows the typical shapes of rate against concentration for these examples.

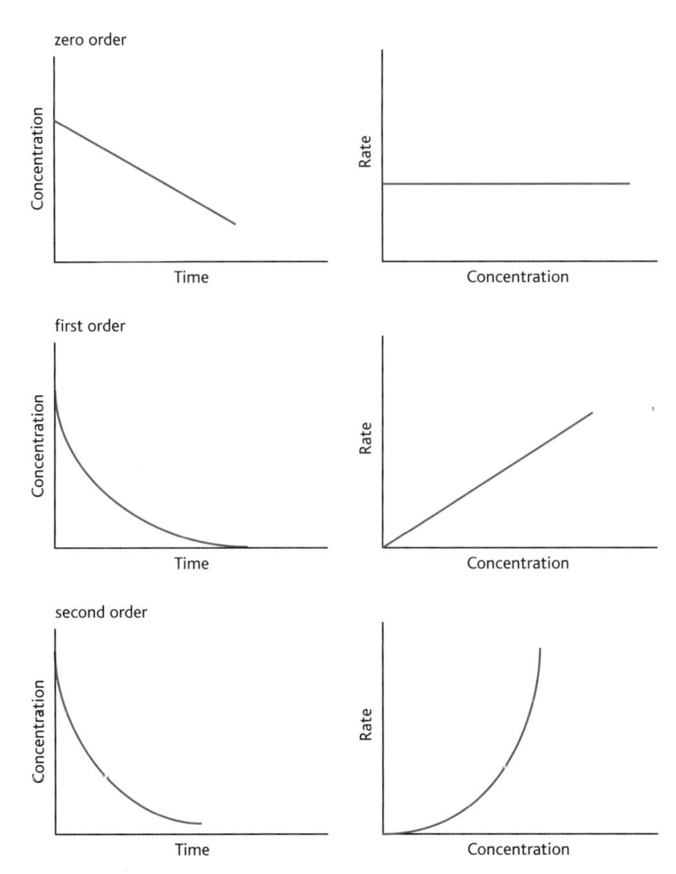

fig. 1.1.12 Typical curves for different orders of reaction.

Questions

1 When plotting a graph of rate against concentration (e.g. **fig.1.1.11**), you have to measure the gradient of a curve at a number of different points. Why would only two points lead to uncertainty?

2 In the gas phase, molecule A decomposes to molecules B and C at high temperatures. A chemist suspects that this reaction is first order with respect to A. In an experiment to explore the kinetics of the reaction, the data in the table were obtained for the decomposition at 800 K.

Time/s	Partial pressure of A/kPa (see page 55)
0	1300
20	1051
40	849
60	685
80	554
100	448
120	361
140	292
160	236
180	191
200	154

a If the reaction is first order with respect to A, write down the rate equation.

b By using a graph, find out if this reaction really is first order with respect to A.

c Calculate the rate constant for the reaction. What are its units?

d What does the rate equation tell you about its mechanism?

Graphical representation of kinetic measurements [4.3a, c, d, f(i)]

Why do the orders of reactions and rate equations matter so much? When chemical engineers design new reactors or chemical plants, they need to know exactly how the rate of the reaction varies with the concentrations of the reactants. This not only informs them about the ideal starting concentrations for the reactants, but also helps them to plan the best possible places to collect the products or to restock the reaction vessels with one or more of the reactants. This is just one reason why it is so important to be able to represent kinetic measurements as accurately as possible.

fig. 1.1.13 Chemical plants cost millions of pounds to build and set up.

The method used to find the order of reaction in the previous section was a method that uses trial and error. This can work satisfactorily if the reaction is first order, or even second order, but if the order is higher or fractional it might take a long time to find the order using this method. A method commonly used involves measuring the rate just at the start of each experiment – the initial rate – and it gives the order of reaction more easily.

Order of reaction from initial rate of reaction

A series of experiments is carried out using different initial concentrations of the reagent under consideration, with every other factor, such as concentration and the temperature, unchanged. It is important that only one variable is changed.

For each experiment and product, a concentration against time graph is plotted (**fig. 1.1.14**). A tangent is drawn to the curve at $t = 0$ and, using a large triangle for maximum accuracy, the gradient is calculated. This is the initial rate of the reaction.

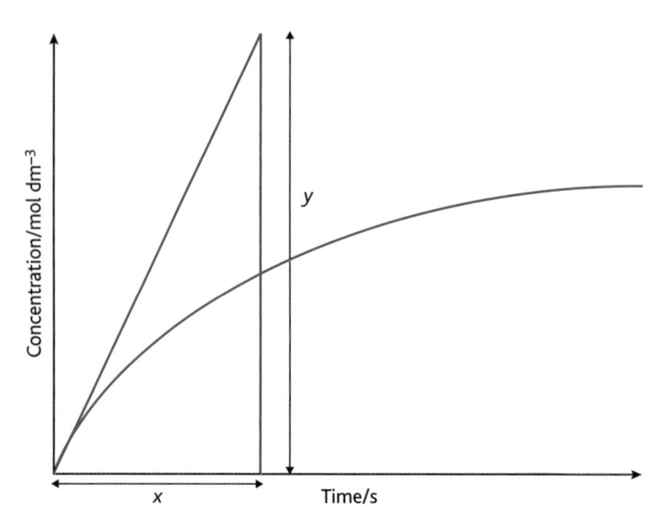

fig. 1.1.14 A graph of concentration against time.

When sufficient initial rate values have been obtained, a graph of initial rate against concentration is drawn (**fig. 1.1.15**).

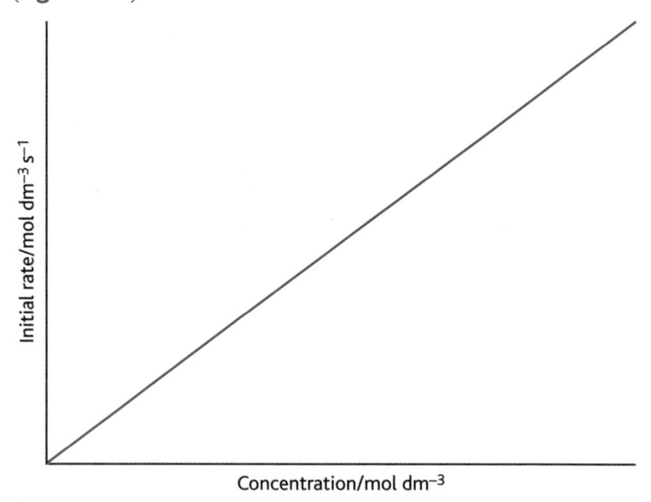

fig. 1.1.15 A graph of initial rate against concentration.

See **fig. 1.1.12** for a reminder of the different shapes and what they mean.

From such data it is possible to work out the order of reaction with respect to each of the reactants.

Bromate(V) and bromide reaction

The following reaction was studied:

$$BrO_3^-(aq) + 5Br^-(aq) + 6H^+(aq) \rightarrow 3Br_2(aq) + 3H_2O(l)$$

Three series of experiments were carried out. The results are shown in **tables 1.1.3–1.1.5**.

Experiment	Initial $[BrO_3^-]$/mol dm^{-3}	Initial $[Br^-]$/mol dm^{-3}	Initial $[H^+]$/mol dm^{-3}	Initial rate/mol dm^{-3}s^{-1}
1	0.10	0.10	0.10	1.2×10^{-3}
2	0.20	0.10	0.10	2.4×10^{-3}
3	0.30	0.10	0.10	3.6×10^{-3}
4	0.40	0.10	0.10	4.8×10^{-3}

table 1.1.3 Series 1: altering the concentration of bromate(V) ions.

From these data, it is clear that rate $\propto$ $[BrO_3^-(aq)]$ because doubling the concentration of $[BrO_3^-]$ doubles the rate. So the reaction is first order with respect to bromate(V) ions.

Experiment	Initial $[BrO_3^-]$/mol dm^{-3}	Initial $[Br^-]$/mol dm^{-3}	Initial $[H^+]$/mol dm^{-3}	Initial rate/mol dm^{-3}s^{-1}
1	0.10	0.10	0.10	1.2×10^{-3}
2	0.10	0.20	0.10	2.4×10^{-3}
3	0.10	0.30	0.10	3.6×10^{-3}
4	0.10	0.40	0.10	4.8×10^{-3}

table 1.1.4 Series 2: altering the concentration of bromide ions.

From these data it can be concluded that rate $\propto$ $[Br^-(aq)]$. So the reaction is first order with respect to bromide ions.

Experiment	Initial $[BrO_3^-]$/mol dm^{-3}	Initial $[Br^-]$/mol dm^{-3}	Initial $[H^+]$/mol dm^{-3}	Initial rate/mol dm^{-3}s^{-1}
1	0.10	0.10	0.10	1.2×10^{-3}
2	0.10	0.10	0.20	4.8×10^{-3}
3	0.10	0.10	0.40	19.2×10^{-3}

table 1.1.5 Series 3: altering the concentration of hydrogen ions.

From these data it can be concluded that rate $\propto$ $[H^+(aq)]^2$ because doubling the concentration of $H^+(aq)$ quadruples the rate of reaction. So the reaction is second order with respect to hydrogen ions.

The rate equation for this reaction is therefore:

$$\text{rate} = k[BrO_3^-(aq)][Br^-(aq)][H^+(aq)]^2$$

Half-life

A convenient way to measure the rate of a reaction, particularly of first order reactions, is to use the **half-life** of the reaction. This is the time taken for half of a reactant to be used up during a reaction process.

Half-life of a first order reaction

On page 17 the decomposition of sulfur dichloride oxide was investigated:

$$SO_2Cl_2(g) \rightarrow SO_2(g) + Cl_2(g)$$

Figure 1.1.16 shows a graph of concentration of $SO_2Cl_2(g)$ against time.

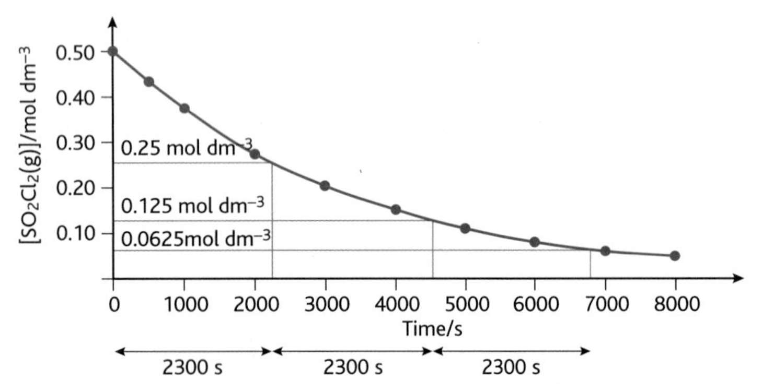

fig. 1.1.16 This graph shows that the reaction has a constant half-life – that is, the time taken for the concentration of SO_2Cl_2 to halve is always constant (at a given temperature). All first order reactions behave like this.

Look at **fig. 1.1.16** carefully. The time it takes the concentration of SO_2Cl_2 to halve is marked on it. Notice that this time is constant – in other words, it takes the same time for the concentration to fall from 0.50 mol dm^{-3} to 0.25 mol dm^{-3} as it takes for it to fall from 0.25 mol dm^{-3} to 0.125 mol dm^{-3}. The time taken for the concentration to halve is known as the half-life ($t_{1/2}$) of the reaction. It is the length of time for half of a given reactant to disappear. If the half-life is short, the reaction is rapid; if the half-life is long, the reaction is slow. The half-life is independent of concentration.

All first order reactions have constant half-lives at a given temperature. You have met the idea of half-life before, in connection with radioactive decay during your AS Chemistry course – radioactive decay is another example of a first order reaction.

Half-life and radioactive decay

Radioactive decay is an example of a first order reaction because its rate is independent of the concentration of the radioactive material. Half-lives vary from a tiny fraction of a second (the decay of polonium-212 has $t_{1/2} = 3 \times 10^{-7}$ s) to millions of years (the decay of uranium-238 has $t_{1/2} = 4.5 \times 10^{9}$ years).

Many radioactive elements have very long half-lives and scientists use these to help to work out the age of rocks in the Earth's crust. For example, some rocks would have originally contained $^{238}_{92}U$ but none of the decay product, $^{206}_{82}Pb$. As time passed and the uranium decayed, the levels of lead began to build up. By analysing the present-day ratio of $^{238}_{92}U$ to $^{206}_{82}Pb$ scientists can estimate the approximate age of the rocks. Current estimates for the age of the Earth based on this method put it at around 4000 million years old.

For working out the age of material that was once living or – in the case of some trees – is still living, scientists use radiocarbon dating. This relies on the half-life of the radioactive isotope carbon-14, which is around 5730 years. This method can produce some very accurate dates, especially when linked to mass spectrometry. However, there are issues with really old material – the technique seems to be effective only over a limited timespan.

Write a short piece of prose to explain, in simple language, the science behind carbon-dating based on counting data, its limitations and why modern methods based on mass spectroscopic measurement of carbon isotope ratios are now used.

Half-life in second order reactions

The half-life of a first order reaction is independent of the initial concentration of the reactants. However, the half-life of a second order reaction does depend on the initial concentrations of the reactants.

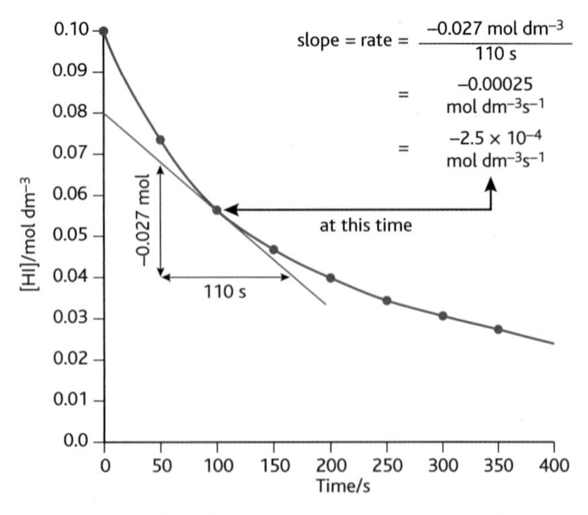

fig. 1.1.17 Graph to show change in concentration of HI with time for the reaction $2HI(g) \rightarrow H_2(g) + I_2(g)$ at 508 °C.

In **fig. 1.1.17**, the initial HI concentration is 0.10 mol dm^{-3} and it takes 125 seconds for the concentration to drop to 0.05 mol dm^{-3} – in other words, the half-life for this first part is 125 seconds. But if you think of 0.05 mol dm^{-3} as the new 'initial concentration', you will see that it takes 250 seconds to fall to 0.025 mol dm^{-3}. So in second order reactions, halving the initial concentration doubles the half-life. In other words, in second order reactions the half-life is inversely proportional to the initial concentration of the reactants. It is related to the rate constant by the following equation:

$$t_{1/2} = \frac{1}{k \times [\text{reactant}]_{\text{initial}}}$$

Graphs of concentration against time and rate against concentration

When you carry out kinetics experiments, characteristic graph shapes are obtained. These shapes are determined by the order of the reaction (see **fig. 1.1.12**). It is useful to be able to recognise them.

Figure 1.1.18 shows concentration against time graphs for zero, first and second order reactions.

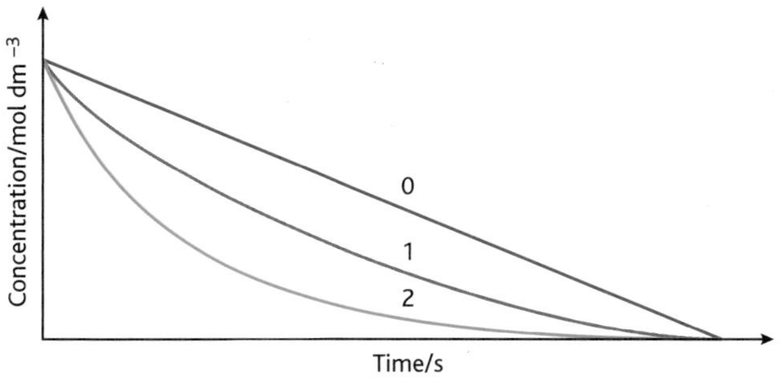

fig. 1.1.18 Graphs of zero, first and second order reactions showing how concentration changes with time.

Figure 1.1.19 shows rate against concentration graphs for zero, first and second order reactions.

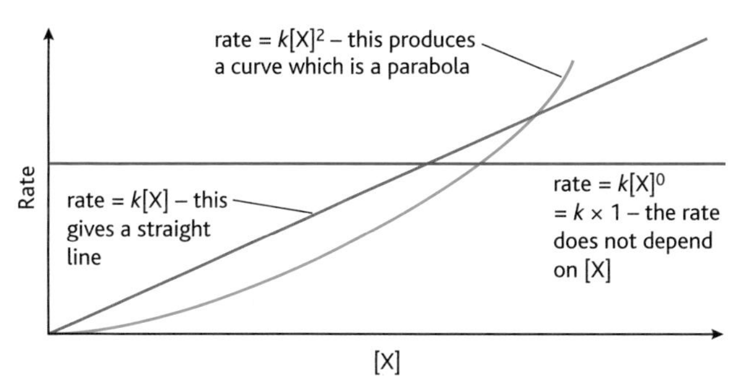

fig. 1.1.19 The variation of reaction rate with [X] for reactions which are zero, first and second order with respect to X.

Questions

1 The equation for the reaction of bromine and methanoic acid is:

$$Br_2(aq) + HCOOH(aq) \rightarrow 2Br^-(aq) + 2H^+(aq) + CO_2(g)$$

a Suggest a way of keeping the concentration of methanoic acid virtually constant.

b The results of an experiment in which the concentration of bromine was monitored throughout a reaction are shown in the table.

Plot a graph of $[Br_2(aq)]$ against time. Show that this reaction is first order by working out three values for half-life.

Time/s	$[Br_2(aq)]$/mol dm^{-3}
0	0.0100
25	0.0090
50	0.0080
75	0.0070
100	0.0065
180	0.0050
240	0.0045
360	0.0030
420	0.0025
480	0.0020

2 The half-life of radioactive iodine-131 is 8.0 days. What fraction of the initial amount of iodine-131 would be present in a patient after 24 days, if none were eliminated through natural body processes?

Activation energy and types of catalysts [4.3a]

You will know from *Edexcel AS Chemistry* chapter 2.6 that substances have to have sufficient energy before they can start reacting – this is called the **activation energy**, E_A.

Many reactions appear not to take place at room temperature – the particles involved don't have sufficient energy when they collide to overcome the activation energy barrier. Raising the temperature can help the molecules to achieve sufficient energy – much depends on the height of the barrier.

Figure 1.1.20 shows the reaction profile for an exothermic reaction. The higher the activation energy barrier, the slower the reaction is likely to be.

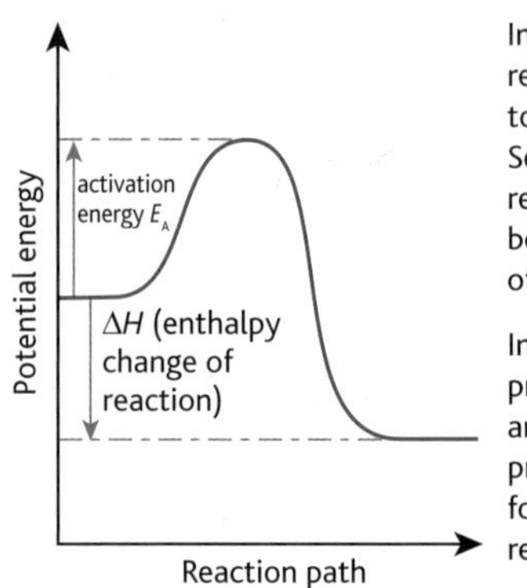

In the upward slope, the reactant molecules are coming together and breaking apart. Separating atoms in the reactant molecules requires bonds to be broken, so this part of the reaction **absorbs** energy.

In the downward slope, the product molecules are forming and moving apart. Producing product molecules involves forming bonds, so this part of the reaction **releases** energy.

fig. 1.1.20 Reaction profile diagram for an exothermic reaction. The progress of the reaction is shown on the horizontal axis, the reaction coordinate.

The effect of using a catalyst is shown in **fig. 1.1.21**. Here the reaction profile shows the energy changes during an uncatalysed reaction (a) and a catalysed reaction (b). You will notice that the effect of the catalyst is to lower the overall activation energy. More colliding particles will possess sufficient energy and so the reaction will be faster.

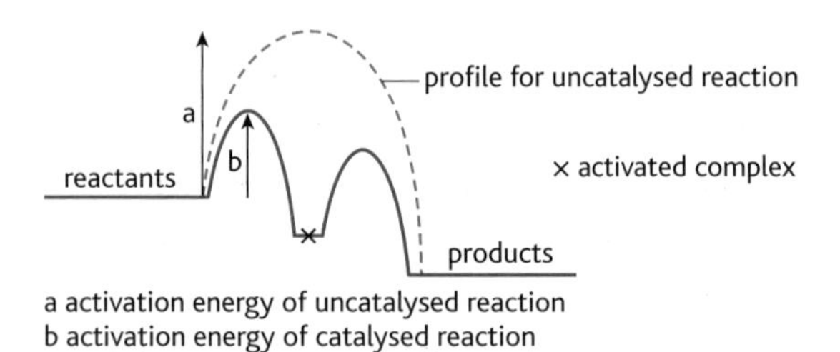

fig. 1.1.21 The effect of using a catalyst.

Catalysts

It has been estimated that 90% of all chemicals produced in industry today use a catalyst at some stage in the manufacturing process.

Catalysts do not make impossible reactions take place – they merely make possible reactions faster. The simple definition of a catalyst is that it is a substance that alters the speed of a reaction without being used up. In practice, catalysts are often used up in secondary reactions.

Catalysts can be divided into two types – homogeneous catalysts and heterogeneous catalysts.

Homogeneous catalysts

A **homogeneous catalyst** is in the same phase (solid, liquid, solution or gas) as the reactants – for example, a gaseous catalyst in a mixture of gases or a liquid catalyst in a mixture of liquids.

Chlorine free radicals act as a homogeneous catalyst in the gas phase. Chlorine free radicals are produced when ultraviolet light from the Sun breaks up chlorine molecules. They are breaking down ozone in the upper atmosphere into oxygen (see *Edexcel AS Chemistry* page 214):

$$Cl\bullet(g) + O_3(g) \rightarrow ClO\bullet(g) + O_2(g)$$

$$ClO\bullet(g) + O_3(g) \rightarrow Cl\bullet(g) + 2O_2(g)$$

Notice that the chlorine free radicals are destroyed, and then regenerated.

The hydrolysis of esters by acid catalysis is an example of homogeneous catalysis with all the reactants, products and the catalyst being dissolved in water:

$$CH_3COOCH_3(aq) + H_2O(l) \rightarrow CH_3COOH(aq) + CH_3OH(aq)$$

Modern examples of homogeneous catalysis include Ziegler–Natta catalysts used to polymerise alkenes (see *Edexcel AS Chemistry* page 134). These catalysts are organometallic compounds containing both a transition metal (e.g. titanium) and an alkyl group.

Heterogeneous catalysts

A **heterogeneous catalyst** is in a different phase from the reactants. For example, in the Haber process the solid iron catalyst catalyses the reaction between two gases, hydrogen and nitrogen.

In a catalytic converter in a car, hot gases from the engine react on the metal surface. The catalysts are used to reduce the amounts of carbon monoxide, unburnt hydrocarbons and nitrogen oxides in the exhaust gases. The first stage is *reduction* – nitrogen oxides are reduced to nitrogen. The second stage is *oxidation* – carbon monoxide and unburnt hydrocarbons are oxidised to form carbon dioxide and water.

Heterogeneous catalysts are frequently transition metals – for example iron, platinum and nickel. The reacting mixtures are usually adsorbed onto the surface of the catalyst, so the catalyst is usually given a very large surface area to maximise its effect on the reaction rate – a large ingot of iron would have little effect compared to the same mass in finely divided form.

Inorganic catalysts such as these are used to catalyse a wide range of different reactions. Although catalysts are not permanently altered during the reactions that they catalyse, they can be poisoned by some impurities and will not work again.

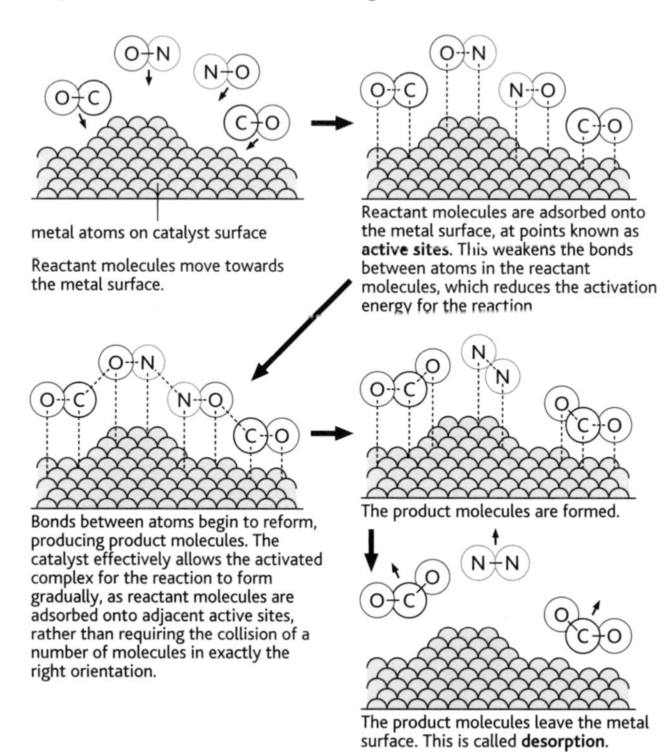

fig. 1.1.22 The mechanism of action of a heterogeneous catalyst.

Figure 1.1.22 shows how a heterogeneous catalyst works in a catalytic converter. Reactant molecules are adsorbed onto the surface of the catalyst and the molecules break down. Reaction takes place on the surface and the product molecules are desorbed from the surface.

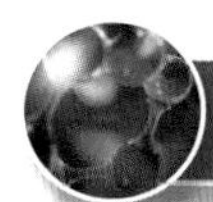

HSW Catalytic converters

When it is three years old, every car has to have a Ministry of Transport test (MOT) to ensure that it is roadworthy. The test has to be repeated every 12 months. Part of the test involves analysing the gases escaping from the exhaust system. In order to reduce harmful emissions, modern cars have been designed to control carefully the amount of fuel they burn. There are sensors mounted before and after the catalytic converter to do this (see over). The main emissions of a modern car engine are shown in **table 1.1.6**.

Gas	How formed
Nitrogen gas	Most passes right through the car engine unchanged
Carbon dioxide	Produced when fuel burns in the engine
Water vapour	Produced when fuel burns in the engine

table 1.1.6

These three gases can be regarded to be acceptable in the atmosphere, although we now realise that carbon dioxide emissions are believed to contribute to global warming. However, because the combustion process is never perfect, smaller amounts of more harmful emissions are also produced in car engines:

- carbon monoxide – a poisonous gas that is colourless and odourless.
- hydrocarbons or volatile organic compounds (VOCs) – produced mostly from unburnt fuel that evaporates. Sunlight breaks these down to form oxidants, which react with oxides of nitrogen to cause ground level ozone, a major component of pollution.
- oxides of nitrogen – contribute to smog and acid rain, and also cause irritation to human mucus membranes.

Most modern cars are equipped with a three-way catalytic converter as part of the exhaust system. 'Three-way' refers to the three regulated emissions it helps to reduce – carbon monoxide, unburnt hydrocarbons and nitrogen oxides.

The first stage of the catalytic converter is the *reduction catalyst*. It uses platinum and rhodium to help to reduce the nitrogen oxide emissions. When such molecules come in contact with the catalyst, the oxygen is removed. Nitrogen atoms bond with other nitrogen atoms, forming nitrogen gas.

The second stage is the *oxidation catalyst*. This reduces the unburnt hydrocarbons and carbon monoxide by oxidising them over a platinum and palladium catalyst. This catalyst aids the reaction of carbon monoxide and hydrocarbons with the remaining oxygen in the exhaust gas to form carbon dioxide and water.

Both catalysts consist of a ceramic structure coated with the metal catalyst. The idea is to create a structure that exposes the maximum surface area of the catalyst to the exhaust stream, while also minimising the amount of catalyst required (they are very expensive).

The third stage in the converter is a *control system* that monitors the exhaust stream using two heated oxygen sensors (also called Lambda sensors) and uses this information to control the fuel injection system. The first of these sensors measures the amount of oxygen in the exhaust gas and this data is used to adjust the composition of fuel and oxygen entering the engine. The second sensor monitors the efficiency of the catalyst in the converter. The goal is to keep the air-to-fuel ratio very close to the 'stoichiometric' point, which is the calculated ideal ratio of air to fuel. Theoretically, at this ratio all the fuel will be burned using all the oxygen in the air. For petrol, the stoichiometric ratio is about 14.7 to 1, meaning that for each kilogram of fuel, 14.7 kilograms of air are burned. The fuel mixture actually deviates from the ideal ratio quite a bit during driving. Sometimes the mixture can be 'lean' (an air-to-fuel ratio higher than 14.7); at other times the mixture can be 'rich' (an air-to-fuel ratio lower than 14.7).

The catalytic converter does a great job of reducing pollution – but there is room for improvement. One of the converter's biggest shortcomings is that it only works at a fairly high temperature. When you start your car cold, the catalytic converter does almost nothing to reduce the pollution in your exhaust. This is an important point to bear in mind when taking your car for its MOT – make sure that it has had a good steady run to heat up the system to working temperature or the car could fail its emissions test.

A simple solution to this problem is to move the catalytic converter closer to the engine. This means that hotter exhaust gases reach the converter and it heats up faster, but this may also reduce the life of the converter by exposing it to extremely high temperatures. Most car makers position the converter under the front passenger seat, far enough from the engine to keep the temperature down to levels that will not harm it. Preheating the catalytic converter using electric resistance heaters could be a good way of reducing emissions. Unfortunately, the 12-volt electrical system on most cars just does not provide enough energy to heat the catalytic converter fast enough.

HSW Tetraethyl lead

Tetraethyl lead (TEL) is an organometallic compound with the formula $(CH_3CH_2)_4Pb$.

It was discovered in 1921 by Thomas Midgley, working for General Motors Research. Due to its extreme toxicity, many early TEL researchers, including Midgley, became poisoned by lead and dozens died.

Tetraethyl lead was once used extensively as an additive in petrol for its ability to increase the octane rating of the fuel. TEL is still the most effective additive for increasing the octane rating of gasoline. A high enough octane rating is required to prevent premature detonation (or 'knocking'). Anti-knock agents allow the use of higher compression ratios for greater engine efficiency. One of the greatest advantages of TEL over other anti-knock agents is the very low concentration needed – typical formulations called for 1 part of TEL to 1260 parts of untreated petrol. Its use in petrol, particularly during the Second World War, was important.

The most important feature of the TEL molecule is the weakness of its four C–Pb bonds. At the temperatures found in internal combustion engines, $(CH_3CH_2)_4Pb$ reacts with oxygen and decomposes completely into carbon dioxide, water and lead:

$$(CH_3CH_2)_4Pb + 13O_2 \rightarrow 8CO_2 + 10H_2O + Pb$$

along with combustible, short-lived ethyl radicals.

The lead can oxidise further to give species such as lead(II) oxide:

$$2Pb + O_2 \rightarrow 2PbO$$

Lead and lead oxide remove radical intermediates in the combustion reactions. This prevents ignition of unburnt fuel during the engine's exhaust stroke. Lead itself is the reactive anti-knock agent.

The lead and lead(II) oxide would accumulate quickly and destroy an engine. For this reason, compounds such as 1,2-dibromoethane are used with TEL. This forms volatile lead(II) bromide, which is lost from the engine through the exhaust.

TEL is no longer used as a petrol additive in most of the world because of the toxicity of lead, and because lead compounds would poison the catalyst inside the converter of a modern car. It is still used as a fuel additive in piston-engined aircraft.

Questions

1 Catalytic converters could not be used if cars used leaded petrol. Suggest why.

2 Why are small quantities of a homogeneous catalyst sufficient for a reaction?

Investigating the activation energy of a reaction [4.3f(v), g]

Activation energy

You know that a certain amount of energy is needed before a reaction can occur. This is called the activation energy. You also know that the rate constant in a reaction is constant only at constant temperature. In this section, we are going to investigate the effect on the rate constant of changing the temperature.

HSW Distribution of particles of different energies in a gas

In the mid-nineteenth century, Ludwig Boltzmann and James Clark Maxwell were working quite independently of each other in Vienna, Austria, and Cambridge, England. The two scientists developed a statistical treatment of the distribution of energy amongst a collection of particles. This led to a greater understanding of the way in which the macroscopic behaviour of matter may be related to the microscopic particles of which it is composed, and to the development of the kinetic theory through a branch of the sciences now called statistical mechanics.

The work of Maxwell and Boltzmann on the distribution of the speeds of particles in a gas underpins much of the understanding of the rate at which chemical reactions occur. Yet Boltzmann's theories on the behaviour of matter were far from accepted at the time. Coming soon after the work of Darwin, many scientists saw Boltzmann's work as threatening the purposeful, God-given workings of the Universe, for if it could be shown that the behaviour of matter on a grand scale could be understood by studying its behaviour on a much smaller scale, what scope was left for the Creator? Stung by the scorn of his fellow scientists, Boltzmann committed suicide.

About 1930, Zartmann and Ko devised an experiment to measure the molecular speeds of gas molecules at different temperatures. They heated tin in an oven and directed the gaseous atoms towards a rotating disc with a slit in it. Any atoms that travelled through the slit hit a second disc behind the first and solidified on it. They found that the tin deposits were not uniform and had different thickness on different parts of the second disk. This meant that there was a spread of velocities and that their distribution was uneven.

Figure 1.1.23 shows the apparatus that Zartmann and Ko used. **Figure 1.1.24** shows the distribution velocities of particles in a gas at different temperatures.

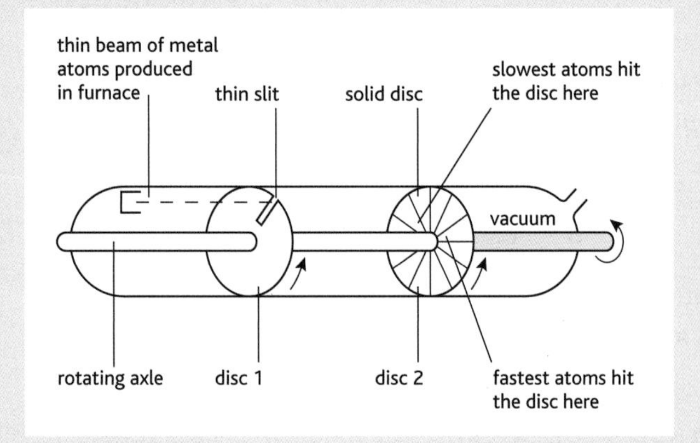

fig. 1.1.23 Zartmann and Ko's apparatus.

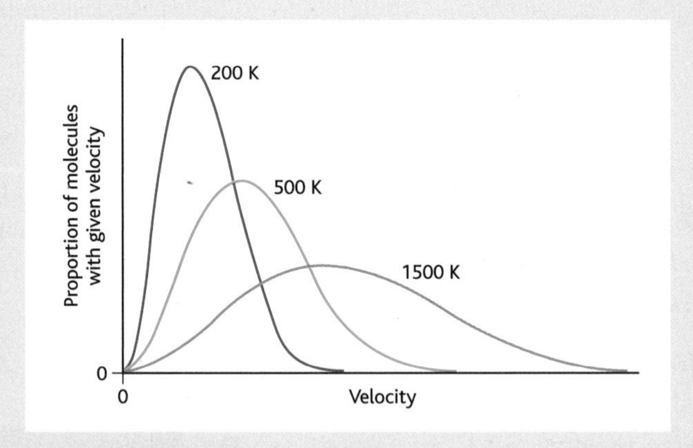

fig. 1.1.24 Zartmann and Ko's results.

Only when these experiments could be carried out could the truth of the work of Maxwell and Boltzmann be verified.

Figure 1.1.25 shows a distribution of the kinetic energies of particles in a sample of gas at a constant temperature. In this graph, E_A represents the activation energy. The area under the curve represents the total number of particles and the shaded area under the curve represents the number of particles with sufficient energy to react – that is, more than the activation energy.

$$\text{fraction of particles with greater than } E_A = \frac{\text{shaded area beneath curve}}{\text{total area beneath curve}}$$

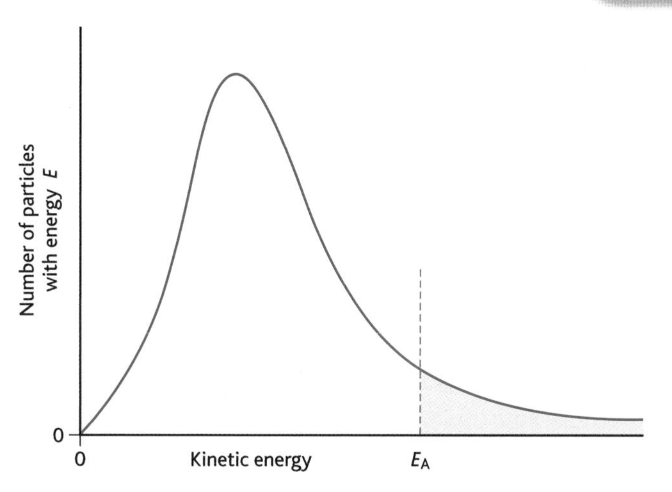

fig. 1.1.25 Distribution of the kinetic energies of particles in a sample of gas.

Using kinetic theory and probability theory, it can be shown that the fraction of molecules with greater energy than E_A J mol^{-1} is given by $e^{E_A/RT}$, where R is the gas constant (8.314 J K^{-1} mol^{-1}), T is the absolute temperature and e is the exponential constant.

From this we can say that at a given temperature:

$$\text{rate of reaction} \propto e^{E_A/RT}$$

But as the rate constant, k, is a measure of the rate of reaction we can write:

$$k \propto e^{E_A/RT} \text{ or } k = Ae^{E_A/RT}$$

A is called the **Arrhenius constant**. The relationship between temperature and rate constant was proposed by the Swedish chemist Svante Arrhenius. He applied natural logarithms throughout:

$$\ln k = -\frac{E_A}{RT} + \text{a constant}$$

A straight line can be represented by $y = mx + c$.
So, from $\ln k = -\frac{E_A}{RT} + \text{a constant}$, plotting a graph of $\ln k$ (that's y) against $\frac{1}{T}$ (that's x) gives a straight line graph, the gradient (m) of which is $-\frac{E_A}{R}$ (see **fig.1.1.26**).

From this value for the gradient, you can work out the value for E_A using the value $R = 8.314$ J K^{-1} mol^{-1}). Note also that this type of graph cuts the y-axis at a certain value – this is the value for the Arrhenius constant, A, for the reaction.

Activation energy

The oxidation of iodide ions by iodate(V) ions can be used to investigate the activation energy by graphical methods.

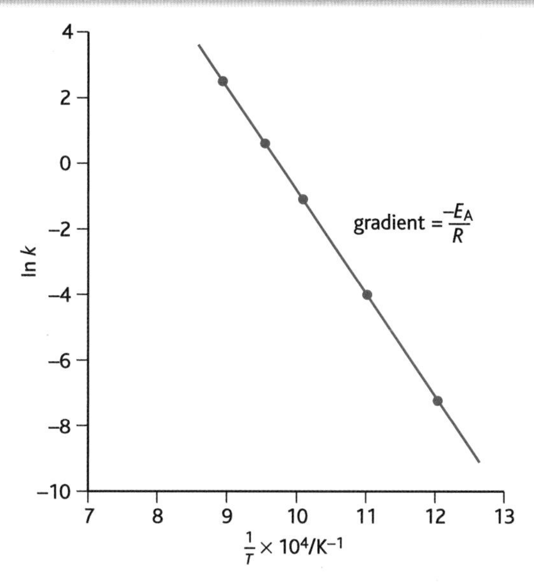

fig. 1.1.26 Working out the value of an activation energy.

Logarithms

You should know there are different types of logarithm:

- logarithms to base 10 (usually called 'logs') can be used to handle very large numbers (e.g. converting hydrogen ion concentrations to pH values)
- natural logarithms to base e (usually called 'ln') – these are used in physical chemistry in kinetics.

Natural logarithms are used in a similar way to logarithms to base 10.

Questions

1 Use the Arrhenius equation to explain:
   a the higher the temperature, the faster is the reaction
   b a reaction with a large activation energy has a slow rate of reaction.

2 The table gives the rate constant for the decomposition of hydrogen peroxide into water and oxygen at different temperatures.

Temperature/K	Rate constant/s^{-1}
295	4.93×10^{-4}
305	1.4×10^{-3}

Estimate the activation energy for the reaction.

Relating a mechanism to the rate-determining step [4.3a, f(iv), h, i, j]

You will know from AS Chemistry that increasing the concentration of a reactant often increases the rate of a chemical reaction. The exact relationship between rate of reaction and concentration can be determined experimentally.

Most chemical reactions take place in a series of steps rather than in a single step. The different steps in a multi-step process have different speeds – some are faster than others. In a multi-step reaction, the slowest step is called the rate-determining step and this determines the speed of the overall reaction.

HSW Supporting a theory

A reaction mechanism worked out for a reaction is only a theory, because in most cases chemists cannot detect or isolate the proposed intermediates. It has to be 'guessed' from the raw data. So to support a model of the mechanism of a particular reaction, the predicted rate law from the mechanism must match the rate law achieved from experimental data. If they don't match, however elegant the mechanism proposed, it is wrong!

The mechanism of substitution reactions

In *Edexcel AS Chemistry* (page 226) there are data to support a discussion of the nucleophilic substitution of two halogenoalkanes – bromoalkanes A and B – with hydroxide ions. These data are shown again in **tables 1.1.7** and **1.1.8**.

In the case of bromoalkane A, the rate of reaction is directly proportional to the concentration of A. So doubling the concentration of A doubles the rate of reaction, and tripling [A] triples the rate of reaction.

The rate of reaction is independent of the concentration of hydroxide ions, so increasing $[OH^-]$ does not make the reaction faster. This means that hydroxide ions do not feature in the rate-determining step.

When 2-iodo-2-methylbutane is refluxed with aqueous potassium hydroxide solution, the first step is the heterolytic fission of the carbon–iodine bond to form a tertiary **carbocation**. Tertiary carbocations are relatively stable. This step is slow and is the rate-determining step.

The second step involves a rapid reaction between the carbocation and hydroxide ions.

The process is summarised in **fig. 1.1.27**. The rate-determining step involves only the 2-iodo-2-methylbutane and so the rate equation does not involve hydroxide ions. This is called an S_N1 reaction (substitution / nucleophilic / first order or unimolecular).

$$CH_3-C(CH_3)(C_2H_5)-I \xrightarrow[\text{rate determining}]{\text{slow}} I^- + C^+(CH_3)_2(C_2H_5) \xrightarrow{+\ :OH^-} HO-C(CH_3)_2(C_2H_5)$$

fig. 1.1.27 The S_N1 substitution reaction mechanism.

Bromoalkane A

Concentration of bromoalkane A /mol dm^{-3}	Concentration of hydroxide ions /mol dm^{-3}	Initial rate /mol dm^{-3}s^{-1}
0.1	0.1	1.11×10^{-5}
0.2	0.1	2.22×10^{-5}
0.3	0.1	3.33×10^{-5}
0.1	0.2	1.11×10^{-5}
0.1	0.3	1.11×10^{-5}

table 1.1.7 Reaction rate data for bromoalkane A.

Bromoalkane B

Concentration of bromoalkane B /mol dm^{-3}	Concentration of hydroxide ions /mol dm^{-3}	Initial rate /mol dm^{-3}s^{-1}
0.1	0.1	1.50×10^{-5}
0.2	0.1	3.00×10^{-5}
0.3	0.1	4.50×10^{-5}
0.1	0.2	3.00×10^{-5}
0.1	0.3	4.50×10^{-5}

table 1.1.8 Reaction rate data for bromoalkane B.

In the case of bromoalkane B, the rate of reaction doubles when the concentration of B doubles (and the concentration of hydroxide ions is unchanged). It also doubles when $[OH^-]$ doubles (and [B] is unchanged). This can be expressed as:

$$\text{rate} \propto [B][OH^-]$$

This can then be represented as a rate equation:

$$\text{rate} = k[B][OH^-]$$

The kinetics of this reaction suggest that the rate-determining step involves both the halogenoalkane and the hydroxide ion.

An example of this mechanism is the reaction of bromomethane with aqueous potassium hydroxide solution. The explanation for this was first given by Christopher Ingold. He suggested that the hydroxide ion joined onto the central carbon atom at the same time as the bromine atom was leaving. Part of the energy required to break the C–Br bond was supplied by the energy released on producing the C–OH bond. Calculations show that the approach of the hydroxide ion along the line of centres of the carbon and bromine atoms is that of lowest energy requirement.

This process is summarised in **fig.1.1.28**. The rate-determining step involves both bromomethane and hydroxide ions. This is called an S_N2 reaction (substitution / nucleophilic / second order or bimolecular).

$$HO^- + H_3C^{\delta+}\text{—}Br^{\delta-} \rightarrow [HO^{\delta-}\cdots CH_3 \cdots Br^{\delta-}] \rightarrow HO\text{—}CH_3 + Br^-$$

fig. 1.1.28 The S_N2 substitution reaction mechanism.

Studying kinetics alone can, in some cases, be insufficient to be sure just which mechanism is being followed – unless the reaction is studied thoroughly under more than one set of conditions.

For example, the hydrolysis of a halogenoalkane R–X in water can lead to confusing results. Water acts as a nucleophilic reagent. If the reaction is S_N2 the rate equation would be:

$$\text{rate} = k[RX][H_2O]$$

But the water is in excess so $[H_2O]$ is effectively constant – so the rate equation would become:

$$\text{rate} = k'[RX]$$

This might lead you to think the reaction was S_N1.

HSW Applications, misunderstandings and risks

It can be difficult to see the relevance of these very theoretical mechanisms, but in fact their application plays a vital role both in many industrial processes and in the pharmaceutical industry. Misunderstanding of the way that a reaction mechanism works can lead to major problems in an industrial process – or to a catastrophic change in a drug. For example, the thalidomide tragedy when thousands of children around the world were born with reduced or missing limbs was in part the result of a lack of understanding of the mechanism by which the drug was made. You will learn more about this in **chapter 1.6**.

Questions

1 The reaction between nitrogen dioxide and carbon monoxide fits the following overall equation:

$$NO_2(g) + CO(g) \rightarrow NO(g) + CO_2(g)$$

a If the rate equation suggested a single-step mechanism, what would be the rate equation?

b The rate equation for this reaction is in fact:

$$\text{rate} = k[NO_2]^2$$

A possible mechanism is:

$$2NO_2(g) \rightarrow NO_3(g) + NO(g) \quad \text{step 1}$$

$$NO_3(g) + CO(g) \rightarrow NO_2(g) + CO_2(g) \quad \text{step 2}$$

Which of these steps do you think is the rate-determining step? Explain your answer.

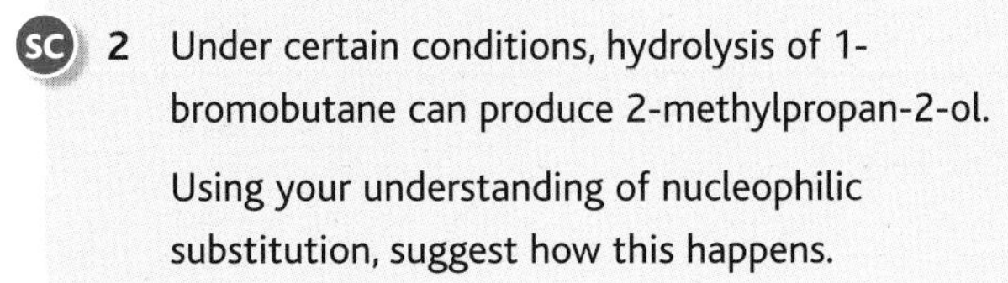

SC 2 Under certain conditions, hydrolysis of 1-bromobutane can produce 2-methylpropan-2-ol.

Using your understanding of nucleophilic substitution, suggest how this happens.

The mechanism of the reaction of iodine with propanone [4.3e]

Iodine and propanone react, in the presence of acid, in a **substitution reaction** to form iodopropanone. The reaction can be represented by:

$$CH_3COCH_3(aq) + I_2(aq) \rightarrow CH_2ICOCH_3(aq) + H^+(aq) + I^-(aq)$$

Kinetic experiments can be carried out using different concentrations of propanone, iodine and hydrogen ions.

HSW

Studying the kinetics of this reaction enables you to understand the mechanism of the reaction. You would never guess that the rate of reaction is determined by the concentration of hydrogen ions and not iodine.

This experiment can be carried out in different ways.

Using a colorimeter

A colorimeter measures the absorption of light during the progress of the experiment. First, the colorimeter has to be calibrated using standard iodine solutions. The reaction mixture containing iodine will be light brown in colour and as the reaction proceeds the solution becomes paler and more light is transmitted (**fig. 1.1.29**).

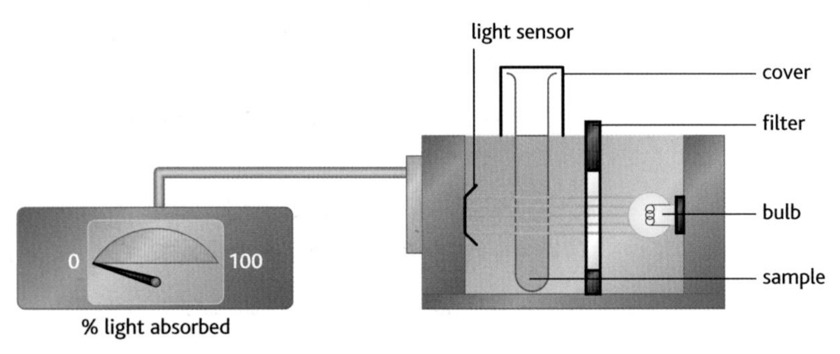

fig. 1.1.29 Using a colorimeter to monitor the progress of a reaction.

Experiment	$[CH_3COCH_3]$ /mol dm^{-3}	$[I_2]$/mol dm^{-3}	$[H^+]$/mol dm^{-3}	Relative rate
1	0.1	0.1	0.1	2
2	0.1	0.2	0.1	2
3	0.1	0.2	0.2	4
4	0.2	0.1	0.1	4
5	0.1	0.1	0.2	4

table 1.1.9 The rate of reaction depends on the concentration of propanone and on that of hydrogen ions, but not on the iodine concentration.

By titration

During the experiment small aliquots (samples) of the reaction mixture are removed with a pipette.

The withdrawn sample is put in a flask and excess sodium hydrogencarbonate is added. This effectively stops the reaction so no further change in iodine concentration occurs during titration. The aliquots are titrated with a standard solution of sodium thiosulfate using starch indicator near the end-point. The measured values for the $[I_2]$ can be used to calculate the concentrations of the other substances involved. **Table 1.1.9** summarises the results of such an experiment. We can conclude that:

- doubling the concentration of iodine has no effect on the relative rate of the reaction (experiments 1 and 2)
- doubling the concentration of propanone doubles the relative rate of the reaction (experiments 1 and 4)
- doubling the concentration of hydrogen ions doubles the relative rate of the reaction (experiments 1, 3 and 5).

Using this we can write the rate equation:

$$\text{rate} = k[CH_3COCH_3][H^+]$$

The reaction is first order with respect to propanone and hydrogen ions, but zero order with respect to iodine. The overall order of the reaction is two. What does this tell us about the mechanism of the reaction?

The reaction involves various steps and the slowest step, the rate-determining step, does not involve iodine. This is why iodine does not appear in the rate equation. Hydrogen ions act as a catalyst – they are regenerated during the reaction.

A possible mechanism for the reaction of iodine and propanone in acid solution is given below.

Step 1

An H^+ ion protonates the oxygen atom in propanone:

$$(CH_3)_2C{=}O + H_3O^+ \rightleftharpoons (CH_3)_2C{=}O^+H + H_2O$$

This is a reversible reaction involving proton transfer (acid–base reaction). Remember that a protonated water molecule, H_3O^+, behaves as an H^+ ion.

Step 2

The electrons in the C=O bond partly shift to form a carbocation – i.e. the positive charge is transferred from the oxygen to the carbon:

$$(CH_3)_2C{=}O^+H \rightleftharpoons (CH_3)_2C^+{-}OH$$

Step 3

This carbocation loses a proton and *slowly* changes into the ***enol*** form. The enol has both alkene and alcohol functional groups and is **isomeric** with the original ketone:

$$(CH_3)_2C^+{-}OH + H_2O \rightleftharpoons CH_3C(OH){=}CH_2 + H_3O^+$$

This involves breaking a strong C–H bond, hence this step has a high activation energy and slow speed. The positive charge on the adjacent carbon of the carbocation facilitates in 'pulling' the C–H bond pair to form the C=C bond and releases the proton to form an H_3O^+ ion.

The rate of formation of the enol thus depends on the concentrations of the ketone and the acid.

Step 4

The iodine molecule acts as an **electrophile** and undergoes a quick **electrophilic addition reaction** (like other alkenes). This produces a protonated iodoketone:

$$CH_3C(OH){=}CH_2 + I_2 \rightarrow CH_3C({=}O^+H){-}CH_2I + I^-$$

Step 5

A water molecule then rapidly removes the proton in another acid–base reaction to form the iodoketone:

$$CH_3C({=}O^+H){-}CH_2I + H_2O \rightarrow CH_3COCH_2I + H_3O^+$$

The whole process is summarised in **fig.1.1.30**.

In this example the slow, rate-determining steps 1, 2 and 3 are the first stage in the reaction and need only CH_3COCH_3 and H^+. The remaining steps happen very quickly. This is also the case when oxygen and hydrogen bromide react together at 700 K. However, the slowest reaction is not always the first.

In the reaction between bromide ions and bromate(V) ions in acid solution (see pages 20–21) the most likely mechanism is that HBr and $HBrO_3$ are made very rapidly before the third, relatively slow, reaction between the two of them takes place. This is the rate-determining reaction, and is followed by two more rapid steps to complete the reaction.

Questions

1 These questions refer to the methods used to carry out the reaction of iodine and propanone in acid solution.

 a Suggest advantages of the method that uses a colorimeter.

 b In the titration method, why is a pipette used to remove aliquots?

 c Why does sodium hydrogencarbonate effectively stop the reaction?

2 Why does I_2 not appear in the rate equation?

fig. 1.1.30 Mechanism for the iodination of propanone.

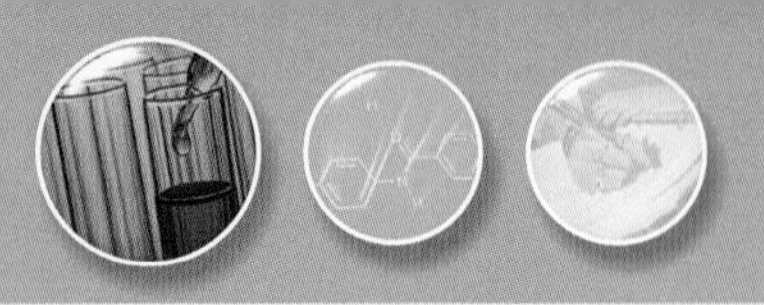

1.2 How far? Entropy

fig. 1.2.1 Fruit goes mouldy, a tidy bedroom becomes a jungle of clothes, dead bodies decay, iron goes rusty – in life as in science, chaos rules!

fig. 1.2.2 Gas jars containing bromine and air, used to demonstrate diffusion.

What is entropy? [4.4b, c, d]

The interaction of energy and matter governs our lives. We take for granted our ability to use the chemicals in batteries to produce sound from an MP3 player. We get into a car and use petrol to travel in hours distances that would have taken our ancestors days. Energy and matter interact on other scales too – from the remnants of the explosion of a star hundreds of light years in diameter, to the nucleus of an atom 10^{-9} mm in diameter. The way in which energy and matter behave affects the entire universe.

The importance of this interaction between energy and matter has not always been recognised. The understanding of this area of science grew out of the Industrial Revolution and the rise of the steam engine as a means of power. Between them, the steam engine and the atom gave rise to a branch of science which is a powerful tool for interpreting what we see around us. This is **thermodynamics** and it can tell us a great deal about why chemical changes take place.

Thermodynamics deals with the laws of heat energy, and its transfer into other types of energy. The first law of thermodynamics tells us that energy can never be created or destroyed. The second law states that **entropy** always increases in the transformation of energy. In everyday understanding, entropy can be thought of as a tendency for everything to move from a state of order to chaos as objects (whether they are atoms or the things in your bedroom) left to themselves mix and randomise themselves as much as they can. Cars rust, hot things cool down – and the universe is on an irreversible journey towards identical nothingness!

Entropy is sometimes described as a measure of the order or disorder in a system. This definition is rather simplistic and has to be used with care. It refers not only to the distribution of particles but also to the ways of distributing the energy of the system in all of the available energy levels.

The units of entropy, S, are $\text{J mol}^{-1}\text{K}^{-1}$.

To help you understand entropy, it helps to consider what happens during **diffusion**.

Diffusion

Start by considering the diffusion of gases. In fig. 1.2.2 there are two gas jars – one contains bromine gas and the other air.

Imagine what happens when the cover slip between the gas jars shown in fig. 1.2.2 is removed. Our experience tells us that, given long enough, we will find the bromine vapour spread evenly throughout the two gas jars – we would certainly *not* expect to find it all in one jar!

In order to convince ourselves of *why* this is so, it is best to look at a much simpler situation where many fewer particles are involved. We can simplify this by imagining starting off with only five bromine particles in the left-hand jar, rather than the 10^{22} or so that there must actually be in the jar in the photograph. Figure 1.2.3 shows the result.

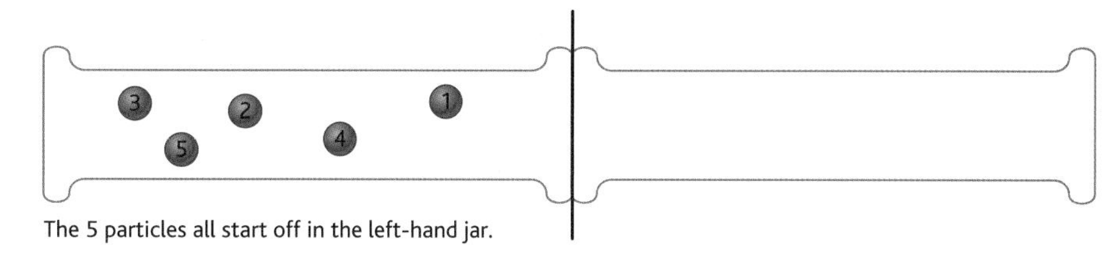

The 5 particles all start off in the left-hand jar.

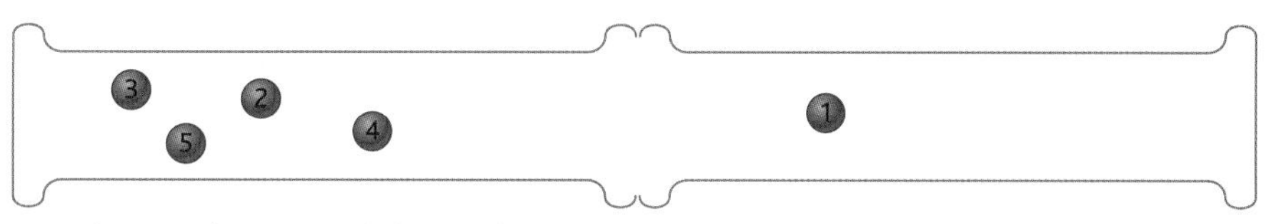

Once the cover slip is removed, the particles are free to move between the jars.

fig. 1.2.3 Diffusion in a gas – one of the ways the particles *might* rearrange. Remember that air molecules are also present but have not been shown.

Each particle has two possible ways of being arranged – in the left-hand jar or in the right-hand jar. If we represent the number of possible arrangements for particle 1 as W_1, for particle 2 as W_2 and so on, then the *total* number of ways, W, that the five particles can be arranged between the two jars is given by:

$$W = W_1 \times W_2 \times W_3 \times W_4 \times W_5$$
$$= 2 \times 2 \times 2 \times 2 \times 2$$
$$= 2^5$$
$$= 32$$

Since only *one* of these 32 ways results in all the particles being in the left-hand gas jar, it would be surprising if this arrangement were to happen very often. Try it yourself – draw 32 joined gas jars and work out every possible arrangement of the five particles!

If we increase the number of particles in the gas jars to 50, W becomes 2^{50}, or about 10^{15}. And for a real situation where there will be something like 10^{22} particles, W has an enormous value – about $2^{10^{22}}$ – and only 2 of these arrangements have all the particles in one jar; the majority have the particles spread more or less evenly between the jars.

To get some idea of what this means, imagine that it is possible to make a note of the position of all the particles in the two jars once every second. Only *once* in every $2^{10^{22}}$ seconds would we be likely to see all the particles in the left-hand jar. The age of the Earth is thought to be something like 10^{17} seconds – so we would have a *very long* wait indeed, and we are quite justified in saying that gases *always* spread out.

In the same way that the spreading out of particles in diffusion represents an increase in entropy because the particles have more disorder, the spreading out of heat energy represents an overall increase of entropy too.

Investigations about the way atoms behave when they interact with energy sources lead scientists to believe that atoms do not deal in energy in 'any old amounts' but only in set quantities or **quanta**. You have already met the idea that energy changes take place when electrons within an atom move between energy levels. However, atoms and molecules can change their energy states in other ways – for example, molecules can vibrate, rotate and translate (move about) as well (fig. 1.2.4). All of these movements involve energy, and the energy for all of these changes comes in quanta.

translational motion: molecules move from one place to another

rotational motion: molecule spins

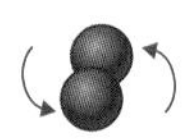

vibrational motion: molecule moves about the same point

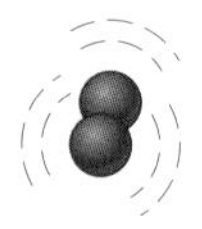

fig. 1.2.4 Ways in which energy can be distributed in a diatomic molecule.

A diatomic molecule, such as hydrogen, will have definite **vibrational energy levels**, which can be imagined rather like the rungs of a ladder. At any given point in time, a single molecule will have a set amount of vibrational energy, which will depend on the number of quanta of energy it possesses. If there are two molecules they may exchange energy quanta, so that their energy levels change (fig. 1.2.5).

(a) Each H_2 molecule has two quanta of vibrational energy.

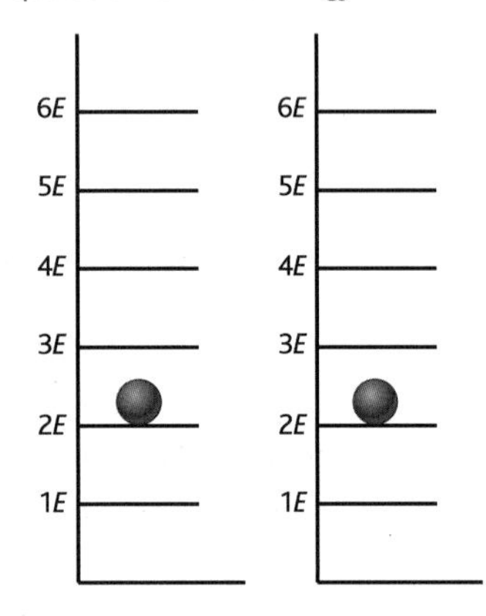

(b) One molecule loses a quantum, the other gains it.

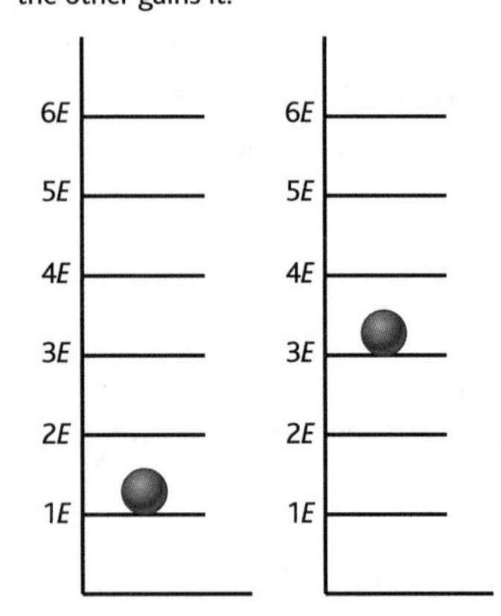

fig. 1.2.5 Model to show the transfer of one quantum of energy from one hydrogen molecule to another.

As you can imagine, there are several ways in which these molecules could share the four quanta of energy – can you work out all five of them? If there are more quanta involved, there are more arrangements possible. And if there are more *molecules* involved, the possibilities get larger again. Sharing energy quanta increases the number of energy states, and so the number of ways in which atoms and molecules can arrange themselves and their energy also increases.

This can be represented mathematically. If W represents the number of ways of arranging the energy, N represents the number of atoms and q represents the number of quanta of energy, then:

$$W = \frac{(N + q - 1)!}{(N - 1)! \times q!}$$

You will appreciate W will be very large with large numbers of particles.

Factorials

Remember that 'three factorial' (3!) means $3 \times 2 \times 1 = 6$. You can work this out on your scientific calculator with the $n!$ button.

If $N = 12$ and $q = 12$:

$$W = \frac{23!}{11! \times 12!}$$

$$= 1\,352\,078$$

HSW Entropy of perfect crystals at absolute zero

There are three laws of thermodynamics. The third law states that the entropy of a perfect crystal at absolute zero (0 K) is zero. This means that in a perfect crystal, at absolute zero, nearly all molecular motion should cease in order to achieve $\Delta S = 0$. A 'perfect' crystal is one in which the internal lattice structure is the same at all times – in other words, it is fixed and non-moving, and the particles are neither rotating or vibrating in their positions. It has neither rotational or vibrational energy. There is only one arrangement in which this order can be attained. Then every particle of the structure is in its proper place.

However, the equation we have used above shows that even when the vibrational quantum number is 0, a molecule still has vibrational energy. This means that no matter how cold the temperature gets, the molecule will vibrate. This is in keeping with the Heisenberg uncertainty principle, which states that both the position and the momentum of a particle cannot be known precisely at any given time.

The closest we can get to an entropy value of zero is a flawless diamond cooled in solid helium.

Changes of entropy with change of state

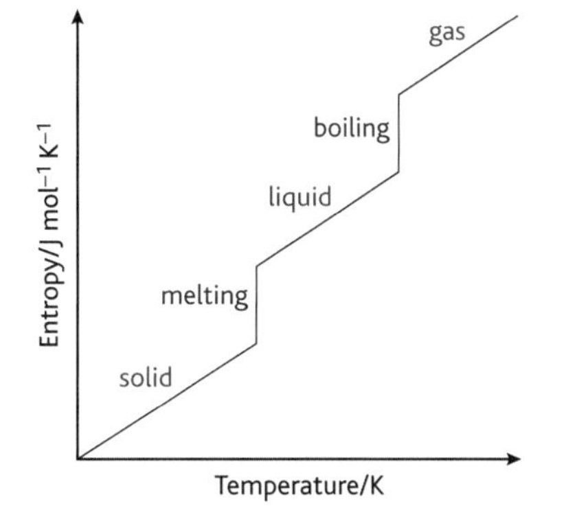

fig. 1.2.6 The changes in entropy that accompany changes in state.

Figure 1.2.6 shows how entropy changes as a solid is changed – first to a liquid, and then to a gas.

When a solid is heated its particles vibrate more and this is shown by the positive slope on the first section of the graph. When melting occurs there is a breakdown in the solid structure and this is accompanied by a rapid increase in entropy. A similar pattern is seen when the liquid is heated – first the energy of the particles increases, and then, at the boiling temperature, the particles break away from each other and there is a further big increase in disorder and entropy. Beyond the boiling temperature the entropy increases steadily. However, it is important to remember that entropy isn't just about the arrangement of the particles – the way energy is *distributed* among them matters as well.

Questions

1 Is it reasonable to expect a cold cup of tea to warm up spontaneously? Explain your answer.

2 Tidying a room reduces its entropy. Does this process contravene the second law of thermodynamics?

3 Arrange the following in ascending order of entropy:
steam at 110 °C; ice at –10 °C; water at 30 °C.

The natural direction of change [4.4e, f]

Spontaneous or not?

A spontaneous reaction is one that will occur on its own, without any external influence. However, it may be fast *or* slow. Spontaneity gives no indication of the rate.

What determines whether a reaction will actually occur spontaneously or not?

Table 1.2.1 gives the **enthalpy** change for three reactions.

Reaction	$\Delta H^\ominus$/kJ mol^{-1}
$Na(s) + ½Cl_2(g) \rightarrow NaCl(s)$	–411.2
$H_2(g) + ½O_2(g) \rightarrow H_2O(g)$	–242.0
$CH_4(g) + 2O_2(g) \rightarrow CO_2(g) + 2H_2O(l)$	–1461.9

table 1.2.1 Enthalpy changes for three exothermic reactions.

All three of these reactions are **exothermic** with a considerable amount of energy being released.

Just as a stone tends to run downhill, *energy tends to run downhill*. The low energy state of the products is more stable than the higher energy state of the reactants – everything always likes to go towards the most stable, or lowest energy, state. You will know from *Edexcel AS Chemistry* (page 200) that it is not as simple as this because of activation energy. Figure1.2.7 shows a simple energy diagram for an exothermic reaction.

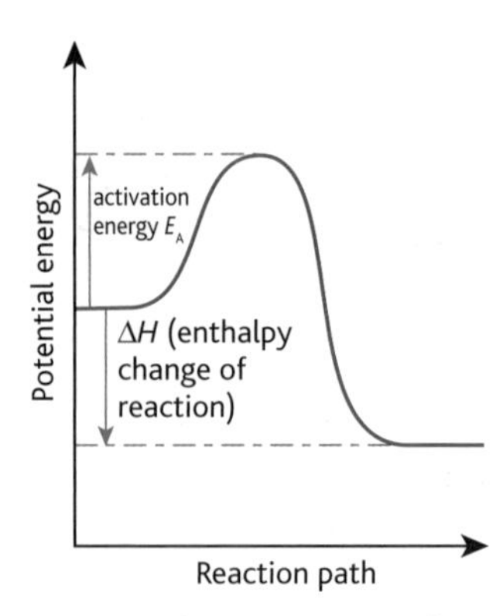

fig. 1.2.7 Before reactants can become products, the activation energy has to be supplied.

There are some **endothermic** reactions that are spontaneous. For example:

$$Br_2(l) + Cl_2(g) \rightarrow 2BrCl(g) \ \Delta H^\ominus = +29.3\,kJ\,mol^{-1}$$

$$H_2O(s) \rightarrow H_2O(l) \ \Delta H^\ominus = +6.01\,kJ\,mol^{-1}$$

You will notice that both of these values are small compared to the values in table 1.2.1. They do not require much energy. If a large amount of energy must be supplied (i.e. a reaction is highly endothermic), the reaction is not likely to be spontaneous.

Therefore, it is clear that whether a reaction is spontaneous or not depends on more than just the enthalpy change. But the enthalpy change is, without a doubt, one of the most important factors determining the spontaneity of a reaction.

The other thing that favours a reaction being spontaneous is an increase in the system's **entropy**.

- All reactions that are highly exothermic *and* lead to greater disorder are spontaneous at all temperatures.
- Any reaction that is neither of these is never spontaneous.

Changes in entropy with change of state

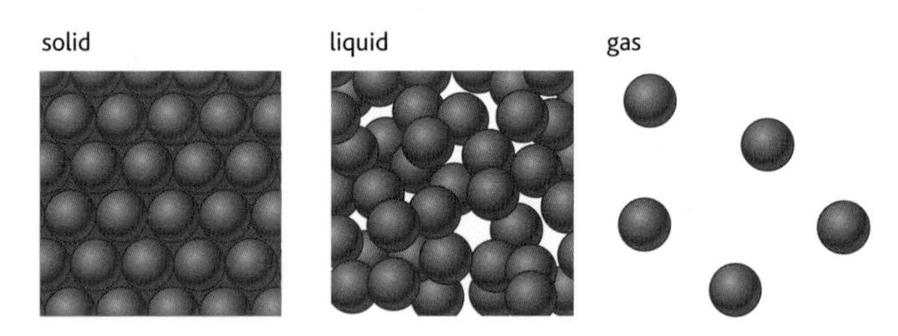

fig. 1.2.8 The arrangement of particles in a crystalline solid, a liquid and a gas.

In a solid, the particles are tightly bound together in fixed positions – the particles have little opportunity to move, apart from vibrations. The standard molar entropy of the solid is low. In diamond, all the carbon atoms are tightly bonded and so little movement of the atoms is possible. In lead, the metallic bonds are not so directional and the atoms can vibrate more. Lead ($S^\ominus = 64.8\,J\,mol^{-1}\,K^{-1}$) has a higher standard molar enthalpy than diamond ($S^\ominus = 2.4\,J\,mol^{-1}\,K^{-1}$).

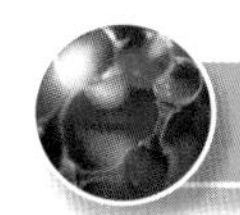

HSW Early ideas of spontaneity

The Danish chemist Julius Thomsen (1826–1909) was the son of a bank auditor. He was born in Copenhagen, Denmark, on 16 February 1826. His academic life centred around Copenhagen. His master's degree was from the University of Copenhagen (1843), where he joined the faculty teaching chemistry, becoming a professor in 1866.

Thomsen carried out many thermochemical measurements and made tables of the amount of heat energy released or absorbed in approximately 3500 chemical reactions. He insisted on carrying out these experiments himself to ensure the consistency of his results. Similar measurements were being made by the French chemist Marcelin Berthelot.

They both regarded the release of heat energy to be the driving force of a chemical reaction. Although we now know that this is not always true, it seemed plausible to them by observing chemical reactions that readily released heat energy (exothermic reactions). Thus, they regarded chemical reactions as either 'spontaneous' or 'not spontaneous' according to their transfer of heat energy. This was called Thomsen's hypothesis.

A spontaneous chemical reaction would be one that moved in only one direction, and would give off heat energy while doing so. Conversely, if such a reaction were to be forced into reverse then it would be necessary to absorb heat energy.

For example, hydrogen and oxygen combine to form water; conversely, water can decompose into hydrogen and oxygen. The formation of water gives off a great deal of heat energy very rapidly, even explosively. Such a reaction would be regarded as spontaneous since, once started, it goes rapidly to completion.

fig. 1.2.9 Julius Thomsen.

Conversely, for water to decompose into hydrogen and oxygen requires an input of energy (e.g. electrical energy). The breakdown of the water molecule would be regarded as not spontaneous, because it does not occur at all unless there is a continuous supply of energy and, once started, the reaction ceases immediately without the energy input.

We now know that this simplified idea is incorrect because:

- there are reactions that are spontaneous but do not give out energy, which are accompanied by a drop in temperature
- there are reversible reactions.

A liquid generally has a higher standard molar entropy than the corresponding solid. This is because the particles can vibrate, rotate and move around in more ways. When the solid melts, there are more ways of distributing the particles and energy. The liquid substance has more disorder and a higher entropy.

When the liquid boils and turns to a gas, the particles become widely spaced and free to move. The system has a higher entropy as a gas because the particles are more disordered.

When a solid such as sodium chloride is dissolved in water, there is an increase in entropy because the ions in the crystal are now free to move throughout the liquid.

Questions

1 In each of the following pairs of substances, choose the one you would expect to have the higher standard molar entropy at 298 K.

a $CO_2(g)$ and $CO_2(s)$

b NaCl(s) and NaCl(aq)

c $H_2O(l)$ and $H_2O(s)$

2 State whether the entropy of the system will increase or decrease during each of the following reactions. Explain your choices.

a $2Na(s) + Cl_2(g) \rightarrow 2NaCl(s)$

b $NaCl(s) + aq \rightarrow NaCl(aq)$

c $CaCO_3 \rightarrow CaO(s) + CO_2(g)$

Changes in entropy during chemical reactions [4.4a, g(i) – (iv)]

Exothermic and endothermic reactions

At one time, chemists believed all spontaneous reactions to be exothermic. However, the fact that some spontaneous reactions occur and take in energy from the surroundings suggests that the truth is more complicated. Looking at whether an enthalpy change is positive or negative and by how much it changes does not enable you to decide if a reaction will take place.

This is because entropy plays a part as well – the natural direction of change is the one that gives an increase in entropy, and that entropy involves both the physical state and the arrangement of particles in a substance. Add in the fact that the entropy of the surroundings, as well as that of the reacting system, plays a part and it becomes even more difficult to predict just what is going to happen.

However, as you will see, chemists do have ways of predicting what is likely to happen – and of making reactions move in the desired direction!

Exothermic or endothermic?

You will probably carry out a series of test tube reactions and be asked to estimate if the possible arrangements of the product particles represent a more ordered or a less ordered system than the original reagents. In other words, if the entropy of the system has increased or decreased.

You should also classify the reactions as exothermic or endothermic. If the temperature of the reaction mixture increases during the reaction it is exothermic, and if it decreases it is endothermic.

Investigating chemical reactions

You can apply some basic principles that suggest the direction of the entropy change in a reaction.

- Ions and molecules usually have higher entropies in solution than they do as solids.
- Gases usually have higher entropies than liquids or solids, so if a gas is produced during a reaction it is likely that the overall entropy will increase.
- When large molecules break down into smaller molecules entropy increases. This is because there are far more ways of arranging several small molecules than one big one.

Dissolving a solid

When ammonium nitrate dissolves in water (fig. 1.2.10), the temperature of the solution decreases – the process is endothermic, taking in energy from the surroundings:

$$NH_4NO_3(s) + aq \rightarrow NH_4^+(aq) + NO_3^-(aq) \qquad \Delta H \text{ is positive}$$

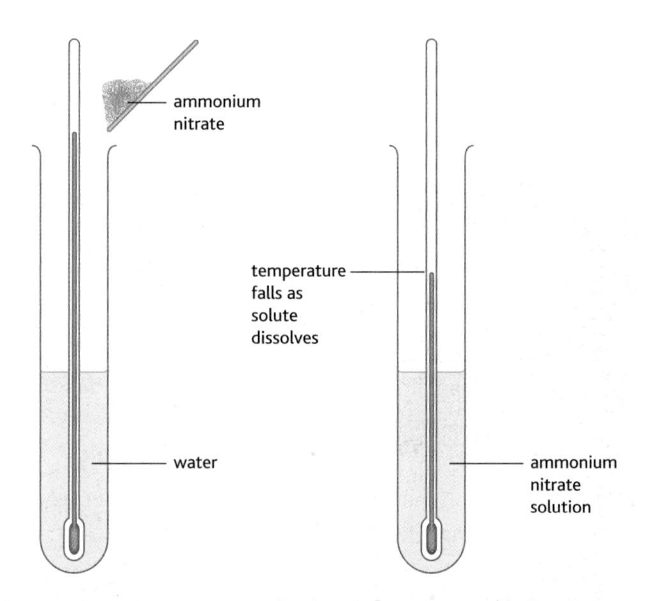

fig. 1.2.10 Ammonium nitrate dissolves in water in an endothermic process and the entropy of the system increases.

When a solid dissolves in a solvent, the level of disorder increases. This is because the particles are in fixed positions in the solid, but become free to move around in solution. However, particularly when ionic solids dissolve, while bonds between the particles are broken, increasing disorder and taking in energy from the surroundings, bonds between the particles and the solvent are made, reducing disorder and releasing energy. So it is hard to *predict* whether the process will be exothermic or endothermic.

Production of gas

The entropy of a gas, with its freely moving particles, is higher than the entropy of corresponding solids and liquids. For example, when ethanoic acid is reacted

with ammonium carbonate, carbon dioxide gas is one of the products:

$$(NH_4)_2CO_3(s) + 2CH_3COOH(aq) \rightarrow 2CH_3COO^-(aq) + 2NH_4^+(aq) + CO_2(g) + H_2O(l)$$
(ΔH is positive)

There is a slight fall in temperature during the reaction telling you that the process is endothermic, taking in energy from the surroundings. However, the particles are well-ordered in the solid, and the disorder increases because a solution and, especially, a gas is formed – so entropy increases during the reaction.

Exothermic reactions that produce a solid

Burn magnesium in air and there is ample evidence of the release of a large amount of energy (fig. 1.2.11):

$$2Mg(s) + O_2(g) \rightarrow 2MgO(s) \quad \Delta H \text{ is negative}$$

In any reaction in which a gas reacts to form a solid, you can expect the entropy of the system to decrease because solids have considerably more order than gases.

fig. 1.2.11 Magnesium burning in air.

Endothermic reactions between two solids

When barium hydroxide and ammonium chloride are mixed, they form a paste and the temperature drops significantly (fig. 1.2.12). This is an endothermic process and ammonia gas can be detected by its smell:

$$2NH_4Cl(s) + Ba(OH)_2.8H_2O(s) \rightarrow BaCl_2.2H_2O(s) + 2NH_3(g) + 8H_2O(l)$$

When two solids are mixed and react together, the entropy change will depend on the physical state of the compounds made and not just on the energy changes in the reaction. Two solids represent a very ordered system with relatively low entropy. If a liquid or, especially, a gas results from the reaction, then the entropy will have increased due to the very disordered arrangement of gas particles.

fig. 1.2.12 The beaker containing the mixture has frozen some water underneath it, enabling the tin to be picked up.

Questions

1 Suggest whether the following processes involve an increase or decrease in entropy of the system and in enthalpy. Explain your answers.

a $NaCl(s) + aq \rightarrow Na^+(aq) + Cl^-(aq)$

b $2H_2(g) + O_2(g) \rightarrow 2H_2O(l)$

c $H_2O(l) \rightarrow H_2O(g)$

d $2Na(s) + Cl_2(g) \rightarrow 2NaCl(s)$

e $NH_4Cl(s) \rightarrow NH_3(g) + HCl(g)$

f $C_3H_8(g) + 5O_2(g) \rightarrow 3CO_2(g) + 4H_2O(g)$

2 a Copy and complete the following table:

Reaction	Enthalpy change	Entropy change of system
Dissolving ammonium nitrate in water	endothermic	increase
Ammonium carbonate reacting with ethanoic acid		
Magnesium burning in air		

b Explain why the enthalpy change of a reaction and its entropy change are not the same.

Calculating entropy changes [4.4h, i, j]

The reaction of sodium and chlorine to form sodium chloride is an exothermic process that involves a decrease in entropy of the system:

$$2Na(s) + Cl_2(g) \rightarrow 2NaCl(s)$$

We know that this is a chemical reaction that definitely does occur, and with the release of a considerable amount of energy, provided that there is a flame to supply the necessary activation energy for the reaction. How can a change in which there is a decrease in entropy of the system be spontaneous?

The answer to this question is that we have considered only the entropy change of the system, ΔS_{sys}. We must also think about the change in entropy of the surroundings, ΔS_{surr}, brought about by energy leaving the system in an exothermic reaction (fig. 1.2.13).

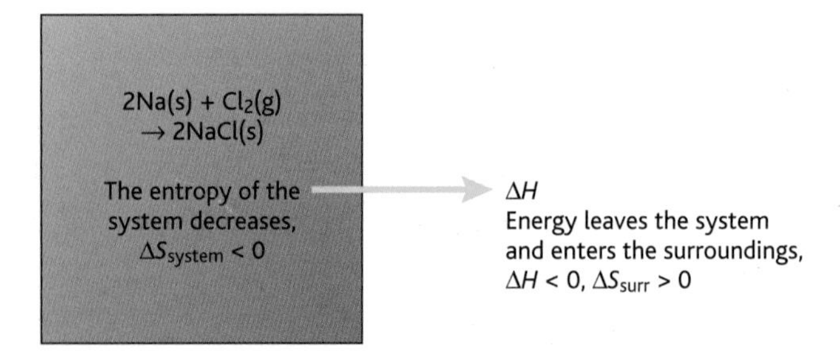

fig. 1.2.13 In this reaction, energy leaves the system. This spreading out of energy brings about an increase in entropy that more than offsets the decrease in the entropy of the system.

Despite the fact that there is a decrease in entropy of the *system* in this reaction, the energy leaving the system produces a substantial increase in entropy of the *surroundings*, because there are more ways of arranging the energy quanta in the rest of the universe than there are of arranging them in the system alone. It is the **total entropy change** which determines whether this reaction is spontaneous or not:

$$\Delta S_{total} = \Delta S_{sys} + \Delta S_{surr}$$

In the case of the reaction of sodium and chlorine, the release of large amounts of heat energy into the universe makes ΔS_{surr} very positive, which is more than enough to offset the negative value of ΔS_{sys}.

Calculating the total entropy change for a reaction

Calculating ΔS_{system}

Finding ΔS_{sys} for a change is generally quite easy – we can look up the standard entropy for the substances on each side of the equation and calculate the difference:

$$\Delta S^{\ominus}_{sys} = \Sigma S^{\ominus}_{products} - \Sigma S^{\ominus}_{reactants}$$

We use data like those given in table 1.2.2.

Substance	$S^{\ominus}$/J mol^{-1}K^{-1}	Substance	$S^{\ominus}$/J mol^{-1}K^{-1}
C(diamond)	2.4	$N_2(g)$	191.4
C(graphite)	5.7	$H_2(g)$	130.6
$O_2(g)$	205.0	Na(s)	51.0
$Cl_2(g)$	223.0	NaCl(s)	72.4
$CO_2(g)$	214.0	$NH_3(g)$	192.0

table 1.2.2 Standard entropies of some substances.

For example, to calculate the standard entropy change for the synthesis of 1 mole of ammonia:

$$N_2(g) + 3H_2(g) \rightarrow 2NH_3(g)$$

$$\Delta S^{\ominus}_{sys} = \Sigma S^{\ominus}_{products} - \Sigma S^{\ominus}_{reactants}$$
$$= (2 \times 192.0) - (191.4 + 3 \times 130.6)$$
$$= -199.2\,\text{J mol}^{-1}\,\text{K}^{-1}$$

There is a reduction in entropy during the reaction. This is to be expected because there are fewer gas molecules in total in the products and therefore less disorder.

However, note that this is the entropy change for the formation of 2 moles of ammonia – so the required answer is half of this, $-99.6\,\text{J mol}^{-1}\,\text{K}^{-1}$.

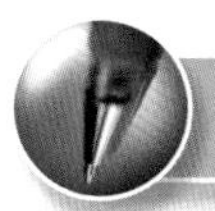

Worked example

Calculate the entropy change in making 1 mole of sodium chloride from its elements:

$Na(s) + \frac{1}{2}Cl_2(g) \rightarrow NaCl(s)$

The entropy change of the system is given by:

$$\Delta S^\ominus_{sys} = \Sigma S^\ominus_{products} - \Sigma S^\ominus_{reactants}$$
$$= 72.4\,J\,mol^{-1}\,K^{-1} - (51.0 + \tfrac{1}{2} \times 223.0)\,J\,mol^{-1}\,K^{-1}$$
$$= -90.1\,J\,mol^{-1}\,K^{-1}$$

Calculating $\Delta S_{surroundings}$

To calculate the entropy change of the surroundings, we need to know the energy transferred to them, which is given by the change in enthalpy, ΔH. The entropy change is calculated from the relationship:

$$\Delta S_{surr} = -\frac{\Delta H}{T}$$

where T is the *absolute* temperature.

The entropy change of the surroundings depends on temperature, with the transfer of a given quantity of energy to surroundings at low temperature producing a greater entropy change than the transfer of the same amount of energy to the surroundings at a higher temperature.

The reason for the difference can be understood if we think about the effect of transferring quanta between the regions of the universe. Transferring 10 quanta to a region already containing 5 quanta causes a large change in the number of ways of arranging quanta, while transferring 10 quanta to a region containing 20 quanta has a much smaller effect. One way of understanding this is to think about the sound of someone shouting. This is barely noticeable in the roar of a football crowd, but in the quiet of a library ...

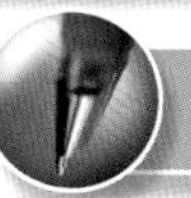

Worked example

The standard enthalpy change for the formation of sodium chloride is $-411\,kJ\,mol^{-1}$. Calculate the entropy change in the surroundings, also under standard conditions.

$$\Delta S^\ominus_{surr} = -\frac{-411\,000\,J\,mol^{-1}}{298\,K}$$
$$= +1379\,J\,mol^{-1}\,K^{-1}$$

Notice that ΔH must be in $J\,mol^{-1}$, because ΔS is in $J\,mol^{-1}\,K^{-1}$.

Calculating ΔS_{total}

When you know the value of the change in entropy of the system, ΔS_{sys}, and that of the surroundings, ΔS_{surr}, you can calculate the total change in entropy, ΔS_{total}, of the reaction involved using:

$$\Delta S_{total} = \Delta S_{sys} + \Delta S_{surr}$$

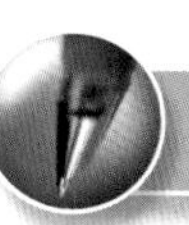

Worked example

Calculate the total entropy change in the formation of 1 mole of sodium chloride from its elements in their standard states:

$Na(s) + \frac{1}{2}Cl_2(g) \rightarrow NaCl(s)$

$$\Delta S^\ominus_{total} = \Delta S^\ominus_{sys} + \Delta S^\ominus_{surr}$$
$$= -90.1\,J\,mol^{-1}\,K^{-1} + 1379\,J\,mol^{-1}\,K^{-1}$$
$$= +1289\,J\,mol^{-1}\,K^{-1}$$

Overall, the change in entropy is positive – so this is a spontaneous process.

Questions

1 Use the data in table 1.2.2 to work out the entropy change for the system during the following reaction:

$C_{graphite}(s) + O_2(g) \rightarrow CO_2(g)$

2 Which substance in the pair $HCl(g)$ / $NH_3(g)$ would you expect to have the higher standard molar entropy at 298 K? Explain your choice.

The feasibility of a reaction, thermodynamic stability and kinetic inertness [4.4k, l, m]

Spontaneous reactions

A **spontaneous reaction** is one which tends to go without being driven by any outside agency.

Figure 1.2.14 shows an analogy for a spontaneous reaction.

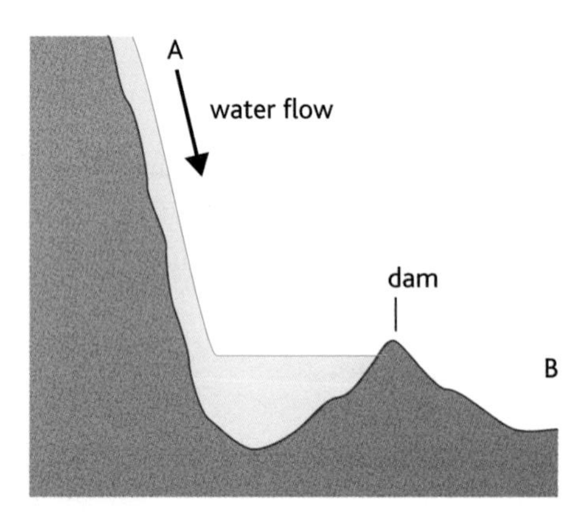

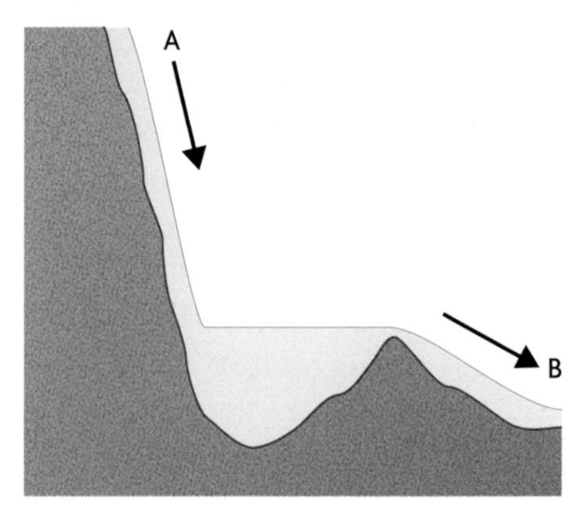

fig. 1.2.14 A spontaneous change.

Water flows downhill naturally, but the rate of flow will depend on various factors. You can think of the water flowing from A to B as the spontaneous reaction. In fig. 1.2.14 the water is held up by a dam and before it can continue its flow towards B the water must rise to the top of the dam. You can think of this as overcoming the activation energy in order to continue.

The reaction of a mixture of methane and air is spontaneous when you consider thermodynamic data, but it does not occur until a source of heat energy is provided to overcome the activation energy. Overall the reaction is exothermic.

We have seen that citric acid and sodium hydrogencarbonate react with effervescence and the release of carbon dioxide gas. The reaction cools the surroundings because it is endothermic – but it also is spontaneous.

Chemists frequently use the value of $\Delta H^\ominus$ for a reaction to predict whether a reaction is likely to happen or not. Exothermic reactions make products that are energetically more stable than the reactants from which they are formed. This suggests that a reaction with a large negative value for $\Delta H^\ominus$ is very likely to happen. Although many spontaneous reactions are strongly exothermic, knowing the value of $\Delta H^\ominus$ alone is not sufficient to predict if a particular reaction is likely to proceed.

There are several reasons for this:

- $\Delta H^\ominus$ tells us about the energy changes that occur in a reaction, but nothing about the kinetic stability of the reactants. The enthalpy change of combustion of petrol is enormous, yet the relative safety of our road transport system relies on the fact that a mixture of petrol and air is kinetically stable.
- $\Delta H^\ominus$ tells us about enthalpy changes under *standard* conditions. Actual enthalpy changes under different conditions of temperature, pressure and concentration are often very different.
- Factors other than enthalpy changes are often involved in chemical changes. These factors are concerned with the way the system and the surroundings are organised.

Entropy, free energy and spontaneous change

The relationships:

$$\Delta S_{total} = \Delta S_{sys} + \Delta S_{surr}$$

and

$$\Delta S_{surr} = -\Delta H / T$$

can be combined to produce the relationship:

$$\Delta S_{total} = \Delta S_{sys} - \Delta H / T$$

Multiplying this relationship by T gives:

$$T\Delta S_{total} = T\Delta S_{sys} - \Delta H$$

or

$$-T\Delta S_{total} = \Delta H - T\Delta S_{sys}$$

The quantity $-T\Delta S_{total}$ is called the **Gibbs free energy** change, or ΔG, named after the American scientist Josiah Willard Gibbs. Under standard conditions, the standard free energy change for a reaction is given by:

$$\Delta G^\ominus = \Delta H^\ominus - T\Delta S^\ominus_{sys}$$

Since ΔS_{total} must be positive for a change to occur spontaneously, it follows that such changes have ΔG less than zero (in other words, negative).

The link between thermodynamic data and spontan…

What does the relationship

$$\Delta G = \Delta H - T\Delta S_{sys}$$

mean when we try to predict spontaneous change? There are four cases to consider.

1 Exothermic changes that are accompanied by an increase in the system's entropy will *always* happen spontaneously:

$\Delta H < 0$ and $\Delta S_{sys} > 0$, and so $-T\Delta S_{sys} < 0$ (remember that $T > 0$ always).

So, $\Delta G < 0$ always.

Example: $C_2H_5OH(l) + 3O_2(g) \rightarrow 2CO_2(g) + 3H_2O(g)$

2 Endothermic changes that are accompanied by a decrease in the system's entropy will *never* happen spontaneously:

$\Delta H > 0$ and $\Delta S_{sys} < 0$, and so $-T\Delta S_{sys} > 0$.

So, $\Delta G > 0$ always.

Example: $CO_2(g) \rightarrow C(s) + O_2(g)$

3 Endothermic changes that are accompanied by an increase in the system's entropy will be spontaneous if the temperature is sufficiently high:

$\Delta H > 0$ and $\Delta S_{sys} > 0$, and so $-T\Delta S_{sys} < 0$.

So, $\Delta G < 0$ if the magnitude of $T\Delta S_{sys}$ > the magnitude of ΔH.

Example: water boiling, $H_2O(l) \rightarrow H_2O(g)$

4 Exothermic changes that are accompanied by a decrease in the system's entropy will be spontaneous if the temperature is sufficiently low:

$\Delta H < 0$ and $\Delta S_{sys} < 0$, and so $-T\Delta S_{sys} > 0$.

So, $\Delta G < 0$ if the magnitude of ΔH > the magnitude of $T\Delta S_{sys}$.

Example: steam condensing, $H_2O(g) \rightarrow H_2O(l)$

These four cases are summarised in **table 1.2.3**.

		ΔH	
		Positive	**Negative**
ΔS_{sys}	**Positive**	spontaneous only at high temperatures	always spontaneous
	Negative	never spontaneous	spontaneous only at low temperatures

table 1.2.3

Reaction of ammonium carbonate and ethanoic acid

The equation for the reaction is:

$$(NH_4)_2CO_3(s) + 2CH_3COOH(aq) \rightarrow 2CH_3COO^-(aq) + 2NH_4^+(aq) + CO_2(g) + H_2O(l)$$

The entropy change is positive. There is more disorder in the products because of the formation of carbon dioxide gas.

Magnesium burning in air

Initially lighting magnesium is not easy, but when it starts the magnesium continues to burn in air spontaneously (see fig. 1.2.11):

$$2Mg(s) + O_2(g) \rightarrow 2MgO(s) \qquad \Delta H^\ominus = -1203.4\,\text{kJ mol}^{-1}$$

There are fewer possible arrangements of energy in the solid magnesium oxide than in the solid magnesium and oxygen gas. We would therefore expect a decrease in the entropy of the system.

We can get the standard entropy values for reactants and products from pages 2–4 and 20–8 in the *Edexcel Data Booklet*. The values at 298 K are:

$$S^\ominus[Mg(s)] = +32.7\,\text{J mol}^{-1}\,\text{K}^{-1}$$

$$S^\ominus[O_2(g)] = +205.0\,\text{J mol}^{-1}\,\text{K}^{-1}$$

$$S^\ominus[MgO(s)] = +26.9\,\text{J mol}^{-1}\,\text{K}^{-1}$$

From this we can work out a value for $\Delta S^\ominus_{sys}$:

$$\begin{aligned}\Delta S^\ominus_{sys} &= \Sigma S^\ominus_{products} - \Sigma S^\ominus_{reactants}\\ &= 2S^\ominus[MgO(s)] - S^\ominus[O_2(g)] - 2S^\ominus[Mg(s)]\\ &= (2 \times +26.9) - (+205.0) - (2 \times +32.7)\\ &= -217\,\text{J mol}^{-1}\,\text{K}^{-1}\end{aligned}$$

This is a negative value for $\Delta S^\ominus_{sys}$, but we must also consider the entropy change in the surroundings.

$$\Delta S_{surr} = -\frac{\Delta H}{T}$$

$$\begin{aligned}\Delta S^\ominus_{surr} &= -\frac{\Delta H^\ominus}{298}\\ &= -\frac{(-1\,203\,400\,\text{J mol}^{-1})}{298\,\text{K}}\\ &= +4040\,\text{J mol}^{-1}\,\text{K}^{-1}\end{aligned}$$

Now we know that $\Delta S^\ominus_{sys} = -217\,\text{J mol}^{-1}\,\text{K}^{-1}$ and $\Delta S^\ominus_{surr} = +4040\,\text{J mol}^{-1}\,\text{K}^{-1}$. So:

$$\begin{aligned}\Delta S^\ominus_{total} &= \Delta S^\ominus_{sys} + \Delta S^\ominus_{surr}\\ &= (-217\,\text{J mol}^{-1}\,\text{K}^{-1}) + (+4040\,\text{J mol}^{-1}\,\text{K}^{-1})\\ &= +3820\,\text{J mol}^{-1}\,\text{K}^{-1}\end{aligned}$$

This value is positive, so we expect the reaction to be spontaneous – confirm it by looking at the signs of the changes and table 1.2.3.

Ammonium chloride and barium hydroxide

The equation for this reaction is:

$$2NH_4Cl(s) + Ba(OH)_2.8H_2O(s) \rightarrow BaCl_2.2H_2O(s) + 2NH_3(g) + 8H_2O(l)$$

Ammonia is a gas, so we would expect an increase of entropy in the reaction.

Using $\Delta S^\ominus$ and $\Delta H^\ominus_f$ values from the *Edexcel Data Booklet*, at 298 K for all reactants and products, we can work out $\Delta S^\ominus_{sys}$ using:

$$\Delta S^\ominus_{total} = \Delta S^\ominus_{sys} + \Delta S^\ominus_{surr}$$

$$\Delta H^\ominus_{reaction} = \Sigma\Delta H^\ominus_f[\text{products}] - \Sigma\Delta H^\ominus_f[\text{reactants}]$$

which gives

$$\Delta S^\ominus_{sys} = +298.6\,\text{J mol}^{-1}\,\text{K}^{-1}$$

$$\Delta H^\ominus_{reaction} = +21.2\,\text{kJ mol}^{-1}$$

The value for $\Delta S^\ominus_{surr}$ can be calculated using:

$$\begin{aligned}\Delta S^\ominus_{surr} &= -\frac{\Delta H^\ominus_{reaction}}{T}\\ &= -\frac{(+21\,200\,\text{J mol}^{-1})}{298\,\text{K}} = -71.1\,\text{J mol}^{-1}\,\text{K}^{-1}\end{aligned}$$

$$\begin{aligned}\text{Now, } \Delta S^\ominus_{total} &= \Delta S^\ominus_{sys} + \Delta S^\ominus_{surr}\\ &= +298.6 + -71.1\,\text{J mol}^{-1}\,\text{K}^{-1}\\ &= +227.5\,\text{J mol}^{-1}\,\text{K}^{-1}\end{aligned}$$

The positive value confirms that the reaction is spontaneous at 298 K.

Research the changes involved in photosynthesis and explain why it is possible for it to happen.

Use the Internet to find out about Ellingham diagrams and use the information to explain why the Bronze Age came before the Iron Age – and why we now live in the 'aluminium age'.

HSW Gibbs – applying thermodynamics to physical chemistry

Josiah Willard Gibbs was one of the greatest American scientists of all time. The seventh in line in his family to be a leading academic, he was a pioneering chemist, physicist and mathematician. His father, also Josiah Willard, was a prominent authority in the field of sacred literature.

fig. 1.2.15 **Josiah Willard Gibbs (1839–1903).**

In 1866, he left Yale University to begin a period of study in Europe, spending a year each in Paris, Berlin and Heidelberg, where he was influenced by Kirchhoff and Helmholtz. At the time, all the leading authorities in thermodynamics lived in Germany. Later he returned to Yale where he spent the rest of his academic career.

Between 1876 and 1878 Gibbs wrote a series of papers that were published together in a monograph titled *On the Equilibrium of Heterogeneous Substances*. This is now considered one of the greatest scientific achievements of the nineteenth century, setting the foundations of physical chemistry. Gibbs applied thermodynamics to interpret, explain and interrelate what had previously been a mass of isolated facts. You can find copies of these papers on the Internet.

In 1901, Gibbs was awarded the highest possible honour given by the international scientific community of his day, granted to only one scientist each year: the Copley Medal of the Royal Society of London.

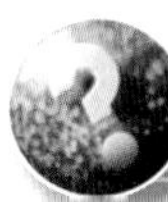

Questions

1 For a reaction, $\Delta H^\ominus = -214\,kJ\,mol^{-1}$ and $\Delta S^\ominus_{sys}$ is $+112\,J\,mol^{-1}\,K^{-1}$. What can you conclude about the spontaneity of the reaction from these data?

2 The equation for the decomposition of zinc carbonate is:

$ZnCO_3(s) \rightarrow ZnO(s) + CO_2(g)$

The values for $\Delta S^\ominus_{sys}$ and $\Delta S^\ominus_{surr}$ are $+175\,J\,mol^{-1}\,K^{-1}$ and $-238\,J\,mol^{-1}\,K^{-1}$, respectively.

a Show that the reaction is not spontaneous at 298 K.

b Calculate the lowest temperature at which the reaction is spontaneous.

Predicting solubility from the enthalpy and entropy of solution [4.4n, o, p]

Introduction

In *Edexcel AS Chemistry* (page 168) we considered the dissolving of an ionic solid in water. We saw that a comparison of the amount of energy required to break up a crystal with the amount of energy released on **hydration** could determine whether the crystal would dissolve or not.

Lattice energy

The amount of energy needed to break up an ionic crystal into separate ions is called its **lattice energy** (pages 84–5 in *Edexcel AS Chemistry*) – strictly defined as the enthalpy of formation of one mole of an ionic compound from gaseous ions under standard conditions:

$$aM^{b+}(g) + bX^{a-}(g) \rightarrow M_aX_b(s)$$

We also found out that the lattice energy cannot be determined directly by experiment, but it can be estimated using Hess's Law in the form of a Born–Haber cycle. It can also be calculated from the electrostatic consideration of its crystal structure.

In *Edexcel AS Chemistry* (page 85), the following trends can be identified:

- as the ionic radii of either the cation or anion increase, the lattice energy decreases
- the solids consisting of divalent ions have much higher lattice energies than solids with monovalent ions.

Enthalpy of hydration

The **enthalpy of hydration,** ΔH_{hyd}, of an ion is the amount of energy released when 1 mole of the gaseous ions dissolve in water under standard conditions to produce a solution of concentration $1.0\,mol\,dm^{-3}$:

$$M^{2+}(g) + aq \rightarrow M^{2+}(aq)$$

where $M^{2+}(aq)$ represents ions surrounded by water molecules and dispersed in the solution. The approximate hydration energies of some typical ions are listed in table 1.2.4 which illustrates the point that as the atomic number increases, so does the ionic size, leading to a decrease in absolute values of enthalpy of hydration.

Ion	ΔH_{hyd} /kJ mol^{-1}	Ion	ΔH_{hyd} /kJ mol^{-1}	Ion	ΔH_{hyd} /kJ mol^{-1}
H^+	−1130	Al^{3+}	−4665	Fe^{3+}	−4430
Li^+	−559	Be^{2+}	−2494	F^-	−483
Na^+	−444	Mg^{2+}	−1921	Cl^-	−340
K^+	−322	Ca^{2+}	−1577	Br^-	−336
Rb^+	−297	Sr^{2+}	−1443	I^-	−295
Cs^+	−276	Ba^{2+}	−1305		

table 1.2.4 Enthalpy of hydration of some common ions.

For a dissolving ionic compound, we can consider that the enthalpy of hydration comes from two components – the enthalpy of hydration of the cation and the enthalpy of hydration of the anion.

So, for the enthalpy of hydration of sodium chloride:

$$\Delta H_{hyd}[NaCl(s)] = \Delta H_{hyd}[Na^+(g)] + \Delta H_{hyd}[Cl^-(g)]$$
$$= (-444\,kJ\,mol^{-1}) + (-340\,kJ\,mol^{-1})$$
$$= -774\,kJ\,mol^{-1}$$

Figure 2.3.14 in *Edexcel AS Chemistry* (page 168) shows the way the water molecules arrange themselves around an anion and a cation.

Enthalpy of solution

The **enthalpy of solution,** $\Delta H^\ominus_{sol}$, of a compound is the enthalpy change when 1 mole of the compound dissolves in a stated amount of water under standard conditions:

$$M_aX_b(s) + aq \rightarrow aM^{b+}(aq) + bX^{a-}(aq)$$

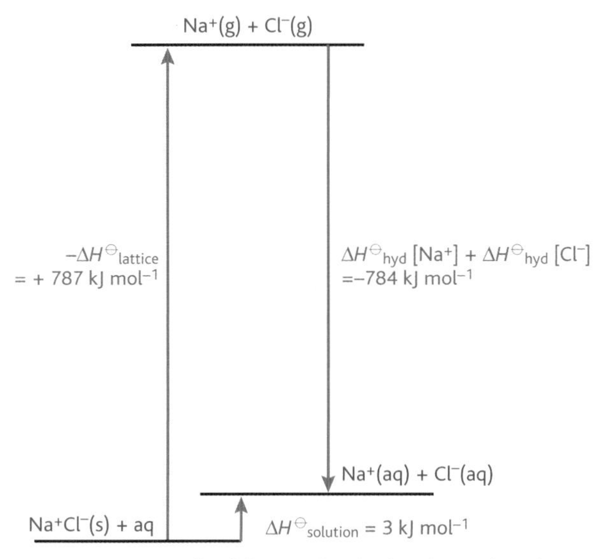

fig. 1.2.16 **An energy level diagram for the dissolving of NaCl in water.**

The enthalpy of solution is the difference between the energy needed to separate the ions from the crystal lattice and the energy given out when the ions are hydrated. We can work out enthalpies of solution from energy level diagrams like fig. 1.2.16. For sodium chloride:

$$\Delta H^\ominus_{sol} = \Delta H^\ominus_{hyd}[Na^+(g)] + \Delta H^\ominus_{hyd}[Cl^-(g)] - \Delta H^\ominus_{latt}[NaCl(s)]$$

$$= (-444\,kJ\,mol^{-1}) + (-340\ kJ\,mol^{-1}) - (-787\,kJ\,mol^{-1})$$

$$= +3\,kJ\,mol^{-1}$$

The result of dissolving sodium chloride in water is a slight evolution of energy but insufficient to be observed.

Solubility of ionic compounds

The different solubilities of Group 2 ionic compounds were observed in *Edexcel AS Chemistry* (page 183). Certainly, when the hydration energy is greater than the lattice energy, the compound will dissolve in water. However, some ionic compounds have a lattice energy greater than hydration energy and yet they still dissolve in water.

The explanation involves considering the entropy change. Dissolving an ionic compound in water leads to an increase in entropy. In order to predict whether an ionic compound will dissolve in water or not, you have to think about a combination of the enthalpy change and the entropy change. These can be combined mathematically to give an important term known as the free energy change, ΔG.

As an approximation – for a reaction to happen, the free energy change ΔG must be negative.

What happens if the enthalpy change is positive – as, for example, when sodium chloride dissolves in water ($+3\,kJ\,mol^{-1}$)?

As long as the entropy change of the system is positive enough, it is possible to have a negative value for free energy change. In the sodium chloride case, you don't need very much of an increase in entropy to outweigh the small enthalpy change of $+3\,kJ\,mol^{-1}$.

Questions

1 Use the data in table 1.2.4 along with the values for the lattice energies of lithium fluoride and lithium chloride, $-1046\,kJ\,mol^{-1}$ and $-861\,kJ\,mol^{-1}$ respectively, to calculate the enthalpy of solution of **a** lithium fluoride; **b** lithium chloride.

2 A $1.0\,mol\,dm^{-3}$ solution of rubidium chloride is made when 1 mole of rubidium chloride is added to water.

$RbCl(s) + aq \rightarrow Rb^+(aq) + Cl^-(aq)$

At 298 K, the enthalpy change on dissolving is $+19\,kJ\,mol^{-1}$.

Standard molar entropies are:

$RbCl(s)$ 95.9 J mol^{-1} K^{-1}

$Rb^+(aq)$ 121.5 J mol^{-1} K^{-1}

$Cl^-(aq)$ 56.7 J mol^{-1} K^{-1}

a Calculate the entropy change in the system, in the surroundings and the total entropy change.

b Why does rubidium chloride dissolve in water when this is an endothermic process?

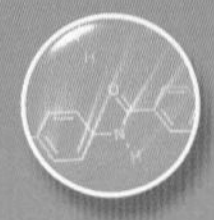

1.3 Equilibria

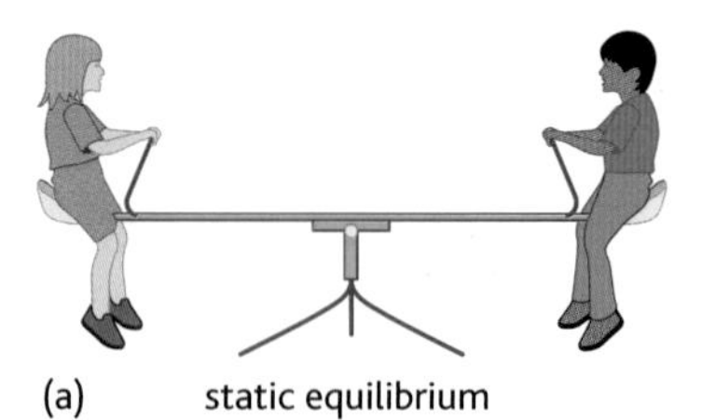

(a) static equilibrium

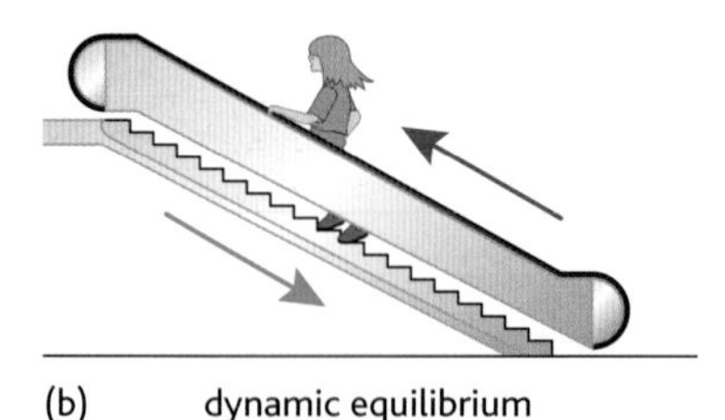

(b) dynamic equilibrium

fig. 1.3.1 Everyday models of static and dynamic equilibrium. However, running up a down escalator is *not* recommended, even to demonstrate a chemical principle!

The idea of an equilibrium constant [4.5a, b, c, e]

Some processes are one way only – boil an egg and the change in the egg white as it cooks cannot be undone. Other processes are reversible – a jelly sets when it is cool, but melts again if it is warmed up. When we are looking at equilibria, we are only concerned with physical processes and chemical reactions that are reversible.

There are many situations in the world around you where two or more factors have contrasting effects on a system (see fig. 1.3.1). When opposing tendencies are balanced, a system is in **equilibrium**. In *Edexcel AS Chemistry* (pages 202–7) you were introduced to the ideas of equilibrium in a chemical context – you are going to revisit those ideas. The concept of equilibrium, and understanding how we can change the position of equilibrium, is very important to chemists both in the laboratory and in industry. You will be looking at this in theory and in the chemical industry.

Static and dynamic equilibrium

There are two types of equilibrium. In a **static equilibrium** all the processes which might disturb the equilibrium have stopped. The everyday example shown in fig. 1.3.1(a) is of two children on a see-saw. They are balancing each other so no movement of the see-saw takes place. On the other hand, in a **dynamic equilibrium** such as shown in fig. 1.3.1(b) things are happening to maintain the equilibrium. The girl is moving up the escalator at the same rate as the escalator is moving down. So long as the girl and the escalator continue to move in opposite directions at the same rate, the dynamic equilibrium will be maintained.

In your AS Chemistry studies, you met the closed container containing liquid and vapour shown in fig. 1.3.2. It shows particles leaving and re-entering a liquid. To an external observer, the system appears to have come to rest, with constant amounts of liquid and vapour. However, examination of the behaviour of the particles in the system shows that they are still moving, which is why this is described as a *dynamic* equilibrium. This sort of simple equilibrium can only take place in a closed container – if the vapour or liquid can leave the system, or some can come into it from outside, then equilibrium can never be reached.

fig. 1.3.2 Particles can leave the liquid (evaporate) if they have enough energy, and particles can leave the vapour and rejoin the liquid (condensation) giving up some of their energy to the other particles in the liquid.

A similar state of equilibrium can be seen when chemicals react together in a reversible reaction.

You can observe equilibrium happening in a very visual way if you observe a mixture of two solutions of iodine. Iodine is soluble in aqueous potassium iodide solution, forming a pale brown solution. It is also soluble in hexane, forming a violet solution. These two solvents do not mix. Figure 1.3.3 shows what happens if you shake iodine with the two solvents.

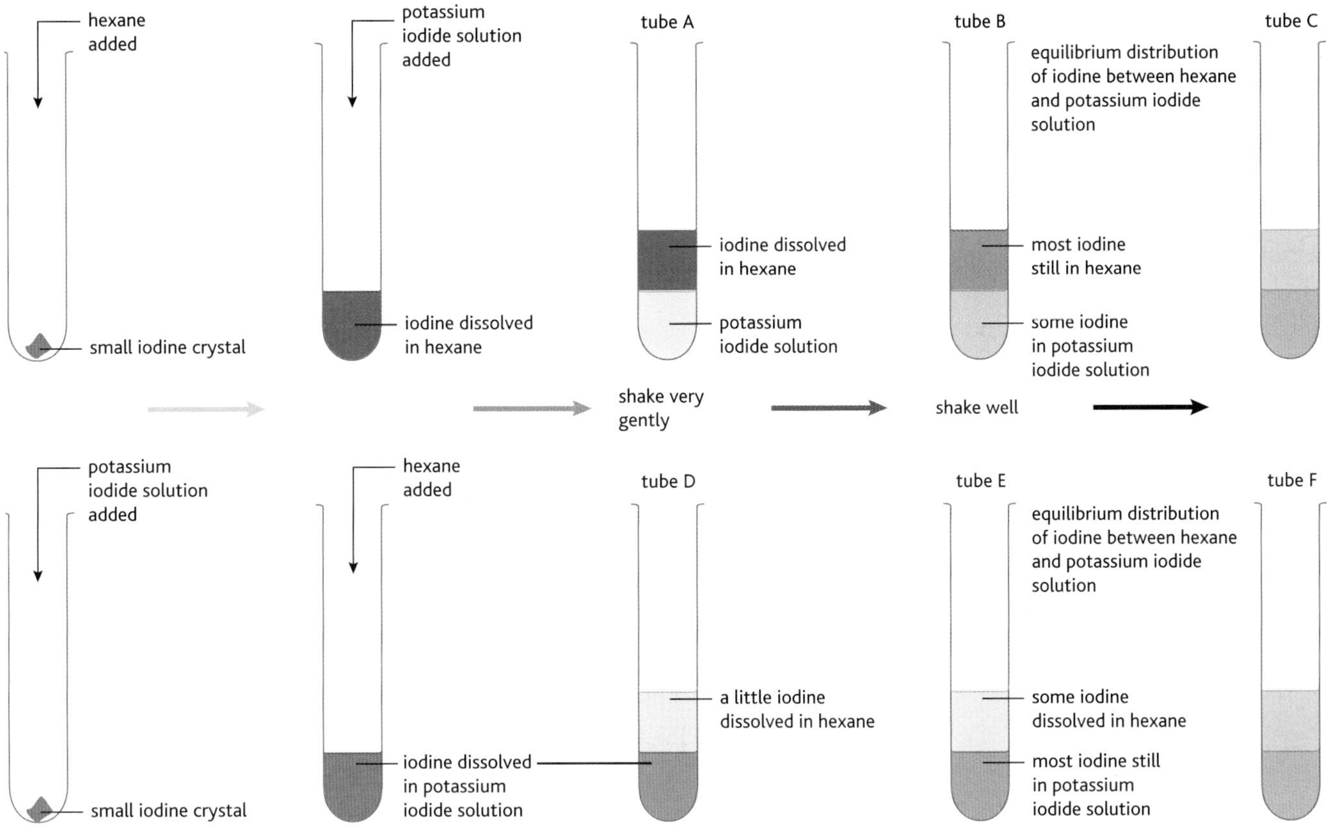

fig. 1.3.3 The distribution of iodine between two solvents.

This simple demonstration shows two important factors of equilibrium processes:

- equilibrium can be approached from either direction – the reactant side or the product side
- at equilibrium, the concentrations of the reactants and the products do not change.

Changing the equilibrium position

So long as everything remains the same, a system will remain in equilibrium. But if one or more of the conditions changes, the equilibrium is lost. What happens when the sealed container in the fig. 1.3.2 example is heated? As the temperature of the liquid in the container rises, its molecules gain more energy and so more of them leave the liquid. The system is no longer in equilibrium, because more molecules are leaving the liquid than are entering it. Eventually, however, there will be a large enough number of molecules in the vapour for the rate of condensation to equal the rate of evaporation again, and so equilibrium is established once more. You can represent this process as

$$\text{energy} + \text{liquid} \rightleftharpoons \text{vapour}$$

in which the double arrows '$\rightleftharpoons$' represent the dynamic equilibrium of particles moving from the liquid to the vapour and vice versa. Transferring energy to the system tends to make the equilibrium shift to the right, so that more molecules leave the liquid and go into the vapour. This disturbs the equilibrium for a time. We shall see many such examples of disturbances to equilibria in this section.

Examples of dynamic equilibria

If a block of ice is put into water at 0 °C and the conditions kept constant, both the ice and the water remain. However, the melting and freezing processes continue at equal rates – the system is in dynamic equilibrium. **Figure 1.3.4** shows particles moving from and to the ice in the dynamic equilibrium.

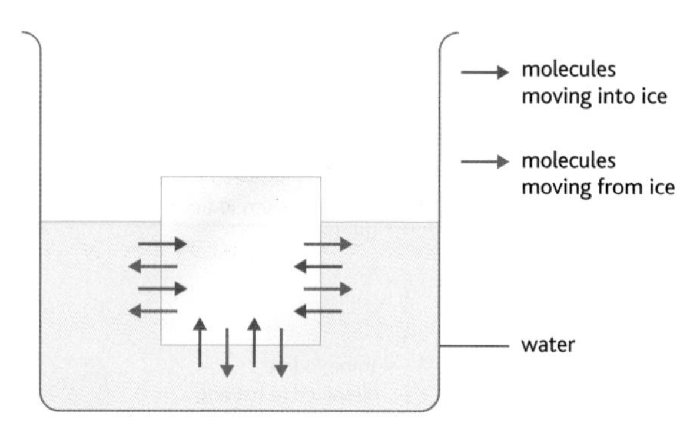

fig. 1.3.4 Ice and water in equilibrium.

When salt is added to water, the salt dissolves to form a solution. As more salt is added, the solution becomes more concentrated, until eventually a saturated solution is formed when no more salt appears to dissolve in the water. However, there is a state of dynamic equilibrium. Sodium ions and chloride ions are leaving the solid and entering the solution at the same rate as sodium ions and chloride ions are entering the crystals from the solution (**fig. 1.3.5**).

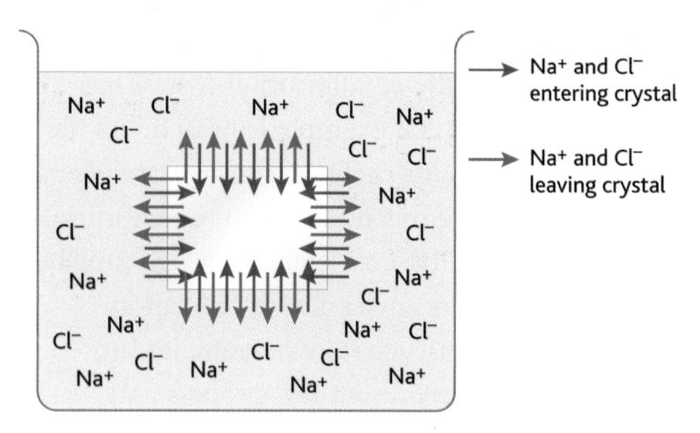

fig. 1.3.5 The same number of ions leave the crystal as re-enter it when the solution is saturated.

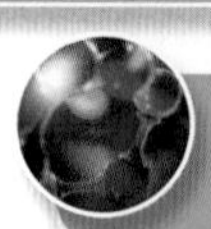

HSW How do we know if a system is in dynamic equilibrium?

If a sample of sodium chloride containing sodium-24 as a tracer is added to a saturated solution of sodium chloride that does not contain sodium-24, then after a period of time sodium-24 will be found in both the crystals and the saturated solution.

In *Edexcel AS Chemistry* (page 203) there is a consideration of the equilibrium reaction between nitrogen dioxide and dinitrogen tetroxide:

$$2NO_2(g) \rightleftharpoons N_2O_4(g)$$

brown   colourless

Under constant conditions of temperature and pressure in a closed container, the system remains in dynamic equilibrium. The colour of the mixture remains unchanged because the concentrations of the reactants and products remain unchanged.

Almost all chemical processes eventually reach a state of dynamic equilibrium like this, although we cannot always see that this is the case. For example, in water vapour at room temperature the equilibrium:

$$2H_2O(g) \rightleftharpoons 2H_2(g) + O_2(g)$$

exists, but because water molecules are so stable there are no easily detectable amounts of hydrogen or oxygen present – effectively, this reaction is one that 'doesn't go'. Similarly, the equilibrium:

$$HNO_3(aq) \rightleftharpoons H^+(aq) + NO_3^-(aq)$$

exists when nitric acid dissolves in water, although in practice the acid is completely ionised so no HNO_3 can be detected – this reaction 'goes to completion'.

The equilibrium law

When a chemical system is in equilibrium, there is a simple relationship between the molar concentrations of the products and reactants. This relationship can be demonstrated if we look at the equilibrium in which hydrogen iodide is formed from hydrogen and iodine:

$$H_2(g) + I_2(g) \rightleftharpoons 2HI(g)$$

If several experiments are set up in which the initial amounts of hydrogen, iodine and hydrogen iodide are different, keeping the temperature in all the experiments the same (in this case, at 700 K) the results shown in **table 1.3.1** are obtained.

Expt.	Initial concentration/mol dm^{-3}			Equilibrium concentration/mol dm^{-3}		
	[H_2(g)]	[I_2(g)]	[HI(g)]	[H_2(g)]	[I_2(g)]	[HI(g)]
1	0.0442	0.0364	0	0.0132	0.0054	0.0620
2	0.0442	0.0288	0	0.0181	0.0027	0.0510
3	0.0334	0.0288	0	0.0093	0.0047	0.0482
4	0	0	0.0518	0.0050	0.0050	0.0366

table 1.3.1 Initial and equilibrium concentrations of hydrogen, iodine and hydrogen iodide in several experiments.

Notice how all the concentrations are different in each experiment, which is not perhaps particularly surprising. What is surprising though is that we can make sense of these figures using a relationship that comes from the balanced equation for the reaction. For each set of experimental results in table 1.3.1, if we take the equilibrium concentration of hydrogen iodide and square it, and then divide this by the product of the equilibrium concentration of hydrogen and the equilibrium concentration of iodine, we get a number – these are shown in table 1.3.2.

Expt. 1	Expt. 2	Expt. 3	Expt. 4
53.9	53.2	53.2	53.6

table 1.3.2 The values for the results in table 1.3.1.

Within experimental error, the numbers in the second row are constant. These quantities have no units – the expression has units of (mol dm^{-3})^2 on the top line and (mol dm^{-3}) × (mol dm^{-3}) on the bottom line, so they cancel out.

Experiments on this equilibrium reaction show that whenever the equilibrium concentrations in a mixture of hydrogen, iodine and hydrogen iodide at a temperature of 700 K are measured carefully, the equilibrium value always works out to be between 53.0 and 54.0.

The relationship:

$$K_c = \frac{[HI(g)]_{eq}{}^2}{[H_2(g)]_{eq}\,[I_2(g)]_{eq}}$$

is called the equilibrium expression for the reaction between hydrogen and iodine, and K_c is called the **equilibrium constant** for the reaction (the 'c' indicates that concentrations were used to calculate it). K_c has a value of 54.0 for this reaction when it is carried out at 700 K. If the value is anything else, then the mixture is not at equilibrium, and the gases will react further until an equilibrium exists.

Predicting the equilibrium law for a reaction

The equilibrium law for a particular reaction can always be predicted from its stoichiometric equation. The general equilibrium for substances A, B, C and D:

$$aA + bB \rightleftharpoons cC + dD$$

has the equilibrium law:

$$K_c = \frac{[C]^c\,[D]^d}{[A]^a\,[B]^b}$$

It is a convention that the concentrations of the products of the reaction are put in the numerator of the equilibrium expression – the concentrations of the reactants appear in the denominator. This means you need to be very careful to identify exactly which reaction you are describing.

For example, at 700 K the equilibrium constant for the reaction between hydrogen and iodine to produce hydrogen iodide is 54.0, as we have found out above. But the equilibrium expression for the dissociation of hydrogen iodide into hydrogen and iodine at the same temperature is:

$$2HI(g) \rightleftharpoons H_2(g) + I_2(g)$$

$$K_c' = \frac{[H_2(g)]\,[I_2(g)]}{[HI(g)]^2}$$

$$= \frac{1}{54} = 0.0185$$

So the equilibrium constant of the backward reaction is the reciprocal of the equilibrium constant for the forward reaction. In other words:

$$K_c' = \frac{1}{K_c}$$

Units and the equilibrium constant

In the case of the hydrogen, iodine and hydrogen iodide reaction, we saw that K_c has no units. This is not the case for many equilibria. For example, in the nitrogen dioxide/dinitrogen tetroxide equilibrium:

HSW Erors and un...certainties

Errors and mistakes are a fact of life – although that one was deliberate! All methods of measurement have limits to how precisely they can measure quantities. Whenever we make measurements using instruments, there is some error or uncertainty in the result. Sometimes the uncertainty may be due to the experimenter (reading a scale wrongly) and sometimes it may be due to the equipment (a wrongly calibrated thermometer). The word 'uncertainty' is used because the problems which arise in making measurements are not always due to mistakes – they may often be due to the limits of the equipment being used, or of the experiment itself.

The important thing in any experiment that involves measuring physical quantities (mass, temperature, pH and so on) is that some attempt should be made to estimate the magnitude of these uncertainties and to show how they are likely to affect the final result.

fig. 1.3.6 **If an entire class carries out an identical practical procedure, they would almost all get slightly different results because of differences in their equipment and their practical skills. Chemists build an allowance for these sorts of uncertainties into their calculations – but try to eliminate errors as far as possible.**

$$2NO_2(g) \rightleftharpoons N_2O_4(g)$$

$$K_c = \frac{[N_2O_4(g)]}{[NO_2(g)]^2}$$

$$= \frac{\text{mol dm}^{-3}}{\text{mol dm}^{-3} \times \text{mol dm}^{-3}}$$

$$= \frac{1}{\text{mol dm}^{-3}}$$

$$= \text{dm}^3\,\text{mol}^{-1}$$

Using the same arguments, we can show that the units of K_c for the equilibrium:

$$N_2(g) + 3H_2(g) \rightleftharpoons 2NH_3(g)$$

are $\text{dm}^6\,\text{mol}^{-2}$.

Questions

1 a Use the reaction shown in fig. 1.3.3 to explain how a dynamic equilibrium works.

b What would you expect to happen if the temperature in which the mixtures were shaken was raised or lowered?

2 Write down equilibrium expressions for the following reactions:

a $Ag^+(aq) + 2NH_3(aq) \rightleftharpoons [Ag(NH_3)_2]^+(aq)$

b $CH_3COOH(l) + C_2H_5OH(l) \rightleftharpoons CH_3COOC_2H_5(l) + H_2O(l)$

c $CH_4(g) + H_2O(g) \rightleftharpoons CH_3OH(g) + H_2(g)$

Give the units for each equilibrium constant.

SC Phosphorus pentachloride decomposes as shown below:

$$PCl_5(g) \rightleftharpoons PCl_3(g) + Cl_2(g)$$

Given that the equilibrium constant for this reaction is 11.5 mol dm^{-3} at 300 K, what is the equilibrium concentration of PCl_3 when the concentration of PCl_5 at equilibrium is 0.07 mol dm^{-3}?

Calculations involving K_c and K_p [4.5e]

Calculations using K_c

An equilibrium constant expressed in terms of concentration only applies when all the reactants and products are in the same phase – e.g. a mixture of liquids or gases. This usually involves only **homogeneous equilibria**.

Worked example

0.256 g of hydrogen iodide was heated at 764 K in a sealed flask of volume 100 cm^3.

$$2HI(g) \rightleftharpoons H_2(g) + I_2(g)$$

When equilibrium was established, the flask was cooled quickly, and it was found that the mixture contained 0.00028 moles of iodine. Calculate a value for K_c.

First we need to work out the number of moles of hydrogen iodide at the start:

$$\text{amount of HI} = \frac{0.256\,g}{128\,g\,mol^{-1}} = 0.002\,mol$$

To work out the amount of HI unreacted you need to realise that twice as many HI molecules are used up as I_2 molecules formed. So:

$$\text{amount of HI} = 0.002\,mol - 0.00056\,mol = 0.00144\,mol$$

The amount of iodine formed is 0.00028 mol, and (from the equation) this must also be the amount of hydrogen formed.

We can now work out the equilibrium concentrations. The volume of the flask is 0.1 dm^3. So:

$$[HI] = \frac{0.00144\,mol}{0.1\,dm^3} = 0.0144\,mol\,dm^{-3}$$

$$[H_2] = [I_2] = \frac{0.00028\,mol}{0.1\,dm^3} = 0.0028\,mol\,dm^{-3}$$

$$K_c = \frac{[H_2]\,[I_2]}{[HI]^2} = \frac{0.0028 \times 0.0028}{0.0144^2} = 0.038$$

Note that the units cancel, so K_c has no units.

Equilibrium constants for reactions involving gases

So far, we have used the concentration of a substance in equilibrium constant calculations. For gases it is often more convenient to use their **partial pressure** – this is the pressure that the gas would have if it occupied the volume alone.

The partial pressure of gas A, p_A, in a mixture of gases A, B, C and D is calculated by:

$$p_A = \frac{\text{number of moles of A}}{\text{total number of moles in the mixture}} \times \text{total pressure}$$

$$= \frac{n_A}{n_A + n_B + n_C + n_D} \times p_{tot}$$

Suppose we have this reaction involving gases:

$$aA(g) + bB(g) \rightleftharpoons cC(g) + dD(g)$$

We can write the equilibrium expression using partial pressures:

$$K_p = \frac{p(C)^c \times p(D)^d}{p(A)^a \times p(B)^b}$$

Worked example

An equilibrium mixture at a pressure of 200 kPa contains 13.5 mol of nitrogen, 3.6 mol hydrogen and 1.0 mol of ammonia. Calculate the equilibrium constant for the reaction.

$$N_2(g) + 3H_2(g) \rightleftharpoons 2NH_3(g)$$

The total number of moles = 13.5 + 3.6 + 1.0 = 18.1

$$p(N_2) = \frac{13.5 \times 200}{18.1} = 149\,kPa$$

$$p(H_2) = \frac{3.6 \times 200}{18.1} = 40\,kPa$$

$$p(NH_3) = \frac{1.0 \times 200}{18.1} = 11\,kPa$$

$$K_p = \frac{p(NH_3)^2}{p(N_2) \times p(H_2)^3} = \frac{11^2}{149 \times 40^3} = 1.27 \times 10^{-5}\,kPa^{-2}$$

Calculations using K_p

Calculations involving equilibrium expressions for gases are carried out in much the same way as for concentrations, working with partial pressures instead.

Worked example

0.256 g of hydrogen iodide was heated at 764 K in a sealed flask. At equilibrium, the pressure in the flask was 127 kPa.

$$2HI(g) \rightleftharpoons H_2(g) + I_2(g)$$

The flask was cooled quickly, and it was found that the mixture contained 0.00028 moles of iodine. Calculate a value for K_p.

First we need to work out the number of moles of hydrogen iodide at the start:

$$\text{amount of HI} = \frac{0.256\,\text{g}}{128\,\text{g mol}^{-1}} = 0.002\,\text{mol}$$

To work out the amount of HI unreacted you need to realise that twice as many HI molecules are used up as I_2 molecules formed. So:

$$\text{amount of HI} = 0.002\,\text{mol} - 0.00056\,\text{mol} = 0.00144\,\text{mol}$$

The amount of iodine formed is 0.00028 mol, and (from the equation) this must also be the amount of hydrogen formed.

We can now work out the partial pressures of the gases:

total number of moles =
0.00144 + 0.00028 + 0.00028 = 0.002

The partial pressures can now be calculated:

$$p(HI) = \frac{0.00144\,\text{mol}}{0.002\,\text{mol}} \times 127\,\text{kPa} = 91.4\,\text{kPa}$$

$$p(H_2) = p(I_2) = \frac{0.00028\,\text{mol}}{0.002\,\text{mol}} \times 127\,\text{kPa} = 17.8\,\text{kPa}$$

$$K_p = \frac{p(H_2) \times p(I_2)}{p(HI)^2}$$

$$= \frac{17.8\,\text{kPa} \times 17.8\,\text{kPa}}{(91.4\,\text{kPa})^2}$$

$$= 0.038$$

Again, note that the units cancel, so K_p has no units.

In some cases, you can work out K_c using concentrations or K_p using partial pressures.

There is a relationship between K_c and K_p:

$$K_p = K_c(RT)^n$$

where R is the gas constant, T is the temperature in kelvin and n is the number of molecules on the right-hand side of the equation minus the number of molecules on the left-hand side.

Heterogeneous equilibria

All of the chemical equilibria met so far have involved only homogeneous systems. This is the case for many reactions – all the reactants and products are gases, or all the substances concerned are in solution.

Equilibria in which more than one phase exists are called **heterogeneous equilibria**. Examples of this type of equilibrium include the reaction of calcium oxide and sulfur dioxide:

$$CaO(s) + SO_2(g) \rightleftharpoons CaSO_3(s)$$

This reaction can reach equilibrium in just the same way as the other reactions we have studied, provided that it is carried out in a sealed container to prevent any sulfur dioxide escaping. For the equilibrium constant we can write:

$$K_c = \frac{[CaSO_3(s)]}{[CaO(s)][SO_2(g)]}$$

However, this expression can be simplified. The expression contains two terms, $[CaSO_3(s)]$ and $[CaO(s)]$, which relate to the concentrations of two solids. If we had a crystal containing 1 mole of calcium oxide, its mass would be 56 g and its volume 16.75 cm^3 – giving a 'concentration' of 3.34 g cm^{-3} – this, of course, is its density, and this is a constant for a substance under steady conditions. So the 'concentration' of a solid is constant, no matter how much of it there is.

So the equilibrium expression contains two terms which are constants. To simplify it, these constants are combined with the other numerical constant in the expression, the equilibrium constant, to give:

$$K_c = 1/[SO_2(g)]$$

Similarly, it could be written:

$$K_p = 1/p(SO_2(g))$$

if it was more convenient to work in terms of pressure.

By using exactly the same arguments, we can show that the same thing is true for heterogeneous equilibria involving liquids. In general, the equilibrium expression for a heterogeneous reaction is written in such a way that the concentrations of solids and liquids are included in the equilibrium constant.

Worked example

The solubility of the silver halides decreases in the order AgCl >AgBr >AgI. Because of the difference in solubility, bromide ions will displace chloride ions from solid silver chloride. The equilibrium constant K_c for the reaction:

$$AgCl(s) + Br^-(aq) \rightleftharpoons AgBr(s) + Cl^-(aq)$$

is 360 at 298 K. If 0.1 mol dm^{-3} $Br^-(aq)$ is added to solid AgCl, what will be the equilibrium concentrations of $Br^-(aq)$ and $Cl^-(aq)$?

From the equation, we know that when 1 mole of Br^- reacts, 1 mole of Cl^- is formed. This means that if we start with 1 mol of $Br^-(aq)$ and x mol of $Cl^-(aq)$ is formed, then the amount of $Br^-(aq)$ at equilibrium will be $(1 - x)$mol. If we take a solution containing 0.1 mol dm^{-3} of Br^- ions and add excess solid silver chloride to it, then the equilibrium concentrations of ions can be calculated as follows:

	$[Br^-(aq)]$/mol dm^{-3}	$[Cl^-(aq)]$/mol dm^{-3}
Initially	0.1	0
At equilibrium	$0.1 - x$	x

Now

$$K_c = \frac{[Cl^-(aq)]}{[Br^-(aq)]}$$

Substituting the value of K_c and the expressions for the ion concentrations gives:

$$360 = \frac{x \text{ mol dm}^{-3}}{(0.1 - x)\text{mol dm}^{-3}}$$

$$360(0.1 - x) = x$$

$$36 - 360x = x$$

$$x = \frac{36}{361}$$

$$= 0.099\,72$$

The concentration of $Cl^-(aq)$ in the equilibrium mixture is 0.099 72 mol dm^{-3}, and the concentration of $Br^-(aq)$ is (0.1 – 0.099 72)mol dm^{-3} = 0.000 28 mol dm^{-3}.

The reaction:

$$CH_3COOC_2H_5(l) + H_2O(l) \rightleftharpoons CH_3COOH(l) + C_2H_5OH(l)$$

has a K_c value of 0.25.

0.20 moles of $CH_3COOC_2H_5$ is mixed with 0.60 moles of water in a stoppered bottle. The mixture is kept at 323 K until equilibrium is reached. How many moles of ethanoic acid will be in the equilibrium mixture?

[Hint: Take the volume of the solution to be V dm^3 and let the number of moles of ethyl ethanoate used up at equilibrium be x moles.]

Questions

1 The reaction between iron and steam is:

$$3Fe(s) + 4H_2O(g) \rightleftharpoons Fe_3O_4(s) + 4H_2(g)$$

- a Is the reaction heterogeneous or homogeneous?
- b Write down an expression for K_c for this reaction.
- c What are the units of K_c?

2 Ethanol can be manufactured by the direct hydration of ethene with steam over a phosphoric acid catalyst:

$$C_2H_4(g) + H_2O(g) \rightleftharpoons C_2H_5OH(g)$$

- a Write down the equilibrium expression for K_p.
- b At 290 °C and 7000 kPa, an equilibrium is established with 1 mole of ethene and 1 mole of steam. The equilibrium mixture contains 0.46 moles of ethanol. Calculate a value for K_p and give its units.

3 The equation for the decomposition of calcium carbonate is:

$$CaCO_3(s) \rightleftharpoons CaO(s) + CO_2(g)$$

Write down the expression for K_p.

4 Write down equilibrium expressions in terms of partial pressure for:

- a $N_2(g) + 3H_2(g) \rightleftharpoons 2NH_3(g)$
- b $PCl_3(g) + Cl_2(g) \rightleftharpoons PCl_5(g)$

More calculations involving K_c and K_p [4.5g]

In the previous two sections, we have found out how to calculate equilibrium constants, both K_c and K_p. This section is about how we can use equilibrium constants.

What does an equilibrium constant tell us?

The equilibrium constant for a reaction, whether we use K_c or K_p, is a measure of *how far* a reaction proceeds to completion. The reaction:

$$2H_2(g) + O_2(g) \rightleftharpoons 2H_2O(g)$$

has $K_c = 9.1 \times 10^{80}\,mol^{-1}\,dm^3$ at 25 °C. So at equilibrium:

$$K_c = \frac{[H_2O(g)]^2}{[H_2(g)]^2[O_2(g)]} = 9.1 \times 10^{80}\,mol^{-1}\,dm^3$$

Now 9.1×10^{80} is a very large number. The only way that we can get the numerator in the expression to be so very much bigger than the denominator is for the concentration of water at equilibrium to be huge compared to the concentrations of hydrogen and oxygen. This means that virtually all of the hydrogen and oxygen in the system will be combined as water molecules at equilibrium – so the reaction effectively goes to completion.

For the reaction:

$$N_2(g) + O_2(g) \rightleftharpoons 2NO(g)$$

K_c at 25 °C is 4.8×10^{-31}. In this case, at equilibrium we have:

$$K_c = \frac{[NO(g)]^2}{[N_2(g)][O_2(g)]} = 4.8 \times 10^{-31}$$

To get this very small number, the denominator must be very many times larger than the numerator, so the concentration of nitrogen monoxide must be tiny compared to the concentrations of nitrogen and oxygen. Fortunately for us, this reaction between nitrogen and oxygen does not occur – at least not at 25 °C.

The relationship between the magnitude of an equilibrium constant and the extent to which a reaction proceeds to completion is summed up in **table 1.3.3**. The examples of equilibria that are given are for standard conditions (25 °C and 1 atm pressure).

Value of K	Extent of reaction	
Very large $K > 1 \times 10^{10}$	Reaction proceeds almost to completion, leaving very small amounts of reactants at equilibrium	$C(s) + O_2(g) \rightleftharpoons CO_2(g)$
around 1	Concentrations of reactants and products are nearly the same at equilibrium	$2BrCl \rightleftharpoons Br_2 + Cl_2$ (in CCl_4 solution)
Very small $K < 1 \times 10^{-10}$	Reaction hardly occurs, producing very small quantities of products at equilibrium	$2HBr(g) \rightleftharpoons H_2(g) + Br_2(g)$

table 1.3.3

When we look at reactions and their equilibrium constants, the term 'position of equilibrium' is often used. When a reaction has a large equilibrium constant, the position of the equilibrium is said to lie to the right-hand side of the equation (lots of the products); for a reaction with a small equilibrium constant the equilibrium lies well to the left (lots of the reactants).

Finally, note that the equilibrium constant tells us *nothing* about the rate of a reaction. Even if the equilibrium constant for a reaction is very large, the rate at which it happens may be extremely slow. The reaction of hydrogen and oxygen to form water is a good example of this. Even though K_c for the reaction is around 9×10^{80}, a mixture of hydrogen and oxygen at room temperature does not spontaneously react to form water.

Questions

1 The table gives values for the reaction shown below at different temperatures:

$2H_2O(g) + C(s) \rightleftharpoons 2H_2(g) + CO_2(g)$

Temperature/K	K_p/kPa
298	1.00×10^{-16}
900	3.72
1300	2.04×10^{2}

a What does this suggest about the conditions of temperature used for this process industrially?

b Is this reaction forming hydrogen and carbon dioxide exothermic or endothermic? Explain your answer.

2 At room temperature and pressure, there is no reaction between nitrogen and oxygen. There is some reaction forming nitrogen monoxide at 1500 K. Suggest the magnitude of the value for K_p at each temperature.

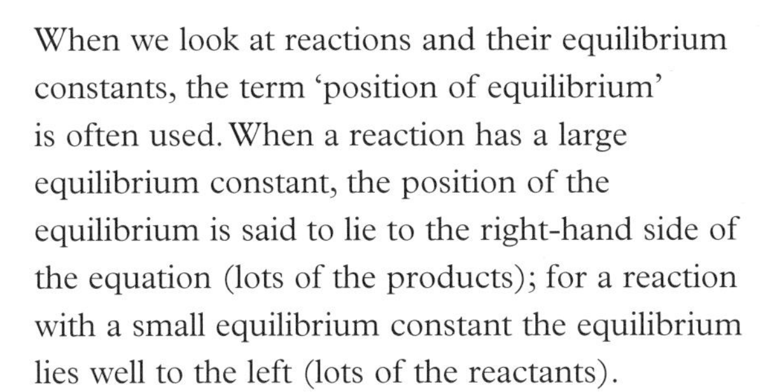

HSW Reactions in the right place

The existence of the equilibrium position for the reaction of nitrogen and oxygen at normal temperatures and pressures means that effectively the gases don't combine, and is good news for us – most of the time! But plants rely on nitrates in the soil to build proteins and grow. One way in which these nitrates are provided is from the lightning in a thunderstorm – the immensely high temperatures push the equilibrium over to a position where nitrogen oxides are formed, which are then washed down into the soil in the rain.

fig. 1.3.7 A lightning strike produces air temperatures of 30 000 K, which certainly changes the equilibrium position for the formation of nitrogen oxides!

The high temperatures and pressures in car engines can also result in a shift in the equilibrium of the reaction between nitrogen and oxygen to form nitrogen oxides. This also contributes to acid rain. Nitrogen dioxide breaks down in a photochemical reaction in sunlight to form nitric oxide and an oxygen atom, which can react with oxygen molecules in the air to form ozone, a major component of the smog found in many major cities in the world.

fig. 1.3.8 The conditions of heat and pressure in car engines encourage the formation of nitrogen oxides – this in turn leads to the formation of photochemical smogs.

One of the biggest problems for scientists is to ensure that the reactions they want to happen take place at the right time, in the right place and at the right speed – and understanding equilibria helps them to do this.

Determination of an equilibrium constant [4.5d]

The value for the equilibrium constant for a reaction can only be found by experiment. This will involve measuring the concentrations of reactants and products at equilibrium at a constant temperature.

Determining an equilibrium constant involves three basic steps:

- Combining known quantities of reactants under carefully measured and controlled conditions and allowing the reaction to come to equilibrium at a fixed temperature.
- Measuring accurately the concentration of one (or more) of the substances in the equilibrium mixture, and from this calculating the equilibrium concentrations of the other substances in the mixture.
- Substituting these equilibrium concentrations in the equilibrium expression and calculating the value.

These steps should be repeated using different initial concentrations to confirm the value of the equilibrium constant. Think carefully about these steps and identify as many potential sources of errors and uncertainties as you can to give you an idea of the problems involved.

You have considered how experimental results (**table 1.3.1** on page 53) are used to determine the equilibrium constant for the reaction between hydrogen and iodine to form hydrogen iodide. Now you are going to consider how this experimental approach works in other reactions.

Hydrolysis of ethyl ethanoate

The equation for this reaction is:

$$CH_3COOC_2H_5(l) + H_2O(l) \rightleftharpoons CH_3COOH(l) + C_2H_5OH(l)$$

Different, known quantities of ethyl ethanoate and water are mixed and left to reach equilibrium. Without a catalyst the reaction may take several years to reach equilibrium. A small quantity of concentrated hydrochloric acid is added to each reaction mixture to act as catalyst. When a system reaches equilibrium, the concentrations of the reactants and products remain constant. The number of moles of ethanoic acid can be found by titration with standard sodium hydroxide solution. (Some of the sodium hydroxide solution will react with the hydrochloric acid catalyst and this should be considered in the calculation, but for simplicity it has been ignored here.)

If the initial numbers of moles of ethyl ethanoate and water were x, and the number of moles of ethanoic acid was y, then at equilibrium, the amount of each substance is:

$$\underset{(x-y)\,\text{mol}}{CH_3COOC_2H_5(l)} + \underset{(x-y)\,\text{mol}}{H_2O(l)} \rightleftharpoons \underset{y\,\text{mol}}{CH_3COOH(l)} + \underset{y\,\text{mol}}{C_2H_5OH(l)}$$

If the volume is $V\,\text{dm}^3$, then the concentrations are $\frac{(x-y)\,\text{mol}}{V\,\text{dm}^3}$, $\frac{(x-y)\,\text{mol}}{V\,\text{dm}^3}$, $\frac{y\,\text{mol}}{V\,\text{dm}^3}$ and $\frac{y\,\text{mol}}{V\,\text{dm}^3}$, respectively.

$$K_c = \frac{[CH_3COOH(l)]\,[C_2H_5OH(l)]}{[CH_3COOC_2H_5(l)]\,[H_2O(l)]}$$

$$= \frac{y/V\,\text{mol dm}^{-3} \times y/V\,\text{mol dm}^{-3}}{(x-y)/V\,\text{mol dm}^{-3} \times (x-y)/V\,\text{mol dm}^{-3}}$$

$$= \frac{y^2}{(x-y)^2}$$

Note that because the number of moles of reactants and products are the same, the volume V cancels out. This is not always the case – and you must use concentrations and not numbers of moles in the calculation.

Table 1.3.4 shows some experimental results. The numbers of moles have been converted into concentrations. In the final column, K_c has been calculated. You can see that the values obtained for K_c from the three experiments are very close together, certainly within experimental error.

Equilibrium concentrations/mol dm^{-3}				
$CH_3COOC_2H_5(l)$	$H_2O(l)$	$CH_3COOH(l)$	$C_2H_5OH(l)$	K_c
13.31	5.27	4.40	4.40	0.276
13.6	7.87	5.47	5.47	0.280
10.0	17.4	7.00	7.00	0.281

table 1.3.4

A spreadsheet is a good way of carrying out these routine calculations.

Measuring K_c

You may carry out a similar experiment with ethanol and ethanoic acid. You can use an Excel spreadsheet to calculate your values for K_c.

Distribution of iodine between two immiscible liquids

Hexane and water are **immiscible** liquids. When added together, they form separate layers with hexane floating on the water. If some iodine is added, it will dissolve in both the hexane layer and the water layer, although it is relatively insoluble in water. This reaction is the same as the one shown in fig. 1.3.3 on page 51 – when the system is in equilibrium, the concentration of iodine in each layer remains constant:

$$I_2(\text{hexane}) \rightleftharpoons I_2(\text{water})$$

An equilibrium exists between iodine in the hexane layer and iodine in the water layer. Iodine molecules pass from the hexane layer to the water layer at the same rate as molecules pass from the water layer to the hexane layer.

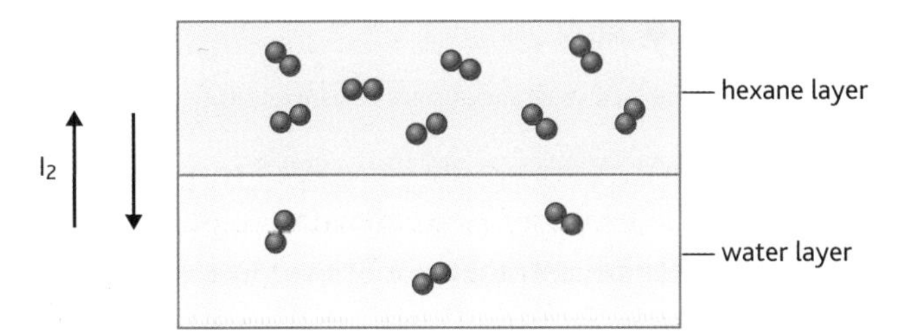

fig. 1.3.9 A dynamic equilibrium is set up as the iodine moves between the two solvents.

A known quantity of iodine is added to a known volume of hexane and water, and the system is allowed to come to equilibrium. The equilibrium concentration of iodine in the water layer can be found by titration with standard sodium thiosulfate solution. Knowing the original amount of iodine used enables the concentration of iodine in the hexane layer to be calculated. The equilibrium constant for this reaction is called the **partition constant**, K_{part}:

$$K_{part} = \frac{\text{concentration of iodine in hexane}}{\text{concentration of iodine in water}}$$

Questions

1 Is it necessary to know the volume of the solution in the reaction of silver ions and iron(II) ions? Explain your answer.

2 This question is about the following reaction:

$$2HI(g) \rightleftharpoons H_2(g) + I_2(g)$$

Hydrogen iodide was heated in a bulb of known volume. When equilibrium was established, the bulb was cooled rapidly to room temperature and the iodine present was determined by titration with standard sodium thiosulfate solution.

Why was there almost no change in the concentration of iodine when the bulb was cooled quickly to room temperature?

3 1.68 moles of $PCl_5(g)$ was mixed with 0.36 moles of $PCl_3(g)$ in a container of volume 2 dm^3. 1.44 moles of $PCl_5(g)$ remained at equilibrium.

Calculate K_c for this reaction, giving the correct units:

$$PCl_5(g) \rightleftharpoons PCl_3(g) + Cl_2(g)$$

4 A series of experiments were carried out heating different quantities of hydrogen and iodine to the same temperature in sealed containers until equilibrium was reached.

$$H_2(g) + I_2(g) \rightleftharpoons 2HI(g)$$

The results are summarised in the table.

	$H_2(g)$/ mol dm^{-3}	$I_2(g)$/ mol dm^{-3}	$HI(g)$/ mol dm^{-3}
a	4.56	0.74	13.54
b	3.56	1.25	15.59
c	2.25	2.34	16.85

Calculate the K_c values.

Relating entropy to equilibrium constants [4.5f, h]

We are going to use ideas from the earlier section on entropy and apply them to equilibrium situations.

fig. 1.3.10 When you see speed skaters like these, the fact that the pressure of the skates on the ice is changing the equilibrium position between water as a solid and a liquid is probably not the first thought that comes into your head – but that is why they can go so fast!

The effect of increasing ΔS on the equilibrium constant

In chapter 1.2, the method for calculating total entropy was introduced:

$$\Delta S_{total} = \Delta S_{sys} + \Delta S_{surr}$$

It was also stated that ΔS_{total} is positive for all spontaneous changes.

However, when dealing with reversible reactions that can reach equilibrium, both the forward and the backward reactions are spontaneous and so ΔS_{total} must be positive in both cases. At first sight, this seems to be impossible. Consider the equilibrium reaction:

$$N_2O_4(g) \rightleftharpoons 2NO_2(g)$$

colourless brown

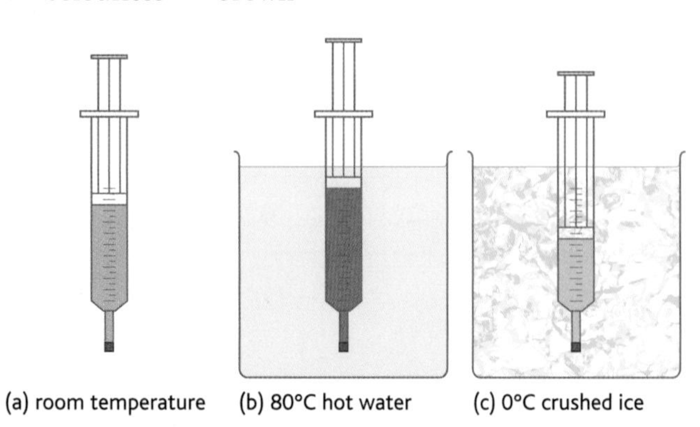

fig. 1.3.11 A mixture of dinitrogen dioxide and nitrogen dioxide in a gas syringe: (a) at room temperature; (b) immersed in hot water; (c) immersed in crushed ice.

The differences in colour show a movement of the position of the equilibrium. In hot water, the mixture becomes browner in colour (fig. 1.3.11(b)). This shows that the equilibrium has moved to the right to produce more nitrogen dioxide. The colour darkens even though the gas mixture expands as it warms up.

In the iced water, the mixture becomes a paler brown in colour – even though its volume contracts (fig. 1.3.11(c)). This shows that the equilibrium has moved to the left to produce more dinitrogen tetroxide.

Now you are going to combine what you know about the equilibrium situation here with an understanding of the entropy changes that are taking place. Consider the changes in molecular disorder taking place as dinitrogen tetroxide dissociates to form nitrogen dioxide. In fig. 1.3.12 you can see the changes in disorder as the molecules of dinitrogen tetroxide split to form more molecules of nitrogen dioxide.

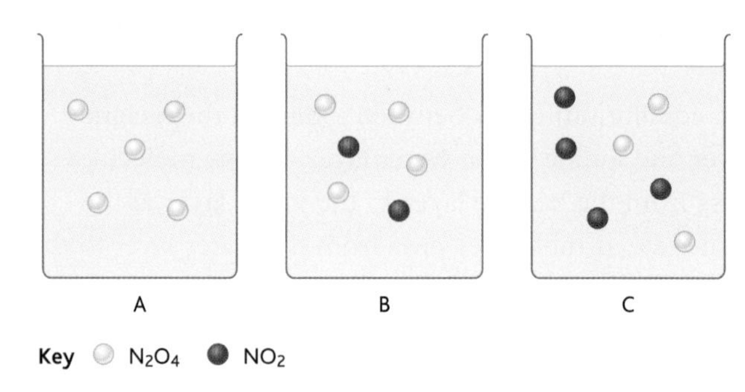

fig. 1.3.12 As the molecules dissociate, entropy increases.

The molecules in A are clearly disordered to a degree, but the disorder increases as we move to B. There are more molecules, and they are of two different types. When we move from B to C, there is a further increase in disorder as more molecules split up, but the percentage increase is less because there are already some nitrogen dioxide molecules there. When equilibrium is reached, the level of disorder stabilises because the rates of association and dissociation are equal. We could apply a similar argument to the backward reaction. The entropy changes in the surroundings also depend on the extent of the reaction.

We can conclude that when a system is in dynamic equilibrium, the total entropy change is zero and

$$\Delta S_{total}[\text{forward reaction}] = \Delta S_{total}[\text{backward reaction}] = 0$$

Earlier we examined the fact that the spontaneous direction of a change is determined by its total entropy change. We have seen here that the direction and extent of a change are determined by the equilibrium constant for that change. You would expect some connection between these two ideas – it turns out that they are related by:

$$\Delta S_{total} = R \ln K$$

where K can be K_c or K_p, and R is the gas constant. Note that this is consistent with the conclusions that we have reached – a high value for K corresponds to a large total entropy change.

Table 1.3.5 shows rough guidelines to the extent of reaction to expect for different values of the total entropy change.

ΔS_{total}/J mol^{-1} K^{-1}	Extent of reaction
+40 to –40	Mixture of reactants and products
>+200	Reaction is complete – all the reactants turned to products
< –200	No evidence of products – reaction does not take place

table 1.3.5

The effect of temperature change on the value of ΔS_{total}

There is little change in ΔS_{system} with change in temperature unless there is a change in state of one of the reactants or products, but there are significant changes to $\Delta S_{surroundings}$.

The entropy change of the surroundings during a chemical reaction is determined by:

$$\Delta S_{surroundings} = -\frac{\Delta H}{T}$$

where ΔH is the enthalpy change and T is the absolute temperature.

For an exothermic reaction, ΔH is negative and so $\Delta S_{surroundings}$ is positive – there is a large increase in entropy of the surroundings.

Consider burning magnesium in oxygen:

$$Mg(s) + \tfrac{1}{2}O_2(g) \rightleftharpoons MgO(s)$$

$$\Delta H^\ominus = -601.7\,kJ\,mol^{-1}$$

$$\Delta S^\ominus_{system} = -216.6\,kJ\,mol^{-1}\,K^{-1}$$

There is a decrease in the entropy of the system because solid magnesium reacts with a gas to form solid magnesium oxide – remember, gases have higher standard molar entropies than solids. At 298 K:

$$\Delta S_{surroundings} = -\frac{(-601.7 \times 1000)\,J\,mol^{-1}}{298\,K}$$

$$= +2019\,J\,mol^{-1}\,K^{-1}$$

$$\Delta S^\ominus_{total} = \Delta S^\ominus_{system} + \Delta S^\ominus_{surroundings}$$

$$= -216.6\,J\,mol^{-1}\,K^{-1} + 2019\,J\,mol^{-1}\,K^{-1}$$

$$= 1802.4\,J\,mol^{-1}\,K^{-1}$$

This value is positive, so the reaction is spontaneous.

Questions

1 The thermal decomposition of zinc carbonate is shown by:

$ZnCO_3(s) \rightarrow ZnO(s) + CO_2(g)$

a Explain why there is an increase in entropy during the reaction.

b Determine if the reaction is spontaneous at 700 K given the following information:

$\Delta S^\ominus_{system} = +175\,J\,mol^{-1}\,K^{-1}$

$\Delta H^\ominus = +71\,kJ\,mol^{-1}$

1.4 Application of rates and equilibrium

How temperature, pressure and catalysis affect an equilibrium constant [4.6a]

A high proportion of the chemicals produced in industry are made in processes involving equilibrium reactions – for example, the Haber process for making ammonia. It is important for manufacturers to control equilibrium reactions to produce the best yields at an economic price.

Examples of industrial reversible reactions

- The Haber process for manufacturing ammonia:

  $N_2(g) + 3H_2(g) \rightleftharpoons 2NH_3(g)$ (see *Edexcel AS Chemistry* page 206)

- The contact process in manufacturing sulfuric acid:

  $2SO_2(g) + O_2(g) \rightleftharpoons 2SO_3(g)$

- Esterification – for example:

  $CH_3COOH(l) + C_2H_5OH(l) \rightleftharpoons CH_3COOC_2H_5(l) + H_2O(l)$

At AS level you had a brief introduction to the effects of changing different factors acting on dynamic equilibria. In this chapter you are going to look in more detail at how some of these factors, including temperature, pressure and catalysts, affect the equilibrium composition and the equilibrium constant, and then apply the ideas to another industrial process – the manufacture of methanol.

Factors that affect equilibria

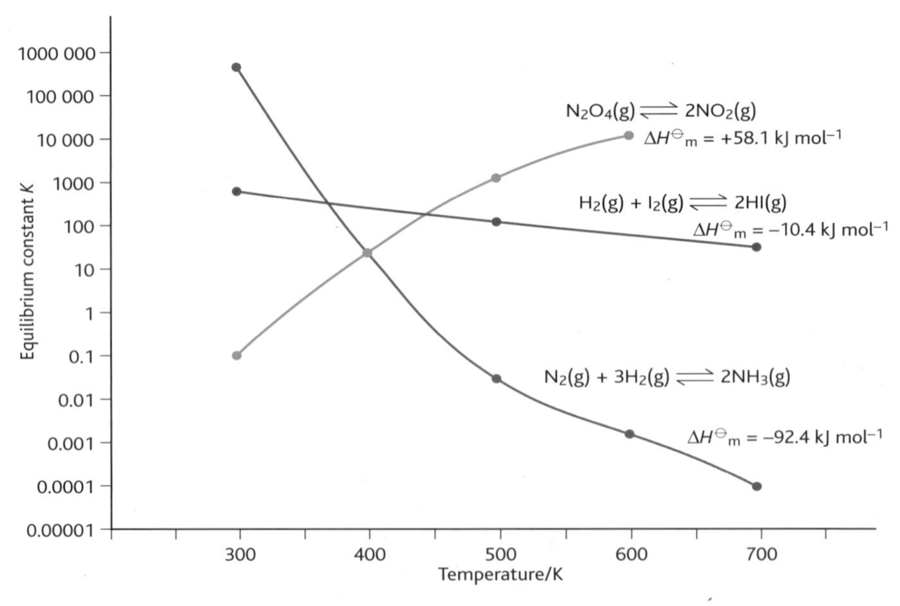

fig. 1.4.1 The effect of changing the temperature on the equilibrium constants for three different reactions.

The effect of temperature changes

The size of the equilibrium constant (K_c or K_p) for a reaction is affected by changes in the temperature. The effect of a temperature change will depend on the energy changes in the reaction concerned (**fig. 1.4.1**). For a reaction that is exothermic in the forward direction, the equilibrium constant *decreases* with *increasing* temperature. If the reaction is endothermic in the forward direction, the equilibrium constant *increases* with *increasing* temperature.

As a consequence, temperature changes also affect equilibrium positions. In general:

- If the temperature is increased, the equilibrium position of an exothermic reaction tends to move to the left (meaning that K decreases). Conversely, the equilibrium position of an endothermic reaction tends to move to the right.
- If the temperature is decreased, the equilibrium position of an exothermic reaction tends to move to the right (meaning that K increases). Conversely, the equilibrium position of an endothermic reaction tends to move to the left.

This is illustrated by the reaction:

$$\underset{\text{blue}}{[Cu(H_2O)_6]^{2+}(aq)} + 4Cl^-(aq) \rightleftharpoons \underset{\text{yellow}}{CuCl_4^{2-}(aq)} + 6H_2O(l)$$

Figure 1.4.2 shows the effect of changing the temperature on this equilibrium.

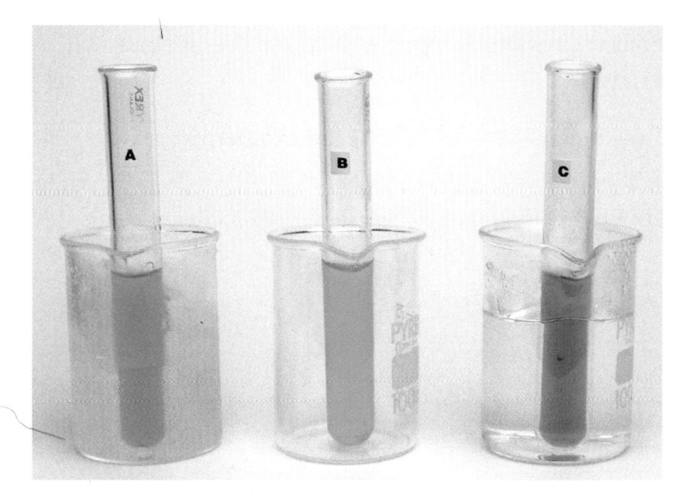

fig. 1.4.2 The tube in the middle is at room temperature, and contains a mixture of $[Cu(H_2O)_6]^{2+}$ and $CuCl_4^{2-}$ ions. Cooling the reaction mixture (left) shifts the equilibrium to the left, producing more of the blue $[Cu(H_2O)_6]^{2+}$ ions. Raising the temperature (right) shifts the equilibrium to the right, producing more of the yellow $CuCl_4^{2-}$ ions. These changes in the equilibrium position show that the reaction is endothermic in the forward direction.

The effect of pressure changes

Increasing the pressure acting on a mixture of gases is usually done by decreasing the volume in which they are reacting. The effect of a change in pressure on the equilibrium position of a mixture of gases depends on the reaction. If there are more moles on the left-hand side of the equation than there are on the right – for example in the production of ammonia – increasing the pressure will move the equilibrium position across to the right. This will tend to oppose the change in pressure by reducing the volume of gas present (in the case of ammonia, 4 moles of reactant gases produce 2 moles of ammonia).

$$N_2(g) + 3H_2(g) \rightleftharpoons 2NH_3(g)$$

A decrease in pressure would have the opposite effect – it would encourage the backward reaction with an increase in the number of moles of gas to maintain the pressure as the volume increases.

In an equilibrium reaction with the same number of moles of gas on both sides of the equation, such as:

$$H_2(g) + I_2(g) \rightleftharpoons 2HI(g)$$

changes in pressure have no effect on the equilibrium position.

When there are different numbers of moles on each side of the equation, the following guidelines apply:

- an increase in pressure, or a decrease in volume, tends to move the reaction in the direction that will decrease the number of moles of gas
- a decrease in pressure, or increase in volume, tends to move the reaction in the direction that will increase the number of moles of gas.

The effect of a catalyst

Catalysts affect the rates of reactions. The effect is the same on both the forward and the backward reactions in an equilibrium. This means that there is no overall effect on the composition of the equilibrium mixture. The only result of a catalyst is that the system reaches equilibrium more quickly – which may well be beneficial.

The effect of changing the reacting quantities

Another factor that can have a big effect on the equilibrium position, and which is very important in industrial chemistry, is adding substances to, or removing them from, the equilibrium mixture. This changes the concentration of the substances in the equilibrium mixture and so the system has to change to restore equilibrium. The general principles are:

- if a substance is added, the equilibrium tends to move in the direction that will use up the added substance
- if a substance is removed, the equilibrium tends to move in the direction that will replace the substance removed.

It should be obvious as to how useful this last point is for enabling chemical industries to increase the yield of the substances they want to manufacture! Now you are going to consider how these general principles are applied in the manufacture of methanol. However, in an industrial setting it is not only the equilibrium mixture which is important. The rate at which the reaction takes place, and the economics of the process, have to be considered as well.

Manufacture of methanol

Today, methane (natural gas) or naphtha from the fractional distillation of crude oil are used to produce methanol. Alternatively coal could be used.

Whichever method is used, the basic chemistry is the same. The first stage is the production of synthesis gas (a mixture of carbon monoxide and hydrogen).

If coal is used, and there is reason to believe that this might be the best raw material in the future, the coal is first heated out of contact with air to produce coke. Then steam is passed over white-hot coke (at 800–1200° C):

$$C(s) + H_2O(g) \rightleftharpoons CO(g) + H_2(g) \quad \Delta H = +131\,kJ\,mol^{-1}$$

Since the forward reaction is endothermic, high temperatures will move the equilibrium to the right producing more carbon monoxide and hydrogen. High temperatures will also, of course, speed up the reaction considerably.

There is an increase in entropy in the forward reaction because there are more gaseous molecules on the right-hand side. Looking at **table 1.2.3** in **chapter 1.2** (page 45), an endothermic reaction with an increase in entropy is only spontaneous at high temperatures. This is consistent with our everyday experience that there is no reaction between coke and water at ordinary temperatures.

The reaction can be carried out at normal pressure since attempting to use higher pressures will only push the equilibrium to the left, reducing the amount of products. Notice that there are 2 moles of gas on the right-hand side but only 1 mole of gas on the left-hand side.

Synthesis gas is then mixed with hydrogen and then passed over a catalyst of copper, zinc oxide and alumina at 250 °C and 50–100 atmospheres.

$$CO(g) + 2H_2(g) \rightleftharpoons CH_3OH(g) \quad \Delta H = -90\,kJ\,mol^{-1}$$

Theoretically, the maximum yield of methanol would be obtained at low temperature (because the forward reaction is exothermic) and by high pressure (because there are 3 moles of gas on the left-hand side and only 1 mole on the right leading to a decrease in entropy). In practice, a sufficient yield is obtained by a moderately high temperature, high pressure and the use of a catalyst.

Looking again at **table 1.2.3**, an exothermic reaction with a decrease in entropy suggests that this reaction is spontaneous only at low temperatures.

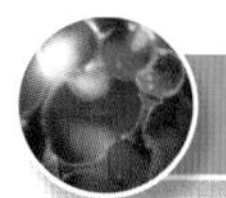

HSW Why is methanol so useful?

Methanol is a most useful fuel because it can be made from carbon or almost any carbon-based material. For example, methanol could be useful as a fuel for commercial vehicles – but the cost of a bus engine adapted to burn methanol is twice the cost of a diesel engine.

fig. 1.4.3 A bus in New York powered by methanol.

Methanol has a much higher octane rating than petrol and it has been used as a racing car fuel for decades. The methanol is usually mixed with petrol – for example, M85 is 85% methanol mixed with petrol.

However, there are problems with using methanol.

- A methanol molecule contains oxygen as well as carbon and hydrogen. One litre of methanol contains only about 53% of the energy of petrol. So fuel tanks would need to be larger and cars would be heavier.
- Methanol has a higher flash point (the temperature at which the vapour ignites) than petrol. This would need modifications to the ignition systems of vehicles to ignite the methanol.
- Methanol corrodes metal components and certain types of rubber. It is necessary to use stainless steel and methanol-resistant rubbery materials.
- Methanol-fuelled vehicles produce less hydrocarbon, carbon monoxide and oxides of nitrogen emissions, but they do produce methanal emissions. A much more expensive exhaust system would be required.

Another way of using methanol in a vehicle is in a methanol fuel cell. Unfortunately these will not operate at room temperature but need a temperature of at least 50 °C to operate.

In a new process, methanol is produced from synthesis gas, comprising of hydrogen and carbon monoxide. The synthesis gas is passed through a liquid phase reactor containing a solid catalyst suspended in methanol. In this innovation, methanol acts both as a product and as a suspension medium for the catalyst. Doing the reaction this way gives a higher conversion to methanol.

Questions

1 There are factors which affect both the equilibrium mixture of a reaction and the rate of the reaction. Why do these factors have to be considered very carefully in an industrial process?

2 If methane is used to make methanol, the first stage in producing synthesis gas is represented by:

$$CH_4(g) + H_2O(g) \rightleftharpoons CO(g) + 3H_2(g) \quad \Delta H = +206\,kJ\,mol^{-1}$$

The synthesis gas produced by this method contains more hydrogen from the same quantity of steam. If carbon dioxide is mixed with the synthesis gas made from methane, a secondary reaction takes place which also produces methanol.

a Suggest, with reasoning, likely conditions for the reaction of methane with steam.

b Write a balanced equation for the reaction of hydrogen with carbon dioxide.

3 a Write the rate equation, in terms of partial pressures, for the reaction of carbon monoxide and hydrogen to produce methanol.

b At 671 K, the value of K_p is 12.07 MPa^{-2}. Would the value of K_p be higher, lower or the same at 771 K?

Choosing conditions for industrial processes [4.6b]

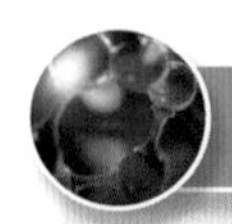

HSW Ammonia production

The Haber process for making ammonia is described on page 206 of *Edexcel AS Chemistry*. Now you have the chemical tools to look at the process in more detail.

Until 1904, nobody had been able to bring about a direct combination of nitrogen and hydrogen to form ammonia without the help of an electrical discharge, although the experiments of Berthelot and Thomson proved that the combination occurred exothermically. With your knowledge of reaction rates and equilibria, you can see that this negative result was due to the slowness of the reaction at low temperatures, and unfavourable equilibrium conditions at high temperatures.

A young chemist called Fritz Haber applied thermodynamic theory about the behaviour of gases to establish industrial requirements for creating reactions, earning him an international reputation as an expert in adapting science to technology.

In 1905 Haber began his groundbreaking work on the synthesis of ammonia. An initial single experiment gave his team hope of finding a technical solution to the problem of producing ammonia from nitrogen and hydrogen. They worked at a temperature of about 1000 °C and normal pressure, using iron as a catalyst. They found that at red heat temperatures and above, and also at higher pressures, traces of ammonia could be formed.

During this work it was also shown experimentally, for the first time, that a real state of equilibrium existed in the system. The equation can be written:

$$\tfrac{1}{2}N_2(g) + \tfrac{3}{2}H_2(g) \rightleftharpoons NH_3(g) \quad \Delta H^{\ominus} = -46.1\,kJ\,mol^{-1}$$

The equilibrium constant for this reaction is:

$$K_p = p(NH_3) / p(N_2)^{1/2} \times p(H_2)^{3/2}$$

This information shows that the formation of ammonia is an exothermic process and that it is accompanied by a reduction in entropy as there are more moles of gas on the left-hand side. Looking at **table 1.2.3** on page 45, it should be clear that this process is likely to be spontaneous only at low temperatures.

It was also clear that the yield of ammonia increased as pressure increased. This was supported by experiments carried out by Haber and his team. They examined the equilibrium under different conditions of pressure (at constant temperature) and temperature (at constant pressure). They also calculated K_p values to four decimal places. The results are shown in **fig. 1.4.4**.

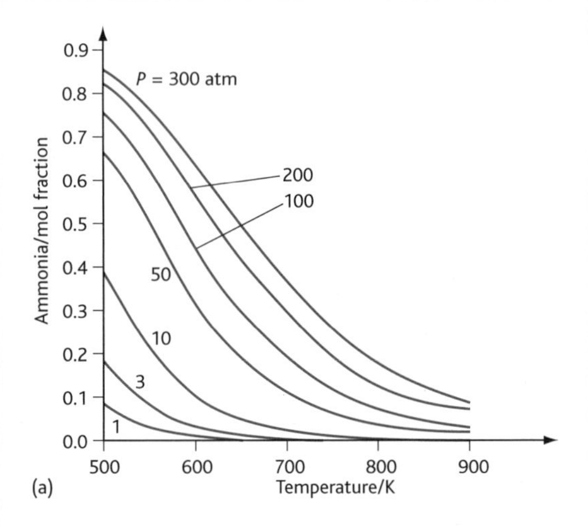

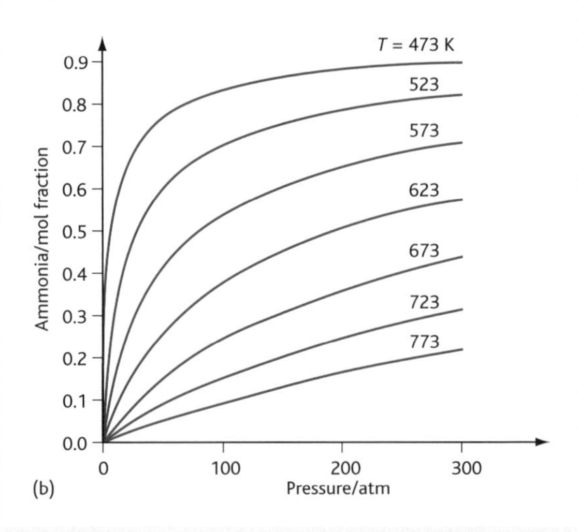

fig. 1.4.4 The mole fraction of ammonia at equilibrium from an initial mixture of nitrogen and hydrogen in a 1 : 3 ratio – (a) at different temperatures at a constant pressure; (b) at different pressures at a constant temperature.

In fact, equilibrium conditions are only part of the story. They don't give any indication about the rate of the reaction. The reaction does not take place at low temperatures because a great deal of energy is required to break the bonds required to turn nitrogen and hydrogen molecules into atoms. A temperature of about 3000 °C is required to break nitrogen molecules, and about 1000 °C to break hydrogen molecules. Unfortunately, raising the temperature to break these bonds will push the equilibrium to the left.

The importance of a suitable catalyst was realised. **Figure 1.4.5** shows the energy profiles for ammonia synthesis with and without an iron catalyst. The use of a catalyst lowers the activation energy, speeding up the reaction sufficiently to obtain ammonia under conditions where the equilibrium conversion is large enough to be useful.

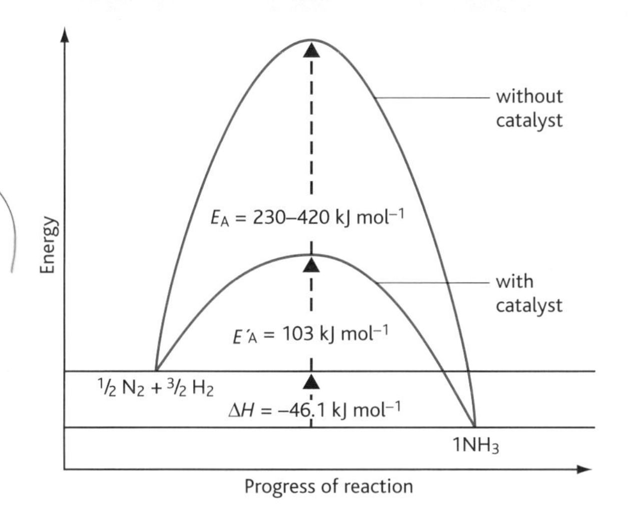

fig. 1.4.5 **Ammonia production with and without a catalyst.**

The nitrogen and hydrogen molecules lose some of their freedom to move when bound to the catalyst surface.

Haber experimented to find suitable catalysts. It was known that iron was suitable, but other transition metals were tried. He found that finely divided osmium could be used, as could uranium. However, these catalysts were too expensive and unstable to use for commercial production. Eventually, iron with a few per cent of alumina and a small amount of potassium was adopted.

Haber recognised that much higher pressures had to be used, which is in fact the real basis for the synthesis of ammonia. He also realised that significant yields of ammonia would not be made in a single pass of nitrogen and hydrogen over the catalyst. He developed the process of recycling unreacted nitrogen and hydrogen after condensation of the ammonia. Haber's ideas represented a significant change from the previous static approach to industrial production and the development of a more dynamic approach. He was using reaction kinetics as well as thermodynamics.

In 1909, Haber established a small-scale laboratory plant using the recycling process. This was capable of producing 80 g of ammonia every hour.

In 1910, construction work began on the first large factory, with an estimated annual output of 30 000 tons of ammonia. Scaling up the process was very much the responsibility of engineer Karl Bosch, whose company Badische Anilin- und Sodafabrik (BASF) had supported Haber's research. Bosch solved some key problems, such as designing containers that could withstand a corrosive process over a period of time. It was, however, not easy and there were a number of explosions in the factories in the early days.

A modern plant produces 1000–1500 tonnes of ammonia per day. The gases move through the plants at speeds of 10 000–20 000 m^3 per 1 m^3 of catalyst per hour. This rapid movement of gas is important in getting a good yield. Improvement of the catalysts (e.g. the use of ruthenium) has enabled reductions to be made in the operating pressure to about 40 atmospheres. The atom economy of the modern Haber process is 100% because all of the reactants are successfully turned into the desired product.

Questions

1 In the contact process for manufacturing sulfuric acid, the key reaction is:

$SO_2(g) + \frac{1}{2}O_2(g) \rightleftharpoons SO_3(g) \qquad \Delta H^\ominus = -98\,\text{kJ mol}^{-1}$

a Write down the expression for K_p.

b Is there an increase or decrease in entropy in the forward reaction? Explain your answer.

c Is this reaction exothermic or endothermic?

d What does this suggest about the spontaneity of the reaction?

e What does this suggest about the most suitable temperature for the process?

f What problems does this cause? Suggest how they are overcome.

The process is carried out using slightly more oxygen than the 2 : 1 ratio suggested by the equation.

g What is the benefit of this? Why is a much greater concentration of oxygen not used?

h The conversion of SO_2 to SO_3 is efficient even with relatively low pressures. What does this mean for the process on an industrial scale?

Controlling reactions for safety, yield, cost and atom economy [4.6c, d]

In any industrial process, the aim is to achieve the maximum yield for the minimum cost. As a result, reaction conditions are balanced to give the best equilibrium position that can be achieved under conditions that are not too expensive to maintain. High pressure needs expensive reaction vessels. High temperatures use a lot of energy and are expensive. So, whenever possible, catalysts are used to increase reaction rates and products are removed to push the equilibrium position towards the desired yield.

Safety is another important issue. It is important for many reasons that an industrial chemical process is as safe as possible, and this is another reason why very high temperatures and pressures are avoided if there is a good alternative route.

Finally, for all industrial reactions economy is the bottom line. If a process does not make money, it won't succeed. This is where the concept of the **atom economy** of an industrial process is so important:

$$\text{atom economy (\%)} = \frac{\text{mass of atoms in desired product}}{\text{mass of atoms in reactants}} \times 100\%$$

A reaction with a good atom economy is very efficient at turning reactants into the desired products with very little waste. In pages 31, 206 and 244–5 of *Edexcel AS Chemistry*, you discovered the importance of this concept and how it was developed.

There are a number of ways in which the atom economy of a process can be improved – they include capturing and recycling any unused reactants, and using more efficient alternative reactions in the process where possible, e.g. the alternative methods of producing synthesis gas for the production of methanol. The Haber process takes account of both thermodynamic and kinetic factors as well as recycling unreacted reactants. These ideas can be applied to other industrial processes – always be on the lookout for them.

HSW Producing urea

Urea, NH_2CONH_2, is manufactured in ever-increasing amounts. 90% of the urea manufactured is used as a slow-acting fertiliser. It is also used to manufacture urea–formaldehyde resins, and in the synthesis of melamine (Formica), adhesives, paints, laminates and moulding compounds.

Urea production involves a two-step process in which ammonia and carbon dioxide react to form ammonium carbamate, which is then dehydrated to urea. Ammonia and carbon dioxide are fed to the synthesis reactor, which operates around 180–210 °C and 150 atmospheres.

Reaction 1: $2NH_3(g) + CO_2(g) \rightleftharpoons NH_2COONH_4(s)$ ΔH –ve

Reaction 2: $NH_2COONH_4(s) \rightleftharpoons NH_2CONH_2(s) + H_2O(g)$ ΔH +ve

Reaction 1 is fast, highly exothermic and goes essentially to completion under normal industrial processing conditions. Reaction 2 is slow, endothermic and usually does not reach equilibrium under processing conditions.

In reaction 1 there is a large decrease in entropy of the system because there are 3 moles of gas on the left-hand side and 1 mole of solid on the right. In reaction 2 there is an increase in system entropy because of the formation of a gas. Looking at **table 1.2.3** on page 45, we can summarise the thermodynamics for these processes:

	ΔH	ΔS	Outcome
Reaction 1	negative	decrease	Spontaneous only at low temperatures
Reaction 2	positive	increase	Spontaneous only at high temperatures

Figure 1.4.6 shows the flow diagram for urea production. The process is designed to maximise the yield from these two reactions and stop the formation of an unwanted compound, biuret:

$$2NH_2CONH_2(s) \rightleftharpoons NH_2CONHCONH_2(s) + NH_3(g)$$

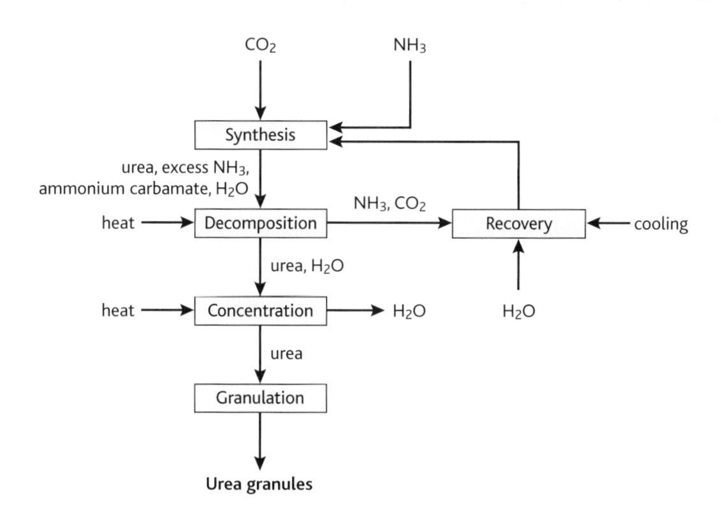

fig. 1.4.6 **The industrial synthesis of urea.**

The formation of biuret reduces the yield of urea. What's more, if it is mixed with the urea used as fertiliser, it will burn the leaves. Another unwanted reaction that can occur during the process is the hydrolysis of urea:

$$NH_2CONH_2(s) + H_2O(l) \rightleftharpoons 2NH_3(g) + CO_2(g)$$

Step 1: Synthesis

This involves two reactor vessels. A mixture of carbon dioxide and ammonia is compressed to 240 atmospheres. An exothermic reaction takes place, and a heat exchanger removes heat and uses it to turn water to steam. About 78% of the carbon dioxide is converted into urea and the liquid is then purified.

The gas from the first reactor and recycled gases go into the second reactor where the pressure is about 50 atmospheres. Here, about 60% of the remaining CO_2 is converted to urea. The solution is then purified in the same way as the liquid from the first reactor.

Step 2: Purification

The major impurities in the mixture at this stage are water, produced during the reaction, and any unused reactants – ammonia, carbon dioxide and ammonium carbamate.

The purification is carried out in three stages to reduce the formation of undesirable products.

The pressure is reduced to 17 atmospheres and the solution is heated. This causes any ammonium carbamate to decompose into ammonia and carbon dioxide. Excess ammonia and carbon dioxide escape from the solution.

Then the pressure is reduced and more carbon dioxide and ammonia escape. Finally, the pressure is reduced to less than 1 atmosphere. Again more carbon dioxide and ammonia escape from the solution. Any ammonia and carbon dioxide remaining are recycled.

Step 3: Concentration

75% of the urea solution is heated under vacuum, which evaporates some of the water, increasing the urea concentration. Some urea crystallises. After evaporating some more water, a concentrated solution of urea is produced.

Step 4: Granulation

Urea is sold as a fertiliser in the form of small granules. These can be formed by spraying the concentrated solution or molten urea onto seed granules. All dust and air from the granulator is removed by a fan into a dust scrubber.

Questions

1 Explain how these reactions maximise the yield while keeping the cost of the process to a minimum and maintaining safety standards.

2 a Which steps of this reaction will help to increase the atom economy of the process?

b Try to find out figures for the atom economy of the manufacture of urea.

c How does the atom economy step in the manufacture of urea compare to the Haber process and why?

3 Urea dust can irritate the skin, eyes and nose. What precautions are taken in the factory to prevent anyone suffering from these effects?

4 Draw the displayed formulae for **a** ammonium carbamate; **b** urea; **c** biuret.

5 Suggest why a sudden reduction of the pressure is not carried out during the purification.

1.5 Acid/base equilibria

What are acids and bases? [4.7a, b, c]

You have probably come across the idea that acids have a 'sharp' taste, and that bases have a 'soapy feel' when rubbed between a damp forefinger and thumb. You may even have tasted an acid like citric acid and felt a base like sodium hydrogencarbonate. However, the 'taste-and-feel' tests for acids and bases are not recommended because many acids and bases can cause considerable harm – citric acid and sodium hydrogencarbonate are fine to treat in this way, but battery acid (sulfuric acid) and sodium hydroxide are very different.

fig. 1.5.1 Acids and bases play an important part in everyday life – both directly and indirectly.

fig. 1.5.2 Arrhenius' theory that aqueous solutions contain electrically charged particles meant that he very nearly failed his PhD – the examiners were extremely critical of his ideas. This work eventually won him a Nobel Prize in 1903!

Theories of acids and bases

The first attempt to define acids and bases was in 1777, when Antoine Lavoisier suggested that acids were compounds that contained oxygen. Shortly after this, it was discovered that hydrochloric acid (HCl) contained no oxygen! The English chemist Sir Humphry Davy then proposed in 1810 that hydrogen, rather than oxygen, was the important element in acids. In 1838, the German chemist Justus von Liebig offered the first really useful definition of an acid – namely, a compound containing hydrogen that can react with a metal to produce hydrogen gas.

The first comprehensive theory of the behaviour of acids and bases in solution was put forward by the Swedish chemist Svante Arrhenius in his PhD thesis published in 1884.

Arrhenius proposed that when acids, bases and salts dissolve in water, they separate partially or completely into charged particles called ions. Because solutions of ions are good electrical conductors, the substances that produce them are called **electrolytes**. Acids were considered to be electrolytes that produce the hydrogen ion (H^+) when they dissolve in water. According to Arrhenius' theory, acids have common properties, despite the obvious differences in their formulae, because they all produce H^+ ions when they dissolve in water:

$$HA(aq) \rightleftharpoons H^+(aq) + A^-(aq)$$

In the same way, Arrhenius' theory explained the common properties of bases as being because they all produce the hydroxide ion in solution:

$$B(aq) + H_2O(aq) \rightleftharpoons BH^+(aq) + OH^-(aq)$$

Use the Internet and/or textbooks to show why Lavoisier's definition based on oxygen was a sound theory, given the knowledge available at the time.

Neutralisation

One of the most important properties of acids and bases is that they can 'destroy' each other when mixed together in the right proportions – a reaction called **neutralisation**. For example, when exactly equal volumes of solutions of hydrochloric acid and sodium hydroxide with exactly the same molar concentrations are mixed together, they form a solution that has no effect on litmus. The reaction between these two chemicals is:

$$HCl(aq) + NaOH(aq) \rightarrow NaCl(aq) + H_2O(l)$$

The reaction between an acid and a base produces a salt (in this case sodium chloride) – we say that the acid neutralises the base, or vice versa.

According to Arrhenius' theory, neutralisation occurs because the number of hydrogen ions from the acid is exactly equal to the number of hydroxide ions from the base, and so the two react completely to form water:

$$H^+(aq) + OH^-(aq) \rightarrow H_2O(aq)$$

Notice that the reaction has been written with a single arrow in the forward direction rather than with equilibrium arrows – as we shall see shortly, the equilibrium constant for this reaction is very large, so it effectively goes to completion.

The oxonium ion

Developments in understanding of the atom in the years following Arrhenius showed that it was almost inconceivable that the hydrogen ion could exist in solution independently. The H^+ ion is simply a proton, with a diameter about 70 000 times smaller than the diameter of a Li^+ ion.

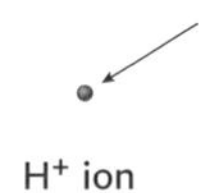

H^+ ion

fig. 1.5.3 If an H^+ ion is represented by a dot 1.0 mm across, a Li^+ ion would need to be represented by a circle 70 metres in diameter!

As a result, it was suggested that the H^+ ion exists in association with a water molecule as the H_3O^+ ion, called the **oxonium ion**:

$$HA(aq) + H_2O(l) \rightarrow H_3O^+(aq) + A^-(aq)$$

When talking about acids and bases, hydrogen ions are often referred to as protons. Strictly speaking, we should always remember that protons do not exist in solution and instead use the term oxonium ion, writing H_3O^+ instead of H^+.

The importance of water in the behaviour of acids was recognised because observations show that substances like hydrogen chloride (HCl) and ethanoic acid (CH_3COOH) do not show acidic properties when dissolved in organic solvents such as methylbenzene. They are non-electrolytes and do not affect dry litmus paper, since the solutions contain no H^+ ions.

Evidence for the H_3O^+ ion

Experimental evidence confirms the existence of the H_3O^+ ion, since the positive ions in an acidic solution carrying an electric current move at a rate which is consistent with them having a size similar to that of a water molecule. If free H^+ ions did exist, the positive ions in acidic solutions would be expected to move much faster than this because of their small size.

The Brønsted–Lowry definition of acids and bases

Although it was useful, the problem with the Arrhenius definition of an acid was that it was limited to situations in which water was present, since it defined acids and bases in terms of the ions which were produced in aqueous solutions. This definition was far too restrictive – there are many reactions which appear to be acid–base reactions that occur in solvents other than water, or even with no solvent at all. One such reaction occurs between hydrogen chloride and ammonia.

In aqueous solution, ammonia (a base) and hydrogen chloride (an acid) react to form a solution of ammonium chloride (a salt). According to the Arrhenius definition, this reaction occurs between the H^+ ions formed by the ionisation of the HCl when it dissolves in the water and the OH^- ions formed when the ammonia dissolves in water, as shown in **fig. 1.5.4**.

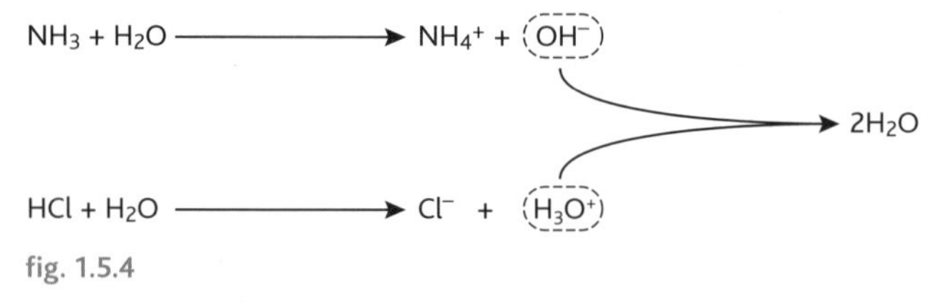

fig. 1.5.4

In fact, the reaction between ammonia and hydrogen chloride does not need water or any other solvent to happen. The white fumes (see **fig. 1.5.5**) are tiny crystals of ammonium chloride, formed as ammonia gas and hydrogen chloride gas react together:

$$NH_3(g) + HCl(g) \rightarrow NH_4Cl(s)$$

fig. 1.5.5 The reaction of ammonia and hydrogen chloride gases to produce a white smoke of ammonium chloride.

This is clearly the same reaction as the one that occurs in water, and therefore it ought to be an acid–base reaction. However, the Arrhenius definition of an acid–base reaction does not allow us to do this, since there is no reaction between H_3O^+ and OH^-.

This problem was recognised in 1923, when the Danish chemist Johannes Nicolaus Brønsted and the English chemist Thomas Martin Lowry independently formulated a more general definition of acids and bases. They proposed that:

- an acid is a proton donor
- a base is a proton acceptor.

Using this definition, the reaction between ammonia and hydrogen chloride can be defined as an acid–base reaction, because the HCl molecule can be seen to act as a proton donor, and the ammonia molecule as a proton acceptor:

$$NH_3(g) + HCl(g) \rightarrow NH_4^+Cl^-(s)$$

Conjugate acids and bases

As a result of the Brønsted–Lowry definition, when an acid dissociates we can consider this to be an acid–base reaction that is in equilibrium:

$$HA(aq) + H_2O(l) \rightleftharpoons H_3O^+(aq) + A^-(aq)$$

In the forward reaction, the HA acts as an acid, donating a proton to a water molecule. Water, in accepting the proton, acts as a base. In the reverse reaction, the H_3O^+ acts as an acid, donating a proton to A^-, which of course acts as a base.

In acid–base reactions it is always possible to find *two* acids (in this case, HA and H_3O^+) and two bases (in this case, H_2O and A^-). In each case, the acid on one side of the equation is formed from the base on the other side – they are called conjugate acid-base pairs. In this example HA and A^- form one such pair, while H_2O and H_3O^+ form the other. HA is the **conjugate acid** of A^-, which means that A^- is the **conjugate base** of HA. H_3O^+ is the conjugate acid of H_2O, so H_2O is the conjugate base of H_3O^+. This is summarised in **fig. 1.5.6**.

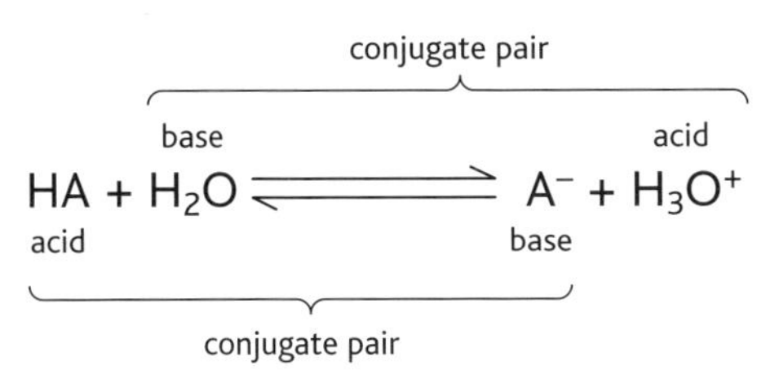

fig. 1.5.6 Conjugate pairs.

Acid or base?

Some substances are able to act as both acids and bases. The most common example of this is water, which acts as an acid when it reacts with substances like ammonia, donating a proton during the reaction:

$$H_2O(l) + NH_3(aq) \rightarrow OH^-(aq) + NH_4^+(aq)$$

However, it behaves as a base when it reacts with substances like hydrogen chloride, accepting a proton in the reaction:

$$H_2O(l) + HCl(aq) \rightarrow H_3O^+(aq) + Cl^-(aq)$$

Strong and weak, acids and bases

We saw at the beginning of this chapter that different acids have different strengths, making it possible to taste an acid like citric acid quite safely, whereas sulfuric acid (the acid in a car's battery) will quickly damage your tongue!

Think about the equilibrium which occurs when a **weak acid** like ethanoic acid dissolves in water:

$$CH_3COOH(aq) + H_2O(l) \rightleftharpoons CH_3COO^-(aq) + H_3O^+(aq)$$

In this example, there are two conjugate acid–base pairs – $CH_3COOH(aq)$ and $CH_3COO^-(aq)$, and $H_3O^+(aq)$ and $H_2O(l)$.

The two acids in this equilibrium, CH_3COOH and H_3O^+, are competing to donate a proton to a base. We can show that this equilibrium lies well over to the left by testing the electrical conductivity of the mixture. This must mean that the oxonium ion is a better proton donor than ethanoic acid – there is much more CH_3COOH than H_3O^+ in the solution. So H_3O^+ is a stronger acid than CH_3COOH. Similarly CH_3COO^- must be a stronger base than H_2O because it is better at accepting protons.

Another point about conjugate acid–base pairs can be made if we look again at what happens when hydrogen chloride – an extremely powerful proton donor – dissolves in water:

$$H_2O(l) + HCl(aq) \rightarrow H_3O^+(aq) + Cl^-(aq)$$

The equilibrium constant for this reaction is so large that we can consider that the reaction effectively goes to completion. As well as telling us that HCl is a very **strong Brønsted acid**, this also tells us that Cl^- is a very **weak Brønsted base** – it cannot even get protons from a good proton donor like H_3O^+. In general:

- strong Brønsted acids have weak conjugate bases
- weak Brønsted acids have strong conjugate bases.

SC Research the Lewis theory of acids and bases, and discuss the relationship between neutralisation and redox.

Questions

1 In each example below, identify two conjugate acid–base pairs.

a $H_2O(l) + HSO_4^-(aq) \rightleftharpoons SO_4^{2-}(aq) + H_3O^+(aq)$

b $CH_3COOH(l) + HClO_4(l) \rightleftharpoons CH_3COOH_2^+(l) + ClO_4^-(l)$

2 How would electrical conductivity indicate the position of the equilibrium in the reaction below?

$$CH_3COOH(aq) + H_2O(l) \rightleftharpoons CH_3COO^-(aq) + H_3O^+(aq)$$

3 Describe how (i) the Arrhenius and (ii) Brønsted-Lowry theories define acids and bases.

A definition for pH and measuring pH for a variety of substances [4.7d, f(i, ii)]

Scientists recognised that some solutions were acidic and some alkaline long before they had a reliable method of measuring just how acidic or alkaline they were. That all changed with the introduction of the pH scale around 100 years ago.

What is pH?

In 0.1 mol dm^{-3} hydrochloric acid, the concentration of H^+ and OH^- ions at 25 °C is 1×10^{-1} mol dm^{-3} and 1×10^{-13} mol dm^{-3}, respectively. So, as you can see, recording the concentrations of H^+ and OH^- ions in aqueous solutions will involve negative indices. This problem was overcome when the Danish chemist Soren Sørensen first introduced the concept of pH in 1909.

HSW Developing the pH scale

The pH scale is as useful today as when it was first developed. How did this useful measure come about? In the nineteenth and early twentieth centuries, a number of scientists were working hard to find a way of indicating the degree of acidity or alkalinity of a solution. The idea of using hydrogen ions as an indicator was becoming accepted, and the rather cumbersome hydrogen electrode was the way things were measured.

Soren Peter Lauritz Sørensen was not really interested in acids and alkalis as such. He started out working in inorganic chemistry but by the early 1900s, when he was in his thirties, he had become deeply involved in the study of amino acids, proteins and enzymes at the Carlsberg Laboratory in Copenhagen. The shapes of protein molecules and the activity of enzymes are heavily dependent on the hydrogen ion concentration of their environment. Sørensen wanted a relatively quick and easy way to measure and refer to the hydrogen ion concentration of his solutions to enable him to control what was happening with his precious proteins. He suggested the notation 'pH', standing for 'power of hydrogen' using the negative logarithm of the concentration of hydrogen ions in solution. The pH scale was basically a useful research tool he came up with to help himself!

Other scientists quickly recognised the potential of this new method. Leonor Michaelis wrote a book on hydrogen ion concentration, and Arnold Beckman developed a simplified pH meter. This prevented the pH scale from disappearing as just a method in Sørensen's experimental procedure, and gave all chemists a tool which is still useful and valid today.

fig. 1.5.7 The development of the pH scale was the result of some creative thinking by Sørensen to help his own protein research – but other chemists loved the idea and their support helped it become completely accepted and widely used.

The definition of pH is the negative logarithm of the hydrogen ion in mol dm^{-3}:

$$pH = -\log[H^+(aq)]$$

One of the advantages of the form of this expression is that it converts negative indices into positive numbers. The other is that is expresses very small numbers in a 'friendly' way. The range of concentration of hydrogen ions in aqueous solutions most commonly encountered is from about 1×10^{-14} mol dm^{-3} and 1 mol dm^{-3}. This gives the range of pH values for these solutions as lying between 0 and 14, since $-\log(1) = 0$ and $-\log(10^{-14}) = 14$.

Logarithms

The pH scale uses logarithms to base 10, and not to base e (natural logarithms). Logarithms to base e are written as 'ln'. Logarithms to base 10 are written as '$\log_{10}$', or just 'log'. You can get pH values using the log button on your scientific calculator.

Worked example

Three solutions of hydrochloric acid have $[H^+(aq)]$ of 0.1, 0.001 and 0.0005 mol dm^{-3} respectively. Calculate the pH of each solution.

If $[H^+] = 1 \times 10^{-1}$ mol dm^{-3}

pH = $-\log(10^{-1})$ = **1**

If $[H^+] = 1 \times 10^{-3}$ mol dm^{-3}

pH = $-\log(10^{-3})$ = **3**

If $[H^+] = 5 \times 10^{-4}$ mol dm^{-3}

pH = $-\log(5 \times 10^{-4}) = -(0.7 + -4)$ = **3.3**

Measuring pH

The pH values of solutions can be determined quickly using a mixture of indicators called universal indicator. The indicator changes show different colours (see **fig. 1.5.8**) depending on the pH of the solution.

fig. 1.5.8 Universal indicator in solutions of various pH values – top row values 1–4, middle row values 5–8, bottom row values 9–12.

It is simple to use universal indicator, but it can only measure pH to the nearest unit and it cannot be used for coloured solutions. More accurate measurements of the pH can be obtained using a pH meter (**fig. 1.5.9**).

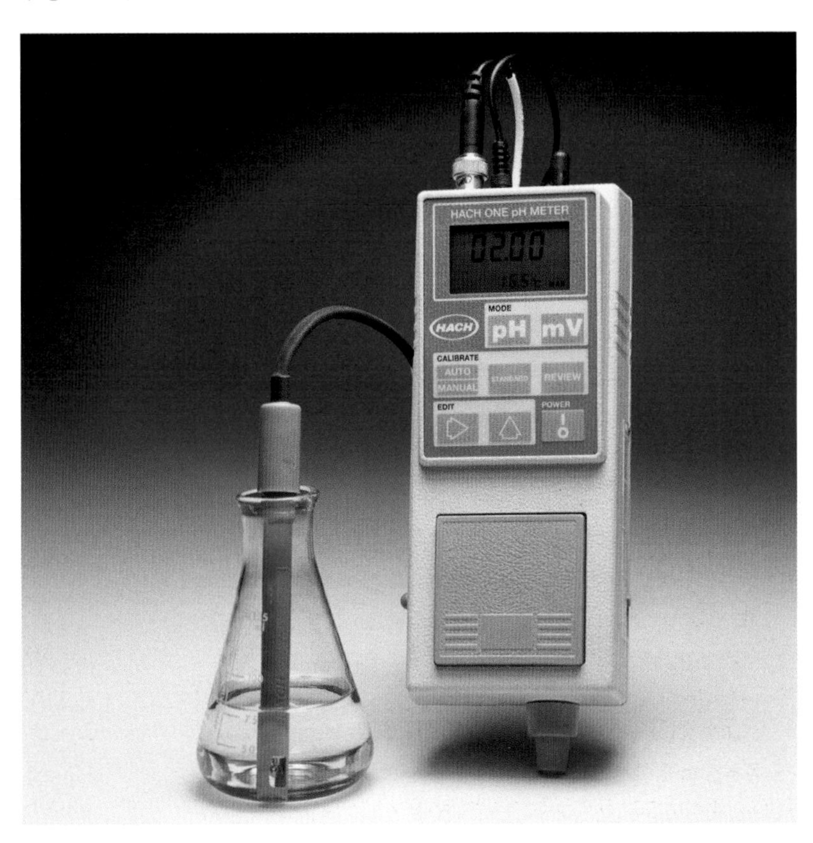

fig. 1.5.9 Digital pH meter in a solution of pH 2.00 indicating a strong acid.

Table 1.5.1 shows the pH values of a range of 0.1 mol dm^{-3} solutions of substances measured by a pH meter.

Substance	A	B	C	D	E	F	G	H	I
pH	1.00	1.00	4.50	5.20	13.00	13.00	8.60	10.30	7.00

table 1.5.1 pH values of ten different substances at the same concentration.

The solutions in the table have the same concentration – however, although A–D are acidic, the solutions do not all have the same pH. Substances A and B are strong acids, in which all the molecules are dissociated into ions. The concentration of hydrogen ions (oxonium ions) in the solutions of A and B is 1×10^{-1} mol dm^{-3}.

Solutions of C and D are also acidic but only some of the acid molecules are dissociated into ions. These substances are called weak acids. They have a higher pH value – and there is more dissociation in C than in D.

Substances E and F are **strong bases**. Here all the molecules are dissociated into ions. The concentration of hydrogen ions and hydroxide ions in solutions E and F will be the same and, hence, their pH values are the same.

Substances G and H are **weak bases** where only some of the molecules are dissociated into ions. There is more dissociation in solution H than solution G.

Solution I is neutral and the concentrations of hydrogen ions and hydroxide ions are equal. This could be pure water or a solution of a salt – but you will find out that salt solutions can be acidic, alkaline or neutral.

Table 1.5.2 shows the pH values of a weak acid and a strong acid at different concentrations.

You can see a simple pattern in the pH values as the strong acid solution is diluted. A 10-fold dilution in the strong acid increases the pH by 1 unit. With a weak acid there is a different pattern. The behaviour of the acid is represented by the equation:

$$HA(aq) + H_2O(l) \rightleftharpoons H_3O^+(aq) + A^-(aq)$$

Strong acid		Weak acid	
Concentration /mol dm^{-3}	pH	Concentration /mol dm^{-3}	pH
0.1	1.00	0.1	2.60
0.01	2.00	0.01	3.40
0.001	3.00	0.001	4.30

table 1.5.2 pH of a weak and a strong acid at different concentrations.

Addition of water increases the dilution as before, tending to decrease hydrogen concentration by a factor of 10 and to increase the pH by 1 unit. However, the addition of water moves the equilibrium to the right, producing more $H^+(aq)$ ions tending to cause the pH to decrease.

HSW pH and the soil

pH is a very important concept for chemists, but it is also a big factor for gardeners, farmers, biologists and horse-breeders – in fact, for anyone who has an interest in the soil. Soil pH varies enormously around the world and from one area of a country to another. The pH of a soil has a major impact on the plants that can grow in it – in turn that affects which animals can live and thrive in a particular area, both above the ground and in the soil. For example, hydrangeas are plants that produce different coloured flowers depending on the pH of the soil in which they are growing. An acid soil with a pH of 4.5–5 produces blue hydrangeas, but as the soil becomes less and less acidic, and even alkaline, the colour becomes pink or red. Note that hydrangea colours are opposite to those for litmus paper! So many plants thrive in one type of soil or other that gardeners often use kits to test soil pH, so they can grow the most suitable plants. Or they can add substances to the soil to make it more favourable for growing a certain crop.

Racehorses thrive on the grass they feed on. In England the average soil pH is around 6.5–7.0, and it contains relatively high levels of calcium and magnesium salts and phosphates. This helps English racehorses to develop strong bones. Kentucky bluegrass feeds some of the best racehorses in the world – and here too the average soil pH is 6.4–7.0. However, in Japan the soil tends to be much more acidic, with a pH of 4.2–5.5, with much lower levels of mineral salts. As a result, Japanese wild grass is not good for racehorses and they tend to have a lower bone density than UK or US horses. Breeders in Japan are treating their paddocks with magnesium oxide and calcium carbonate to try to lower the acidity levels, and they add salt supplements to the feed to increase bone strength.

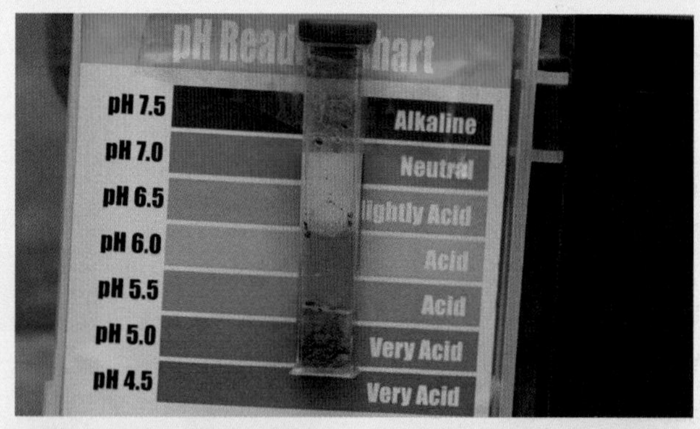

fig. 1.5.10 Soil pH affects many things – from the plants in your garden to the performance of racehorses.

The ionic product of water, K_w

In water, there is an equilibrium between the water molecules and the ions they dissociate into. This can be written as:

$$2H_2O(l) \rightleftharpoons H_3O^+(aq) + OH^-(aq)$$

or more simply

$$H_2O(l) \rightleftharpoons H^+(aq) + OH^-(aq)$$

We can write an expression for the equilibrium constant:

$$K_c = \frac{[H^+(aq)][OH^-(aq)]}{[H_2O(l)]}$$

This can be rewritten as:

$$K_c \times [H_2O(l)] = [H^+(aq)][OH^-(aq)]$$

Now, because the extent of dissociation into ions is small, it is reasonable to take $[H_2O(l)]$ as a constant. This means that $K_c \times [H_2O(l)]$ is a constant – it is called the ionic product of water, K_w.

At 298 K:

$$K_w = [H^+(aq)][OH^-(aq)] = 1 \times 10^{-14}\,mol^2\,dm^{-6}$$

Worked example

Three solutions of sodium hydroxide have $[OH^-(aq)]$ of 0.1, 0.001 and 0.0005 mol dm^{-3}, respectively. Calculate the pH of each solution.

$K_w = [H^+][OH^-] = 10^{-14}\,mol^2\,dm^{-6}$ at 25 °C

so $[H^+] = \dfrac{10^{-14}\,mol^2\,dm^{-6}}{[OH^-]\,mol\,dm^{-3}}$

If $[OH^-] = 1 \times 10^{-1}\,mol\,dm^{-3}$

then $[H^+] = 1 \times 10^{-13}\,mol\,dm^{-3}$

and pH $= -\log(10^{-13}) =$ **13**

If $[OH^-] = 1 \times 10^{-3}\,mol\,dm^{-3}$

then $[H^+] = 1 \times 10^{-11}\,mol\,dm^{-3}$

and pH $= -\log(10^{-11}) =$ **11**

If $[OH^-] = 5 \times 10^{-4}\,mol\,dm^{-3}$

then $[H^+] = 2 \times 10^{-11}\,mol\,dm^{-3}$

and pH $= -\log(2 \times 10^{-11})$

$= -(0.3 + -11) =$ **10.7**

So, in pure water at 25 °C the concentrations of both H^+ and OH^- ions are $1 \times 10^{-7}\,mol\,dm^{-3}$.

Under these conditions, water has a pH of 7. However, K_w is an equilibrium constant and its value will change with temperature. As the equilibrium constant changes, so does the pH. **Table 1.5.3** shows the effect of changing temperature on the pH of pure water.

Temperature/°C	K_w /mol^2 dm^{-6}	pH
0	0.114×10^{-14}	7.47
10	0.293×10^{-14}	7.27
20	0.681×10^{-14}	7.08
25	1.008×10^{-14}	7.00
30	1.471×10^{-14}	6.92
40	2.916×10^{-14}	6.77
50	5.476×10^{-14}	6.63
100	5.130×10^{-13}	6.14

table 1.5.3 The effect of changing temperature on K_w and pH.

Questions

1 Calculate the pH of:

a 0.1 mol dm^{-3} nitric acid, HNO_3

b 0.2 mol dm^{-3} potassium hydroxide, KOH

c 0.05 mol dm^{-3} sulfuric acid, H_2SO_4

2 a Explain the difference between a strong acid and a concentrated acid.

b Is it possible to have a concentrated solution of a weak acid?

K_a, K_w and strong and weak acids and bases [4.7d, e, h]

You will know from **chapter 1.3** how we can calculate a value for the equilibrium constant, K_c, from the concentrations of the reactants and products for a system in equilibrium at constant temperature.

The dissociation of weak acids and bases into ions are equilibrium processes and the equilibrium constants for the dissociation of an acid and a base are represented by K_a and K_b, respectively.

Conjugate acids and bases

In any acid–base reaction there is a competition for protons. For example:

$$\underset{\text{acid 1}}{NH_4^+(aq)} + \underset{\text{base 2}}{H_2O(l)} \rightleftharpoons \underset{\text{base 1}}{NH_3(aq)} + \underset{\text{acid 2}}{H_3O^+(aq)}$$

$NH_4^+(aq)$ and $NH_3(aq)$ are called a conjugate acid–base pair. Moving from one to another involves either a loss or a gain of a proton. Similarly, $H_3O^+(aq)$ and $H_2O(l)$ are also a conjugate acid–base pair.

K_a for acids

The strength of an acid or a base is concerned with how good it is at donating or accepting protons. For an acid HA in aqueous solution the following equilibrium exists:

$$HA(aq) \rightleftharpoons H^+(aq) + A^-(aq)$$

A measure of the strength of an acid is the value of the equilibrium constant for its dissociation into ions – for the equilibrium above this is:

$$K_a = \frac{[H^+(aq)][A^-(aq)]}{[HA(aq)]}$$

The equilibrium constant for this reaction, K_a, is known as the **dissociation constant** of the acid. K_a provides a direct measure of the strength of the acid because it measures its extent of dissociation – the larger the value of K_a the more dissociated the acid, and the greater its strength as a proton donor. Because it is an equilibrium constant, the value of K_a is unaffected by changes in concentration, but it is affected by temperature.

Table 1.5.4 shows the values of K_a for some acids. As the strength of the acid increases, the value of K_a also increases. Notice too that the strength of the conjugate base of the acid–base pair (A^-) increases as the strength of its conjugate acid decreases. So, nitric acid is a strong acid but its conjugate base (the nitrate ion) is very weak indeed. In contrast, water is a very weak acid but the hydroxide ion is a strong base.

Acid	Equilibrium Acid Base	K_a at 25 °C /mol dm^{-3}
Sulfuric	$H_2SO_4 \rightleftharpoons H^+ + HSO_4^-$	Very large
Nitric	$HNO_3 \rightleftharpoons H^+ + NO_3^-$	40
Trichloroethanoic	$CCl_3COOH \rightleftharpoons H^+ + CCl_3CO_2^-$	0.23
Sulfurous	$H_2SO_3 \rightleftharpoons H^+ + HSO_3^-$	0.015
Hydrofluoric	$HF \rightleftharpoons H^+ + F^-$	0.00056
Methanoic	$HCOOH \rightleftharpoons H^+ + HCO_2^-$	0.00016
Ethanoic	$CH_3COOH \rightleftharpoons H^+ + CH_3CO_2^-$	1.7×10^{-5}
Carbonic	$H_2CO_3 \rightleftharpoons H^+ + HCO_3^-$	4.5×10^{-7}
Hydrogen sulfide	$H_2S \rightleftharpoons H^+ + HS^-$	8.9×10^{-8}
Ammonium ion	$NH_4^+ \rightleftharpoons H^+ + NH_3$	5.6×10^{-10}

table 1.5.4 The values of K_a for some acids. Values for a strong acid like sulfuric or nitric acids are rarely quoted – they are large and we assume that the ionisation is complete.

When quoting hydrogen ion concentrations involving negative indices, we found it useful to use logarithms to produce pH values. We use a similar strategy with K_a and K_b values. Instead of quoting the value of K_a for an acid, we can instead give its pK_a, where:

$$pK_a = -\log K_a$$

For example, K_a for ethanoic acid is 1.7×10^{-5} mol dm^{-3}, so:

$$pK_a = -\log(1.7 \times 10^{-5}) = 4.8$$

The stronger an acid, the *larger* its K_a and the *smaller* its pK_a.

The pK_a of an acid is also useful in another way. From the relationship between pK_a and K_a above, we can write (omitting the (aq) state symbols for clarity):

$$pK_a = -\log \frac{[H^+][A^-]}{[HA]} = -(\log [H^+] + \log \frac{[A^-]}{[HA]})$$

So:

$$pK_a = pH - \log \frac{[A^-]}{[HA]}$$

This is significant because it means that pK_a = pH when:

$$\log \frac{[A^-]}{[HA]} = 0$$

In other words, when:

$$\frac{[A^-]}{[HA]} = 1 \text{ (because } \log 1 = 0)$$

In other words, an acid is exactly 50% dissociated at a pH equal to its pK_a. This fits exactly with the Brønsted–Lowry theory because, by definition, weaker acids have stronger conjugate bases, which will tend to be protonated at higher pH values. Stronger acids have weaker conjugate bases which will not tend to be protonated except at lower pH values.

In the same way as pK_a can be quoted for acids, we can quote values of pK_b for bases – a weak base will have a *small* K_b, and consequently a *large* pK_b.

Calculating pH from K_a and K_b

It is straightforward to calculate the pH of a solution containing an acid from the acid's dissociation constant.

Strong acid

In a solution of a strong acid, like hydrochloric acid, we assume that the acid is 100% dissociated, so that [H$^+$(aq)] is equal to the concentration of the acid in the solution. For example, in a solution containing 1 mol dm^{-3} of HCl, we assume that [H$^+$(aq)] = 1 mol dm^{-3}. The pH of the solution can then be found from:

$$pH = -\log[H^+(aq)] = -\log 1 = 0$$

Weak acid

We must use the concentration and the value of K_a to calculate the pH of a solution. The K_a of methanoic acid is 1.6×10^{-4} mol dm^{-3}. To calculate the pH of a solution containing 0.5 mol dm^{-3} of methanoic acid, we let [H$^+$(aq)] = x mol dm^{-3}. We can then write down the concentrations of undissociated methanoic acid molecules and methanoate ions as follows:

	$HCOOH(aq)$	$\rightleftharpoons H^+(aq)$	$+ HCOO^-(aq)$
Initial concentration (mol dm^{-3})	0.5	0	0
Equilibrium concentration (mol dm^{-3})	$0.5 - x$	x	x

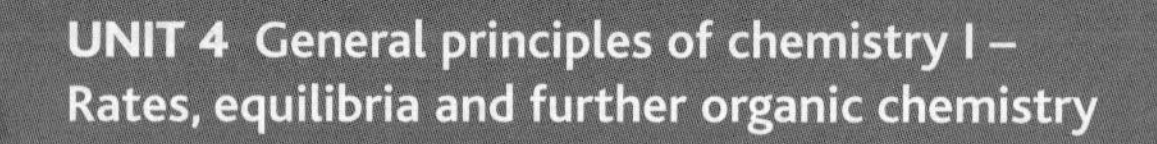

We can now write the relationship between these concentrations and K_a:

$$K_a = \frac{[H^+(aq)][A^-(aq)]}{[HA(aq)]}$$

$$= \frac{x \times x}{0.5 - x}$$

Now the value of x is small for this acid because it is only weakly ionised, so we can assume that x is very small compared to $0.5\,mol\,dm^{-3}$, and can write:

$$K_a = \frac{x\,mol\,dm^{-3} \times x\,mol\,dm^{-3}}{0.5\,mol\,dm^{-3}}$$

Substituting for K_a:

$$1.6 \times 10^{-4}\,mol\,dm^{-3} = \frac{(x\,mol\,dm^{-3})^2}{0.5\,mol\,dm^{-3}}$$

Rearranging this gives:

$$(x\,mol\,dm^{-3})^2 = 1.6 \times 10^{-4}\,mol\,dm^{-3} \times 0.5\,mol\,dm^{-3}$$

$$x = \sqrt{1.6 \times 10^{-4}\,mol\,dm^{-3} \times 0.5\,mol\,dm^{-3}}$$

$$= 8.9 \times 10^{-3}\,mol\,dm^{-3}$$

Now this is the value of $[H^+(aq)]$. The pH can be found from:

$$pH = -\log[H^+(aq)]$$

$$= -\log(8.9 \times 10^{-3})$$

$$= 2.05$$

An assumption about the size of the ion concentration in relation to the size of the solution concentration was made in the example above. Investigate the difference in the calculated value for pH if this assumption was not made.

Strong base

To find the pH of a solution of a base, we must first find the concentration of OH^- ions, and then use the ionic product of water, K_w, to find the concentration of H^+ ions.

We can assume that, like a strong acid, a strong base is completely ionised in aqueous solution. This will be true whether the base produces OH^- ions directly as it dissolves in water (as in the case of sodium hydroxide) or they are produced as a result of the reaction of the base with water – as in the case of potassium oxide, for which

$$K_2O(s) + H_2O(l) \rightarrow 2K^+(aq) + 2OH^-(aq)$$

In a solution containing $0.2\,mol\,dm^{-3}$ of sodium hydroxide, $[OH^-(aq)]$ can be taken to be $0.2\,mol\,dm^{-3}$. Using the relationship for K_w:

$$K_w = [H^+(aq)][OH^-(aq)] = 1.0 \times 10^{-14}\,mol^2\,dm^{-6}$$

we can substitute for $[OH^-(aq)]$:

$$[H^+(aq)] \times 0.2\,mol\,dm^{-3} = 1.0 \times 10^{-14}\,mol^2\,dm^{-6}$$

So:

$$[H^+(aq)] = \frac{1.0 \times 10^{-14}\,mol^2\,dm^{-6}}{0.2\,mol\,dm^{-3}}$$

$$= 5.0 \times 10^{-14}\,mol\,dm^{-3}$$

The pH can be found from:

$$pH = -\log[H^+(aq)]$$

$$= -\log(5.0 \times 10^{-14})$$

$$= 13.30$$

Weak base

Calculating the pH of a solution of a weak base requires the use of K_b – the method is very similar to that used with K_a to find the pH of a solution of a weak acid. For example, ammonia is a weak base:

$$NH_3(aq) + H_2O(l) \rightleftharpoons NH_4^+(aq) + OH^-(aq)$$

It has a K_b of $1.8 \times 10^{-5}\,mol\,dm^{-3}$. To calculate the pH of a $0.1\,mol\,dm^{-3}$ solution of ammonia we let $[OH^-(aq)] = x$. We can then write down the concentrations of unprotonated ammonia and ammonium ions as follows, ignoring the concentration of water which, as usual, can be regarded as being constant.

	$NH_3(aq) + H_2O(l) \rightleftharpoons$	$NH_4^+(aq)$ +	$OH^-(aq)$
Initial concentration ($mol\,dm^{-3}$)	0.1	0	0
Equilibrium concentration ($mol\,dm^{-3}$)	$0.1 - x$	x	x

We can now write down the relationship between these concentrations and K_b:

$$K_b = \frac{[NH_4^+(aq)][OH^-(aq)]}{[NH_3(aq)]}$$

$$= \frac{x^2}{(0.1 - x)\,mol\,dm^{-3}}$$

Now the value of K_b is small because ammonia is a weak base, so we can assume that x is much smaller than $0.1\,mol\,dm^{-3}$, and we can write:

$$K_b = \frac{x^2}{0.1\,mol\,dm^{-3}}$$

Substituting for K_b:

$$1.8 \times 10^{-5}\,mol\,dm^{-3} = \frac{x^2}{0.1\,mol\,dm^{-3}}$$

Rearranging this gives:

$$x^2 = 1.8 \times 10^{-5}\,mol\,dm^{-3} \times 0.1\,mol\,dm^{-3}$$

$$x = \sqrt{1.8 \times 10^{-5}\,mol\,dm^{-3} \times 0.1\,mol\,dm^{-3}}$$

$$= 1.34 \times 10^{-3}\,mol\,dm^{-3}$$

This is the value of the hydroxide ion concentration, $[OH^-(aq)]$. But:

$$[H^+(aq)][OH^-(aq)] = 1.0 \times 10^{-14}\,mol^2\,dm^{-6}$$

$$[H^+(aq)] = \frac{1.0 \times 10^{-14}\,mol^2\,dm^{-6}}{1.34 \times 10^{-3}\,mol\,dm^{-3}}$$

$$= 7.46 \times 10^{-12}\,mol\,dm^{-3}$$

The pH can be found from:

$$pH = -\log[H^+(aq)]$$

$$= -\log(7.46 \times 10^{-12})$$

$$= 11.12$$

Questions

1 Use the ionic product of water to show that the pH of water is 7.0 at 298 K.

2 Calculate the hydrogen ion concentration, and then the pH, of:

a $0.01\,mol\,dm^{-3}$ hydrochloric acid

b $0.01\,mol\,dm^{-3}$ ethanoic acid

c $0.5\,mol\,dm^{-3}$ ammonia solution.

K_a for ethanoic acid is $1.7 \times 10^{-5}\,mol\,dm^{-3}$; K_b for ammonia is $1.8 \times 10^{-5}\,mol\,dm^{-3}$.

3 Vinegar is a dilute aqueous solution of ethanoic acid with small amounts of other components. Calculate the pH of a sample of bottled vinegar that is $0.667\,mol\,dm^{-3}$ ethanoic acid, assuming that none of the other components affect the acidity of the solution. K_a for ethanoic acid is $1.7 \times 10^{-5}\,mol\,dm^{-3}$.

Determination of K_a for a weak acid [4.7g]

We are going to consider a simple method for finding a value for K_a at 298 K for a weak acid such as benzenecarboxylic acid (benzoic acid), C_6H_5COOH – a monobasic acid:

$$C_6H_5COOH(aq) \rightleftharpoons H^+(aq) + C_6H_5COO^-(aq)$$

$$K_a = \frac{[H^+(aq)][C_6H_5COO^-(aq)]}{[C_6H_5COOH(aq)]}$$

Method

A small quantity of the acid is weighed out and dissolved in a small amount of water. It is not a very soluble substance and warm water could be used and the solution allowed to cool. It is then diluted with water in a volumetric flask to make up a standard solution.

We know the mass of the acid used, the volume of the solution, and the molar mass of the acid (122 g), so we will be able to work out the concentration of the acid in $mol\,dm^{-3}$.

Next, we ensure the solution is mixed completely and take a sample of it. We measure its pH using a calibrated pH meter.

We know that $pH = -\log[H^+(aq)]$, so we can work out the hydrogen ion concentration.

The value of K_a can then be found using the equilibrium expression above.

Sample results

Mass of benzenecarboxylic acid used = 0.50 g

Number of moles used = $\frac{0.50\,g}{122\,g\,mol^{-1}} = 0.0041$ mol

Volume of solution = $250\,cm^3 = 0.25\,dm^3$

Concentration of benzenecarboxylic acid solution = $\frac{0.0041\,mol}{0.25\,dm^3}$

$= 0.0164\,mol\,dm^{-3}$

pH of the solution = 3.0

Remembering that $[H^+(aq)] = [C_6H_5COO^-(aq)]$, so
$[H^+(aq)] = 1.0 \times 10^{-3}\,mol\,dm^{-3}$

$$K_a = \frac{[H^+(aq)][C_6H_5COO^-(aq)]}{[C_6H_5COOH(aq)]}$$

$$= \frac{(1.0 \times 10^{-3}\,mol\,dm^{-3})^2}{0.0164\,mol\,dm^{-3}}$$

$$= 6.1 \times 10^{-5}\,mol\,dm^{-3}$$

The value given in the Data Booklet for benzenecarboxylic acid is $6.3 \times 10^{-5}\,mol\,dm^{-3}$ at 298 K.

Our result is quite close to this – but that does not necessarily mean that we have a very accurate experiment. It may be that there are a number of errors and, by chance, they cancel out.

What assumption has been made?

The principal assumption concerns the concentration of benzenecarboxylic acid. We assumed that the concentration of the acid solution that we produced is the actual concentration of benzenecarboxylic acid in the solution. In other words, that no dissociation took place. However, a very small amount of dissociation must have happened to move the pH of the solution from that of water to a value of 3.0.

Evaluation of the experiment

When carrying out evaluation, a chemist looks at the quality of the data and probably repeats an experiment such as this with different masses or different volumes of water. They would also consider the apparatus used and the method of carrying out the experiment.

- There is only one set of results here and so it is difficult to assess the quality of the data. The fact that a value is obtained that is close to that in the Data Booklet is promising.
- We are assuming that pure benzenecarboxylic acid and pure deionised water were used.
- Note that only 0.50 g of the acid was weighed out, and that this is a relatively small amount. However, with a sophisticated balance, capable of reading to at least the nearest 0.01 g, this is achievable.
- A volumetric flask enables a standard solution to be prepared accurately, but the solution must be mixed thoroughly.
- The method of measuring the pH of the acidic solution is clearly important. Even slight differences in pH make a considerable difference to the value of K_a obtained. If only coloured indicators were available, then it would be very difficult to get an accurate and reliable result. With a correctly calibrated pH meter more accurate results are obtainable.

Questions

1 Suppose, in the sample calculation, that a different pH meter was used, and it gave a reading of 2.7 rather than 3.0. What would the value for K_a be now?

2 A student carrying out this experiment thought she could get better results by carrying out repeats. She considered:

  a measuring the pH of the same sample from the volumetric flask three times using the same pH meter

  b measuring the pH of the same sample from the volumetric flask three times using different pH meters

  c measuring the pH of different samples from the volumetric flask using the same pH meter.

Comment on the suitability of each of these possibilities.

pH changes during acid–base titrations [4.7i]

In this section we look at how pH changes during an acid–base titration. You might expect that there will be a gradual change in pH as you add a dilute acid solution to a dilute alkali solution – but this is not so. The shape of the graph of pH against volume of acid added depends on whether the acid and base involved are strong or weak.

When you carry out a simple acid–base titration, you usually use an indicator to show when the acid and alkali are mixed in exactly the right proportions to neutralise each other.

- The **end-point** of the titration is when the appropriate colour change takes place.
- The **equivalence point** is when the two reactants are mixed in exactly the proportions indicated by the equation. Sometimes this is called the **neutral point** but, as you will see, this is not always so.

Titration of 25.00 cm^3 sodium hydroxide (0.10 mol dm^{-3}) solution with hydrochloric acid (0.10 mol dm^{-3}) would require 25.00 cm^3 of hydrochloric acid because there is a 1 : 1 relationship according to the equation:

$$NaOH(aq) + HCl(aq) \rightarrow NaCl(aq) + H_2O(l)$$

An aqueous solution of sodium chloride is what is present at the equivalence point, when all the acid and base have neutralised. In this case, at the equivalence point the pH is 7 – exactly neutral.

However, if sodium hydroxide and ethanoic acid are titrated, then at the equivalence point, where an aqueous solution of sodium ethanoate is present, the solution is not neutral but slightly alkaline. Similarly, if ammonia solution is titrated with hydrochloric acid, then at the equivalence point an aqueous solution of ammonium chloride is present and it is found that this is slightly acidic.

It is wise not to use the term neutral point, but use *equivalence point*. The term end-point refers to when the indicator changes colour – often this is at the equivalence point.

Titration of a strong acid and a strong base

Figure 1.5.11 shows the graph of pH against volume of acid added for the titration of 1.00 mol dm^{-3} hydrochloric acid with 1.00 mol dm^{-3} sodium hydroxide. In this, and the examples that follow, the alkaline solution is pipetted into the conical flask and the acid added from a burette.

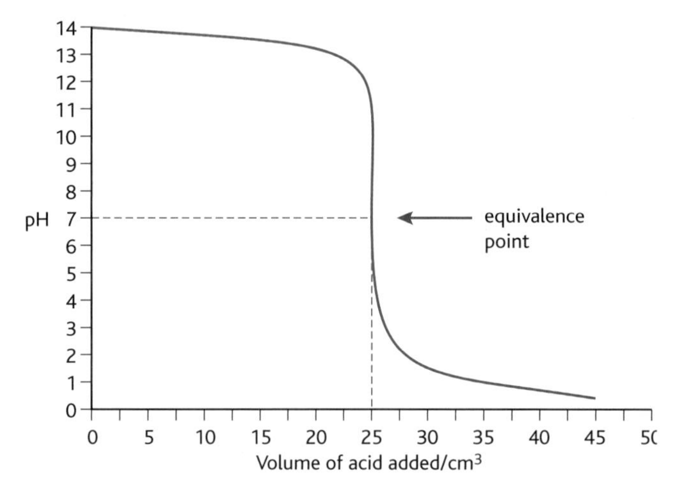

fig. 1.5.11 Change in pH for strong acid–strong base titration.

There is only a very small change in pH for the addition of about the first 22 cm^3 of acid. As the equivalence point is approached, there is a rapid change in the pH.

If you calculate the values, the pH falls all the way from 11.3 when you have added 24.9 cm^3 to 2.7 when you have added 25.1 cm^3. Beyond this, the change in pH value is again very gradual.

Titration of a weak acid and a strong base

For example, the titration of 1.00 mol dm^{-3} ethanoic acid with 1.00 mol dm^{-3} sodium hydroxide.

The curve (**fig. 1.5.12**) will be exactly the same as that for a strong acid and strong base up to the equivalence point.

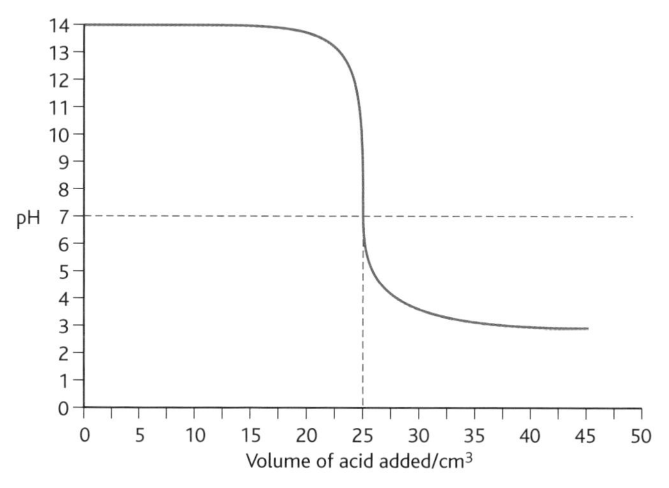

fig. 1.5.12 Change in pH for weak acid–strong base titration.

Beyond the equivalence point, an aqueous solution of ethanoic acid and sodium ethanoate remains. This is a buffer solution (see page 92) and its pH does not change very much when further acid is added. The equivalence point here is not pH 7, but between pH 8 and 9.

Titration of a strong acid and a weak base

For example, the titration of 1.00 mol dm^{-3} hydrochloric acid with 1.00 mol dm^{-3} ammonia solution – see **fig. 1.5.13**.

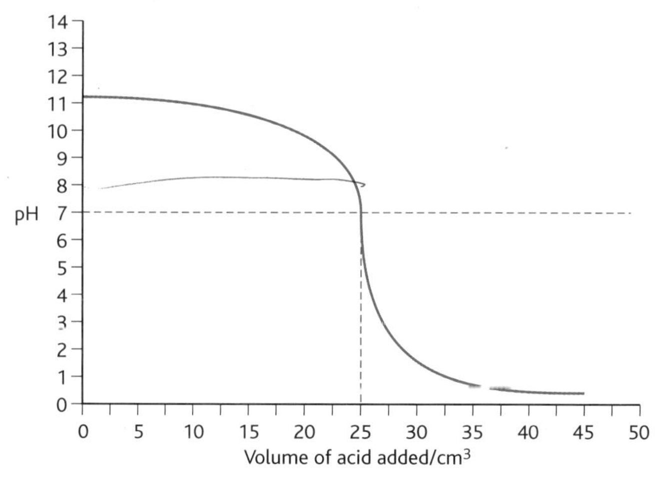

fig. 1.5.13 Change in pH for strong acid–weak base titration.

The equivalence point (when 25.00 cm^3 of the reactants have mixed) is at about pH 5. The equivalence point still falls on the steepest part of the curve.

Titration of a weak acid with a weak base

For example, the titration of 1.00 mol dm^{-3} ethanoic acid with 1.00 mol dm^{-3} ammonia solution – see **fig. 1.5.14**.

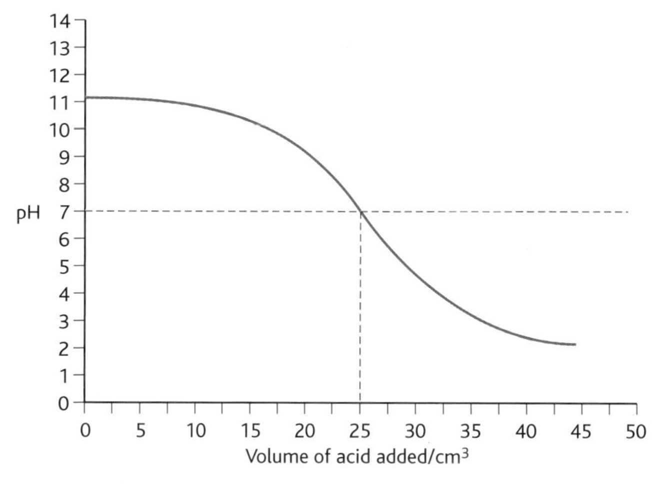

fig. 1.5.14 Change in pH for weak acid–weak base titration.

This curve is similar to the earlier ones. For example, up to the equivalence point the graph resembles the first part of the strong acid–weak base graph, and after the equivalence point the graph resembles the last part of the weak acid–strong base graph. There is one important difference – at no stage is there a steep portion of the graph, as can be seen in the other three cases. There is merely a point of inflexion.

Questions

1 Why is it difficult to get the end-point of a titration with hydrochloric acid and sodium hydroxide when only a single titration is carried out?

2 Sketch the shape of the graph you would expect if a solution of a strong acid is placed in the conical flask and a solution of a strong base is added from a burette.

Choosing suitable indicators [4.7j]

Earlier you looked at titration curves, which show how pH changes as more acid is added in four different cases:

- strong acid–strong base
- weak acid–strong base
- strong acid–weak base
- weak acid–weak base.

These four curves will help us to understand why different indicators are needed for different titrations.

Acid–base indicators

Acid–base indicators are generally weak acids or weak bases. For an indicator that is a weak acid, the dissociation into ions can be represented as:

$$HIn(aq) \rightleftharpoons H^+(aq) + In^-(aq)$$

Either the weak acid, HIn(aq), and/or its conjugate base, In^-(aq), is coloured. If the pH increases (hydrogen ions are removed) the equilibrium moves to the right to produce more In^-, which causes a change in colour. For example, the indicator phenolphthalein is colourless in its protonated form and pink in its unprotonated form. Below a pH of 9, phenolphthalein is colourless. Adding hydroxide ions removes protons from the right-hand side of the equilibrium, pulling it to the right. This causes more of the unprotonated form of the indicator to be produced, and causes the solution to turn pink.

Figure 1.5.15 shows the colours of some common indicators and the pH range over which they change – due to the limits of sensitivity of the human eye, this range is normally about 2 pH units.

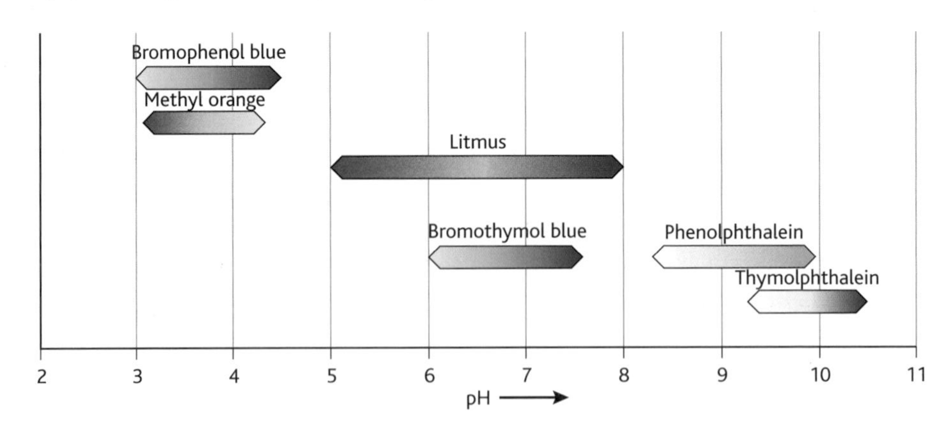

fig. 1.5.15 The colour range and pH range of some indicators.

Indicators and end-points

The end-point of a titration occurs when the two solutions have reacted exactly. The ideal indicator for a given acid–base titration is one that is in the middle of its colour change at the pH of the equivalence point of the titration. This pH depends on the acid and base being titrated.

The reason for this can be understood if we think carefully about the equilibrium constant for the dissociation of the indicator, K_{in}, given by:

$$K_{in} = \frac{[H^+(aq)][In^-(aq)]}{[HIn(aq)]}$$

In the middle of the colour change of the indicator, the two forms of the indicator, HIn and In^-, are present in equal amounts, so $[HIn(aq)] = [In^-(aq)]$. The expression above then simplifies to:

$$K_{in} = [H^+(aq)]$$

or $pK_{in} = pH$

In other words, pK_{in} for an indicator should be equal (or as close as possible) to the pH of the equivalence point of the titration.

Choosing an indicator for a titration

A good indicator must show a dramatic colour change in order to make it easy to detect the end-point of a titration. One of the best indicators from this point of view is phenolphthalein, with a colour change from colourless (acid) to pink (alkali). In contrast, the colour change of methyl orange (red to yellow) may be difficult to see in dilute solutions.

At the equivalence point of a titration, the pH must change sharply by several units if the end-point is to coincide with the equivalence point. The pH change during a titration is greatly dependent on the acid and base being titrated, and indicators must be selected accordingly:

- *Strong acid–strong base:* there is a long vertical portion on the titration curve (**fig.1.5.11**) from about pH 10 to pH 4. Any indicator that changes colour within this range is suitable – e.g. methyl orange, bromothymol blue or phenolphthalein.
- *Weak acid–strong base:* the vertical portion (**fig.1.5.12**) is between about pH 6.5 to pH 11. An indicator that changes colour within this range would be suitable – e.g. phenolphthalein.
- *Strong acid–weak base:* the vertical portion (**fig.1.5.13**) is between about pH 3 and pH 7.5. Any indicator that changes within this range is suitable – e.g. methyl orange or bromothymol blue.
- *Weak acid–weak base:* there is no vertical portion on this curve (**fig.1.5.14**) so no indicator would change colour dramatically. This titration cannot be completed satisfactorily with an indicator. Another method (e.g. monitoring pH with a pH meter) would have to be used.

Questions

1 **Figure 1.5.16** shows the graph of the pH change when dilute hydrochloric acid is titrated with sodium carbonate solution. The equation for the overall reaction is:

$Na_2CO_3(aq) + 2HCl(aq) \rightarrow 2NaCl(aq) + CO_2(g) + H_2O(l)$

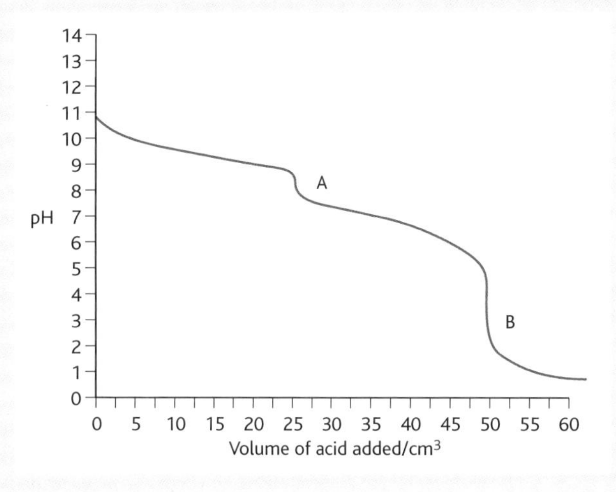

fig. 1.5.16

a Suggest suitable indicators to detect the changes at A and B.

b Why would the change at A be more difficult to detect than the one at B?

c Suggest equations for the reactions taking place at each stage.

2 The table gives some acid–base indicators, their colour change and the pH range where they change colour.

Indicator	Colour change	pH range
Congo red	violet → red	3.0–5.0
Phenol red	yellow → red	6.8–8.4
Thymol blue	yellow → blue	8.0–9.6

a Which of these indicators could be used to detect the end-points in the following titrations?
(i) strong acid–strong base; (ii) weak acid–strong base; (iii) strong acid–weak base.

b Suggest pK_{in} values at 298 K for congo red and thymol blue.

Finding K_a for a weak acid from a pH titration [4.7h, l]

The titration method can be used to calculate K_a for a weak acid. Earlier we found that if half of an acid has been neutralised then:

$$pK_a = pH$$

This method is more accurate and reliable that the method used earlier in this chapter.

Finding K_a

You will probably use one of the methods below to find the value of K_a for a weak acid. You will probably use a pH meter to measure the pH values of solutions. Without a pH meter you may have to resort to using wide-range universal indicator and then narrow-range universal indicator. This takes much longer and relies on colour judgement. You might compare the value you get with the value given in a data book.

Method 1

This experiment involves carrying out a titration starting with the weak acid in the conical flask and a standard solution of sodium hydroxide in the burette. You measure the pH of the solution after each addition of the acid, and plot a graph of pH (*y*-axis) against the added volume of sodium hydroxide solution.

From your graph you can find the volume of sodium hydroxide needed to neutralise the acid (the equivalence point). You can then find the half-equivalence point (i.e. when half the total volume of sodium hydroxide has been added) and estimate the pH at that point.

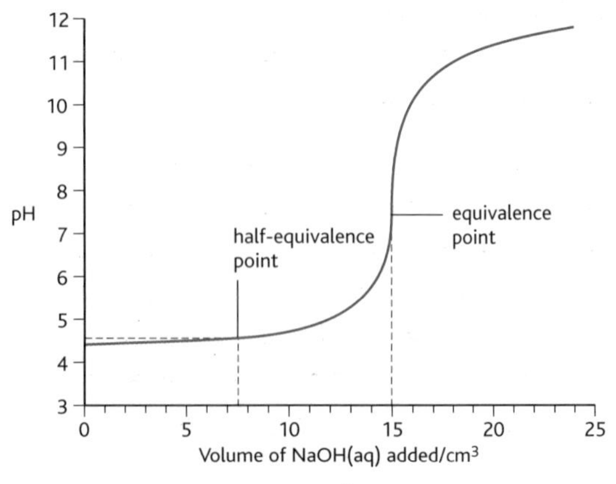

fig. 1.5.17 Titration curve and the half-equivalence point.

Figure 1.5.17 shows a graph of the results from such an experiment. 15.00 cm^3 of sodium hydroxide neutralised the weak acid. The equivalence point was 15.00 cm^3 and so the half-equivalence point was 7.50 cm^3.

When 7.50 cm^3 of sodium hydroxide has been added, half the acid has reacted. At this point the concentrations of acid molecules and conjugate base ions are equal. From **fig. 1.5.17**, the pH is 4.6 at this point. So:

$$pK_a = pH$$
$$= 4.6$$
$$K_a = 2.52 \times 10^{-5}\,mol\,dm^{-3}$$

Method 2

This is sometimes called the 'half volume method' because it involves titrating until half of the molecules of the weak acid have reacted. The concentrations of weak acid molecules and its conjugate base ions are then equal. Of course, you can't actually titrate to this point in practice, so this method is used to reach it.

You prepare a standard solution of known concentration of the weak acid and measure out two portions of equal volume. You titrate the first portion with standard sodium hydroxide solution using phenolphthalein as indicator.

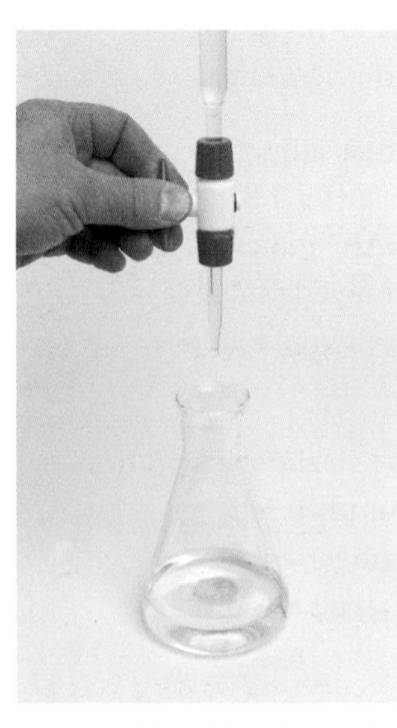

fig. 1.5.18 Phenolphthalein used as the indicator in an acid–base titration.

You stop adding acid when the solution just goes pink and stays pink – this is the end-point. Now you add the second portion of the standard weak acid solution and mix thoroughly. The pH of the solution is measured using a calibrated pH meter. This solution now has equal concentrations of weak acid molecules and conjugate base ions. The value of K_a is calculated as in the first method.

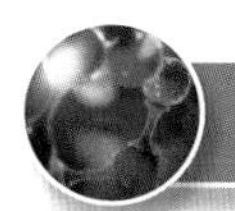

HSW Weak acids and TB

Sometimes it can be difficult to see the relevance of chemistry such as pK_a values beyond the school laboratory. In fact, they are important in many areas of active research. For example, in 2003 the Journal of Antimicrobial Chemotherapy published research into the impact of weak acids on *Mycobacterium tuberculosis*, the bacterium that causes tuberculosis (TB). This is a disease affecting millions of people around the world, so anything which can make treatment more effective would be welcome.

The research team found that the weak acid pyrazinoic acid (pK_a = 2.9), one of the current drugs used against TB, was very effective against the bacteria, particularly at a pH of 4–5.

The team decided to look at the effect of a range of other weak acids at different pH values on two similar bacteria, only one of which causes TB. It was vital for the group to understand the role of pH and the pK_a values of the different acids they used for their results to be meaningful. Moderating the pH of the drugs may make them more effective or reduce the dosage needed, making treatment more affordable. Some of their results are summarised in the table. 'MIC' stands for minimum inhibitory concentration – the lowest concentration at which the acid stops the bacteria growing.

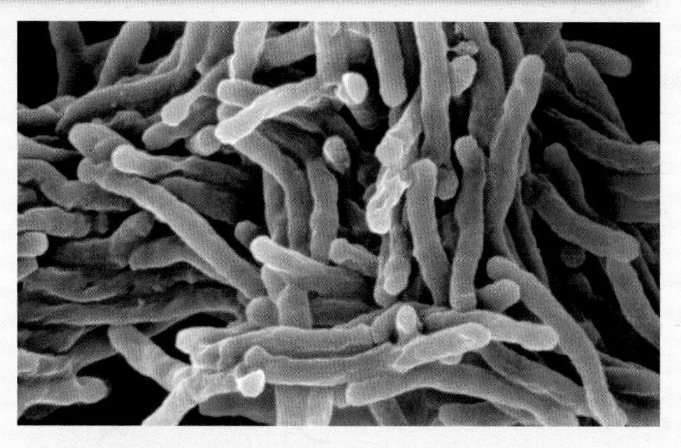

fig. 1.5.19 Mycobacterium causes around 1.7 million deaths a year. Some scientists hope to use weak acids to help produce more effective treatments against this deadly microorganism.

Weak acid	*Mycobacterium tuberculosis* (causes TB)		*Mycobacterium smegmatis* (doesn't cause TB)		pK_a
	MIC (mg dm^{-3}) at pH 6.8	MIC (mg dm^{-3}) at pH 5.5	MIC (mg dm^{-3}) at pH 6.8	MIC (mg dm^{-3}) at pH 5.5	
Benzoic acid	111	37	>333	>111	4.20
Salicylic acid	50–100	10–20	1000	333	3.00
Acetyl salicylic acid (aspirin)	111	37	1000	333	3.49
Mefenamic acid	33	11	1000	333	4.20
Nicotinic acid	>500	200	>1000	1000	4.96

Questions

1 Consider the two methods of measuring K_a described in this section.

a Which would you expect to give the most reliable result as the methods are written?

b What could you do to get better results from the one that appears to be less good?

2 a In what way is the acid–base titration described in method 2 different from most acid–base titrations using phenolphthalein?

b Suggest why the burette should be washed out thoroughly before it is put away.

3 Using the data in the table on inhibition of *Mycobacterium* in the HSW box, plot bar charts of the MIC values against the pH values for the two different bacteria. What do these charts and the data in the table show you about the effect of pH and pK_a on the antimicrobial activity of the different weak acids used to try to stop the bacteria from growing?

An introduction to buffer solutions [4.7k, l]

What is a buffer solution?

A small change in the pH of a system can often have large, sometimes unwanted, results. For example, a small amount of lemon juice added to milk or cream causes large changes to the structure of the proteins, leading to 'curdling'. Much more seriously, if the pH of your arterial blood were to change from 7.4 (its normal value) to 7.00 or to 8.00, you would die.

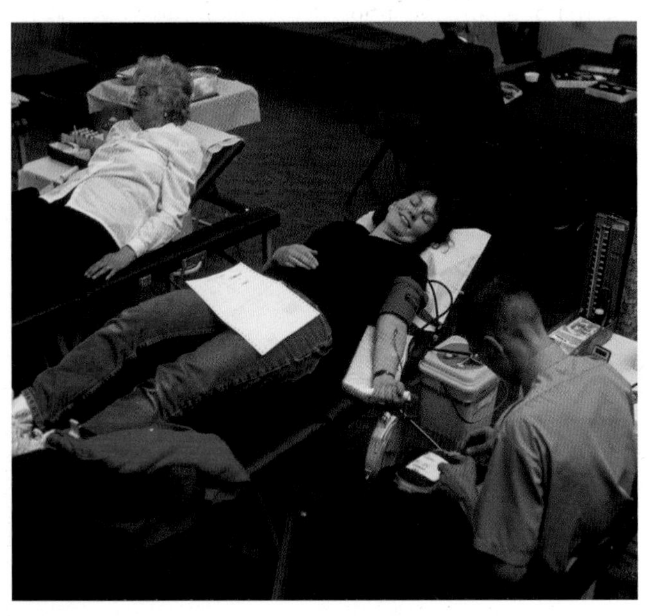

fig. 1.5.20 When people donate blood, a special buffer solution is added to make sure that the pH of the blood does not change, damaging the cells, while it is being stored.

Natural systems have mechanisms that help to prevent large changes in pH happening, and chemists have developed similar ways of stabilising the pH of solutions in which reactions are occurring. Natural and artificial solutions that contain solutes which together resist changes in pH are known as buffer solutions – they are said to be *buffered* against changes in pH. The principles of buffer solutions can be understood using the ideas of equilibria.

A buffer solution usually consists of two solutes dissolved in water. One of these is a weak acid and the other is its conjugate base – in other words, a weak acid and one of its salts.

If we represent the weak acid as HA, the equilibrium concerned can be written:

$$HA(aq) \rightleftharpoons H^+(aq) + A^-(aq)$$

If we represent the salt as MA, this will be fully dissociated:

$$MA(s) \rightarrow M^+(aq) + A^-(aq)$$

From these two equations, you can see that the buffer solution contains a large amount of A^-. This pushes the acid–base equilibrium to the left, so the solution also contains a large amount of HA. The solution thus has a reservoir of the weak acid (HA), and a reservoir of its conjugate base (A^-). **Figure 1.5.21** shows how this helps to prevent changes in pH. However, the pH value will change if a *large* excess of acid or base is added.

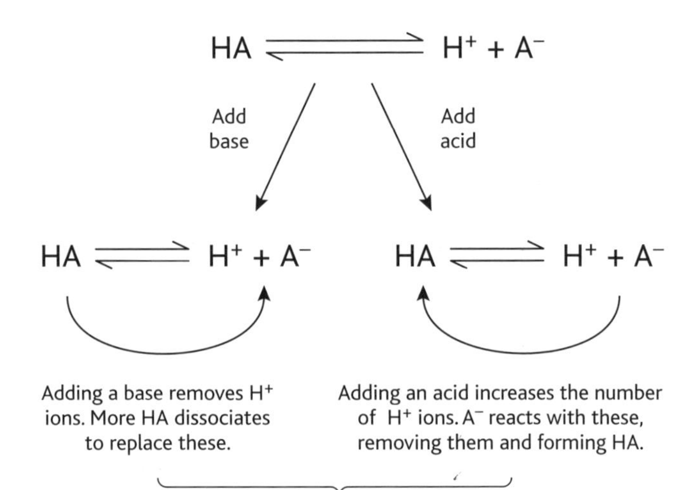

fig. 1.5.21 How a buffer solution works.

As **fig. 1.5.21** shows, the stability of the pH of a buffer solution is due to two factors:

- a reservoir of HA, which supplies H^+ ions if they are removed from the solution
- a reservoir of A^-, which reacts with any H^+ ions added removing them from the solution.

A buffer solution consisting of a weak acid and one of its salts is an acidic buffer solution, with a pH below 7. For example, a solution containing ethanoic acid (the weak acid) and sodium ethanoate (supplying its strong conjugate base) will act as a buffer solution at a pH less than 7. A buffer solution above pH 7 – an alkaline buffer solution – can be made by mixing a weak base and one of its salts. For example, a solution containing ammonia (a weak base) and ammonium chloride (supplying its strong conjugate acid) will act as a buffer solution at a pH greater than 7.

The buffering action of a mixture of ammonia and ammonium chloride can be explained by the equations:

$$NH_3(aq) + H_2O(l) \rightleftharpoons NH_4^+(aq) + OH^-(aq)$$

$$NH_4Cl(aq) \rightarrow NH_4^+(aq) + Cl^-(aq)$$

The buffer solution contains a large concentration of NH_4^+ ions – mainly from the ammonium chloride, which is completely dissociated into ions in aqueous solution. It also has a high concentration of NH_3 molecules because of the tendency of the first reaction to lie well to the left-hand side.

If hydroxide ions are added to this buffer solution, the equilibrium in the first equation moves further to the left. This effectively removes the added OH^- ions and restores the original pH value – it also removes NH_4^+ ions, but there is a large reservoir of these due to the second reaction. If hydrogen ions are added, they remove OH^- ions. The equilibrium in the first equation moves to the right to produce more OH^- ions and restore the pH. If a very large excess of hydrogen ions or hydroxide ions are added then the pH will change – there is a limit to the buffering process.

The pH of a buffer solution

Consider a buffer solution made by dissolving a salt, MA, in a solution of a weak acid, HA. We know that the dissociation constant of the acid, K_a is given by:

$$K_a = \frac{[H^+(aq)][A^-(aq)]}{[HA(aq)]}$$

This relationship can be rearranged so that $[H^+(aq)]$ is on the left-hand side of the equation:

$$[H^+(aq)] = K_a \times \frac{[HA(aq)]}{[A^-(aq)]}$$

Strictly speaking, the concentrations appearing in this expression are *equilibrium* concentrations. However, remember that we are dealing with a solution of a weak acid plus a large reservoir of its conjugate base (A^-). Because of this, the equilibrium:

$$HA(aq) \rightleftharpoons H^+(aq) + A^-(aq)$$

will lie *well* to the left-hand side, and the acid will be hardly dissociated at all. The concentration of the acid at equilibrium will, therefore, effectively be equal to the initial concentration of the acid. In the same way, the concentration of the ion A^- will effectively be the concentration supplied by the salt. We can represent these simplifying assumptions as:

$$[HA(aq)]_{eq} \approx [HA(aq)]_{initial} = [acid]$$

$$[A^-(aq)]_{eq} \approx [A^-(aq)]_{initial} = [base]$$

Substituting these simplifications into the equation for $[H^+(aq)]$ above:

$$-[H^+(aq)] = -K_a \times \frac{[acid]}{[base]}$$

$$-\log[H^+] = -\log K_a - \log\frac{[acid]}{[base]}$$

$$pH = pK_a - \log\frac{[acid]}{[base]}$$

Note that $\frac{[acid]}{[base]}$ is the ratio of the acid to the base. So, the pH of a buffer solution depends on this *ratio* and it will be affected very little by dilution – because diluting it will decrease both concentrations by a similar amount.

It will also be useful to realise that when [acid] = [base] then the relationship simplifies to:

$$[H^+(aq)] = K_a$$

or $pH = pK_a$

Worked example

A buffer solution is made by adding 2.05 g of sodium ethanoate to 1000 cm^3 of 0.09 mol dm^{-3} ethanoic acid. What is the pH of the solution produced? (K_a for ethanoic acid is 1.7×10^{-5} mol dm^{-3}).

$$[H^+(aq)] = K_a \times \frac{[acid]}{[base]}$$

In this case, [acid] = 0.09 mol dm^{-3}

The molar mass of sodium ethanoate is 86 g. So the amount of the salt used is:

$$\frac{2.05\,g}{82\,g\,mol^{-1}} = 0.025\,mol$$

This is dissolved in 1000 cm^3 of solution so [base] = 0.025 mol dm^{-3}.

Therefore:

$$[H^+(aq)] = 1.7 \times 10^{-5}\,mol\,dm^{-3} \times \frac{0.09\,mol\,dm^{-3}}{0.025\,mol\,dm^{-3}}$$

$$= 6.12 \times 10^{-5}\,mol\,dm^{-3}$$

So, the pH of the buffer solution is given by:

$$pH = -\log[H^+(aq)]$$

$$= -\log(6.12 \times 10^{-5})$$

$$= 4.21$$

The effectiveness of a buffer solution

In order to consider how effective a buffer is at stabilising the pH of a solution, consider the solution in the last worked example. What happens if 1 cm^3 of 1.0 mol dm^{-3} sodium hydroxide solution is added to the 1000 cm^3 buffer solution?

The initial pH of the buffer solution is 4.2. When the sodium hydroxide is added, the concentration of acid falls and the concentration of ethanoate anion rises, due to the reaction:

$$OH^-(aq) + CH_3COOH(aq) \rightarrow CH_3COO^-(aq) + H_2O(l)$$

Now the amount of OH^- in 1.0 cm^3 of sodium hydroxide (1.0 mol dm^{-3}) is given by:

$$\text{amount of } OH^- = \frac{1}{1000}\ dm^3 \times 1.0\ mol\ dm^{-3}$$
$$= 0.001\ mol$$

The solution initially contains 0.09 mol of acid. Since 1 mol of OH^- removes 1 mol of CH_3COOH, the new concentration of acid is given by:

$$[CH_3COOH(aq)] = (0.09 - 0.001)\ mol\ dm^{-3}$$
$$= 0.089\ mol\ dm^{-3}$$

Similarly, the concentration of CH_3COO^- rises from its initial value of 0.025 mol dm^{-3}:

$$[CH_3COO^-(aq)] = (0.025 + 0.001)\ mol\ dm^{-3}$$
$$= 0.026\ mol\ dm^{-3}$$

Note that, in both calculations, we have neglected the slight increase in volume of the solution due to adding the 1 cm^3 of sodium hydroxide solution.

So, the new concentration of hydrogen ions in the solution is given by:

$$[H^+(aq)] = 1.7 \times 10^{-5}\ mol\ dm^{-3} \times \frac{0.089\ mol\ dm^{-3}}{0.026\ mol\ dm^{-3}}$$
$$= 5.8 \times 10^{-5}\ mol\ dm^{-3}$$
$$pH = -\log[H^+(aq)]$$
$$= -\log(5.8 \times 10^{-5})$$
$$= 4.24$$

Compared with the worked example, which used the assumptions, this is an increase in pH of just 0.03.

Worked example

For emphasis, let's work out the change in pH if the same amount of the same sodium hydroxide solution was added to a solution of hydrochloric acid, a strong acid, with a pH of 4.21.

From the previous worked example, we know that $[H^+(aq)]$ in a solution with a pH of 4.21 is 6.12×10^{-5} mol dm^{-3}. When 0.001 mol of NaOH are added to 1 dm^3 of a hydrochloric acid solution containing 6.12×10^{-5} mol dm^{-3} of H^+ ions, all these ions will be removed so the final concentration of OH^- ions is given by:

$$[OH^-(aq)] = (1.00 \times 10^{-3}) - (6.12 \times 10^{-5})\ mol\ dm^{-3}$$
$$= 9.39 \times 10^{-4}\ mol\ dm^{-3}$$

The new value of $[H^+(aq)]$ can be found using the ionic product of water:

$$K_w = [H^+(aq)][OH^-(aq)] = 1.00 \times 10^{-14}\ mol^2\ dm^{-6}$$
$$1.00 \times 10^{-14}\ mol^2\ dm^{-6} = [H^+(aq)] \times 9.39 \times 10^{-4}\ mol\ dm^{-3}$$
$$[H^+(aq)] = \frac{1.00 \times 10^{-14}\ mol^2\ dm^{-6}}{9.39 \times 10^{-4}\ mol\ dm^{-3}}$$
$$= 1.065 \times 10^{-11}\ mol\ dm^{-3}$$

The pH of the solution is given by:

$$pH = -\log[H^+(aq)]$$
$$= -\log(1.065 \times 10^{-11})$$
$$= 10.97$$

Compared with the worked example on the buffer solution, this is a pH change of nearly 7 units!

SC Calculate the pH of a buffer solution containing equal volumes of 0.1 mol dm^{-3} ethanoic acid and:

a 0.1 mol dm^{-3} sodium ethanoate

b 0.2 mol dm^{-3} sodium ethanoate.

K_a for ethanoic acid = 1.7×10^{-5} mol dm^{-3}.

Buffer solutions and pH curves

Figure 1.5.22 shows part of the pH curve when titrating a weak acid (ethanoic acid) with a strong base (sodium hydroxide). In the portion marked 'buffer range', the change in pH as sodium hydroxide is added is gradual. Over this range, there are significant concentrations of ethanoic acid molecules and ethanoate ions – and so there is buffer action.

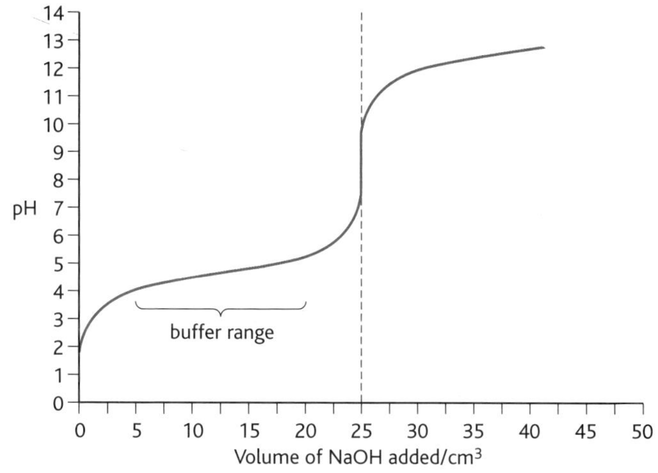

fig. 1.5.22 Typical pH curve for a buffer solution.

HSW Buffers and the cosmetics industry

Producing cosmetics and shampoos is a multi-billion pound industry around the world. People want to look their best – and it is the work of chemists to help! One of the many problems to be solved is the fact that the human skin is naturally slightly acidic. This is vital for keeping skin healthy because it provides the perfect environment for the survival of useful microorganisms – and yet prevents disease-causing organisms from growing. The ideal skin pH is around 5.5.

fig. 1.5.23 Developing cosmetics to enhance the appearance of human skin involves complex chemistry – and lots of buffers!

Skin cosmetics have to be even more complex because people want products which soften, protect against the sun and are anti-ageing – all in one package! Some of the ingredients are stable for only a certain time at different pH values. One way in which manufacturers have overcome this problem is to develop products containing two separate emulsions, containing different ingredients at different pH values. They are only mixed at the point at which the cream is about to be rubbed into the skin. To do this, different buffers are selected for the two different layers according to the pH desired in the emulsion. Each buffer is chosen so that its pK_a value is close to the desired pH value (usually within ±1 pH unit). The best buffer is made when the pK_a of an acid/base ingredient is equal to the pH of the buffer.

There are some common buffer systems used in the cosmetic world. When an acidic pH is required (pH 3–5) the buffer is chosen from glycolic acid, lactic acid, citric acid, ethanoic acid or succinic acid – along with the corresponding salt of course. When a neutral pH buffer is needed (pH 6–8) the buffer system chosen usually contains phosphate monobasic and phosphate dibasic salts. For an alkaline pH (pH 8–10) the buffer system usually contains carbonate ions and hydrogencarbonate ions.

On top of all this, cosmetics chemists must make sure that when different products are used together then the resulting mixture is suitable for the skin – cosmetics chemistry is very complex and challenging!

Questions

1 Given that the pK_a for ethanoic acid is 4.8, calculate the ratio of ethanoate ions to ethanoic acid molecules in a buffer solution with a pH of 5.4.

2 Calculate the pH of a buffer solution containing 12.2 g of benzenecarboxylic acid (C_6H_5COOH) and 7.2 g of sodium benzenecarboxylate (C_6H_5COONa) in 1000 cm^3 solution. pK_a for benzenecarboxylic acid is 4.2.

3 Investigate the pK_a of the different buffers that are used in the cosmetics industry, and explain why those particular chemicals might be chosen.

Buffers in biological systems [4.7m]

In this section we are going to consider the importance of buffers inside the human body and in the food we eat. It is very important to have the correct pH conditions in both cases.

Buffers in the body

The human body operates within a narrow range of pH values. For example, the pH of arterial blood plasma is 7.4 – if its pH were to change significantly, particularly if it became more acidic, it would affect the way the whole body functioned. The body has developed ways of preventing large pH changes and many of these involve buffers.

All the cells in the body need oxygen to function. This is used in respiration to provide energy for the cells. Carbon dioxide is formed as a waste product. Carbon dioxide dissolves in body fluids to form carbonic acid – and this leads to a lowering of pH.

Oxygen is carried in the blood bound to haemoglobin, a protein containing iron. The key equilibrium for this process is:

$$HHb(aq) + O_2(g) \rightleftharpoons H^+(aq) + HbO_2^-(aq)$$

The presence of H^+ on the right-hand side of the equilibrium means that the equilibrium position is very sensitive to the hydrogen ion concentration in the fluid. If the pH falls, $[H^+]$ increases and the equilibrium adjusts to the left displacing oxygen from the haemoglobin. If the pH rises, $[H^+]$ decreases and the equilibrium adjusts to the right causing more oxygen to bind to haemoglobin. Both of these are potentially life-threatening if the pH changes are not controlled.

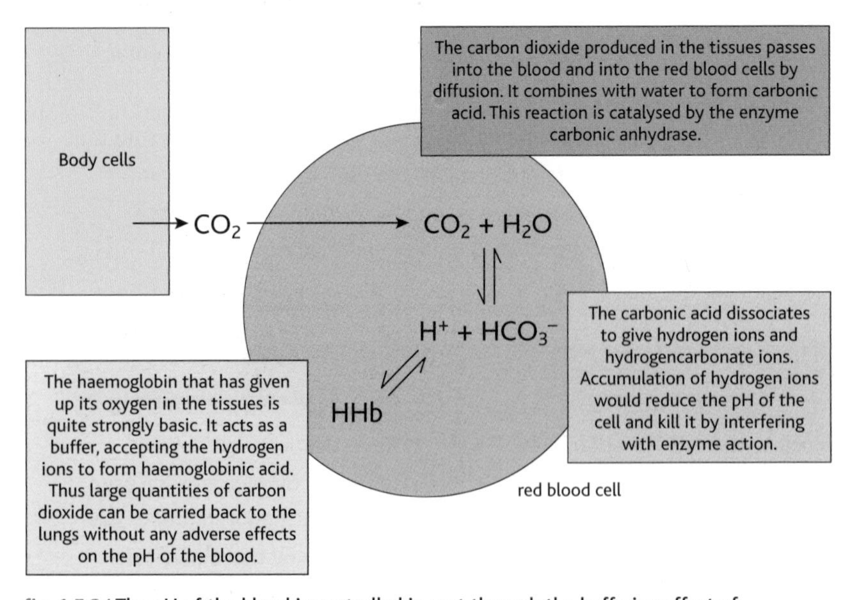

fig. 1.5.24 The pH of the blood is controlled in part through the buffering effect of haemoglobin.

Carbon dioxide produced in body tissues represents the greatest threat to the stability of the blood plasma pH under normal circumstances. **Figure 1.5.24** shows how the ability of haemoglobin to act as a buffer helps to control the pH of the blood, absorbing hydrogen ions produced by carbon dioxide.

The maintenance of a steady blood pH is too important for there to be reliance on only one control system. If there are excessive changes in blood chemistry then the kidneys kick in to prevent the pH from shooting up. They excrete hydrogen ions in the urine should the pH fall by too much; they retain hydrogencarbonate ions if the pH value falls. This means that the pH of the urine is variable, with a normal range of 4.5 to 8.5. A fall in blood pH also stimulates the kidneys to produce ammonia, which combines with hydrogen ions and is excreted in the urine as ammonium salts.

Through this combination of blood buffer systems and the kidneys, the pH of the blood can be maintained within a narrow range, helping the body to stay in balance.

Buffers in food

Food spoilage can be caused by a combination of various factors – such as light, oxygen, heat, humidity and/or many kinds of microorganisms (bacteria, yeasts and fungi). We try to reduce spoilage by keeping certain foods in the dark, in airtight containers and/or in refrigerators.

Spoilage of food by microorganisms depends greatly on the pH value of the food. Most microorganisms thrive

when the pH of their surroundings is approximately neutral (pH 6.6–7.5). The metabolism of these microorganisms is then greatest and they can multiply fast. Most bacteria can survive at pH values as low as 4.4 and as high as 9.0 – only specialised bacteria can survive outside this range.

One important factor in the spoilage of food is its buffer *capacity* – this describes the amount of acid or base required to change the pH of the food significantly. The more protein there is in a food, the higher is its buffer capacity, because amino acids have both acidic and basic characteristics. This means that it takes longer for the pH of the food to change sufficiently for bacteria to stop multiplying. Bacteria and moulds can also produce waste products which act as poisons or toxins, causing harmful ill-effects.

Many processed foods, such as jams, contain buffer systems (such as citric acid and sodium citrate) which help to maintain the pH within a range where growth of microoganisms is very slow or non-existent.

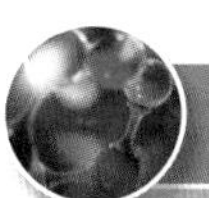

HSW Buffers in the ocean

The oceans of the world act as an enormous buffer system for the carbon dioxide in the atmosphere. When carbon dioxide dissolves in water it forms carbonic acid – and this dissociates to form hydrogen ions and hydrogencarbonate ions:

$$CO_2(g) + H_2O(l) \rightleftharpoons H_2CO_3(aq) \rightleftharpoons H^+(aq) + HCO_3^-(aq)$$

In turn, the hydrogencarbonate ions can dissociate to form carbonate ions and hydrogen ions:

$$HCO_3^-(aq) \rightleftharpoons H^+(aq) + CO_3^{2-}(aq)$$

These carbonate ions are used by many different organisms to combine with calcium ions to form calcium carbonate, which is the basis of many shells and of coral reefs.

The average pH of the oceans is around 7.8–8.0, so the equilibrium positions of the reactions are pushed towards the formation of hydrogencarbonate ions (see the top-right reaction above). Normally, the upper layers of the oceans are supersaturated with carbonate ions. This is good news for tiny organisms like coral polyps that absorb carbonate ions and combine them with calcium ions to make calcium carbonate for their outer protective skeleton.

fig. 1.5.25 Coral reefs are home to thousands of species of animals and plants.

Millions of these organisms build up over hundreds of years to form what we call coral reefs. However, atmospheric carbon dioxide levels are rising, due to human activities at least in part. When carbon dioxide levels increase, more carbonate is needed to buffer the ocean solution. As the carbonate ion concentration in the ocean falls it becomes harder for the living organisms to build their shells or skeletons. If the concentration of carbonate ions drops below saturation point, the calcium carbonate skeletons of the corals begin to dissolve. Scientists are working hard to find out exactly what is happening to the coral reefs around the world, trying to decide how best to reverse the increase in atmospheric carbon dioxide concentration.

Questions

1 Suggest why food such as onions can be preserved by pickling in vinegar.

2 When you exercise vigorously, the blood supply to your active muscles increases. One reason is to supply the tissues with oxygen and food. Why else do you think the increase in blood supply is important?

3 Investigate the carbon dioxide buffering systems in the ocean further and write equilibrium reactions for the processes.

4 Some people say that the increased amount of carbon dioxide in the atmosphere is causing 'acidification' of the oceans – others say that acidification is the wrong term to use. Explain both points of view.

Examzone

You are now ready to try the first Examzone test for Unit 4 (Examzone Unit 4 Test 1) on page 256, which tests you on what you have learnt in the first five chapters of this book.

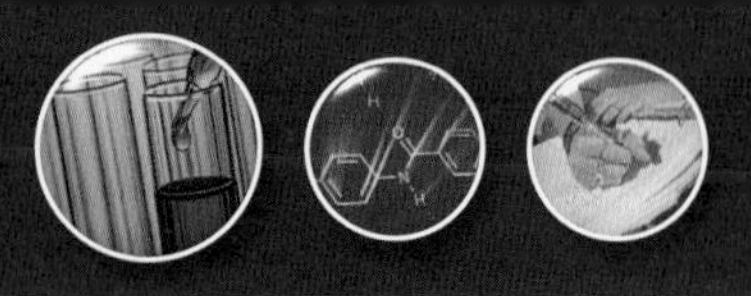

1.6 Further organic chemistry

Isomerism and chirality [4.8.1a, b]

Isomerism was introduced in chapter 1.7 of *Edexcel AS Chemistry*. **Structural isomerism** occurs when the atoms in a molecule can be bonded in different ways to produce distinctly different compounds. **Figure 1.6.1** shows some examples.

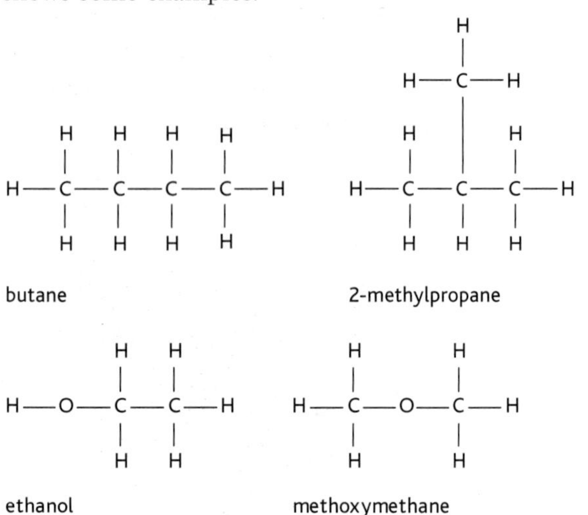

fig. 1.6.1 Examples of structural isomerism.

Another form of isomerism is **stereoisomerism**. Stereoisomers have the same groups of atoms, but in different spatial arrangements. **Geometric isomerism** was introduced in *Edexcel AS Chemistry* on pages 126–128. It arises due to the lack of free rotation of groups around a bond – frequently a double bond. Geometric isomers have components arranged differently with respect to the double bond. The existence of *cis* (*Z*) and *trans* (*E*) isomers is shown in **fig. 1.6.2**.

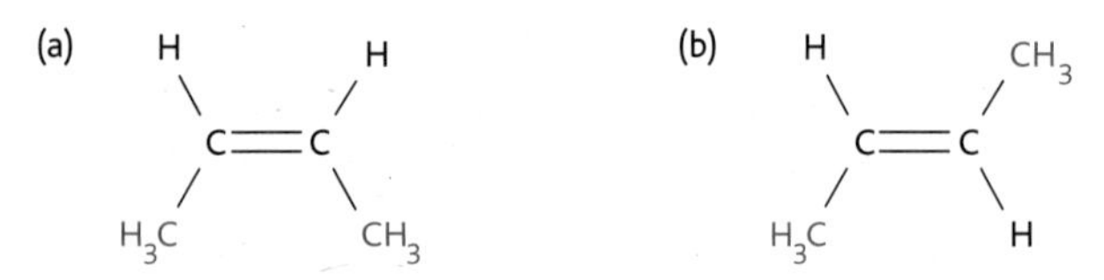

fig. 1.6.2 Geometric isomers of but-2-ene: (a) *cis*-but-2-ene (*Z*); (b) *trans*-but-2-ene (*E*).

Optical isomerism

Optical isomerism is another type of stereoisomerism. If two objects are **mirror images** of each other and they are not superimposable then they are said to be **chiral**. Chirality is a property exclusive to asymmetrical objects. The simplest type of chiral molecule is where four different groups are attached to one carbon atom, which acts as a **chiral centre**. This carbon atom is sometimes called an asymmetric carbon atom and indicated with an asterisk. The word 'chiral' comes from the ancient Greek word for hand.

We use a convention to show three-dimensional molecules in two-dimensional diagrams – this is illustrated in **fig. 1.6.3**.

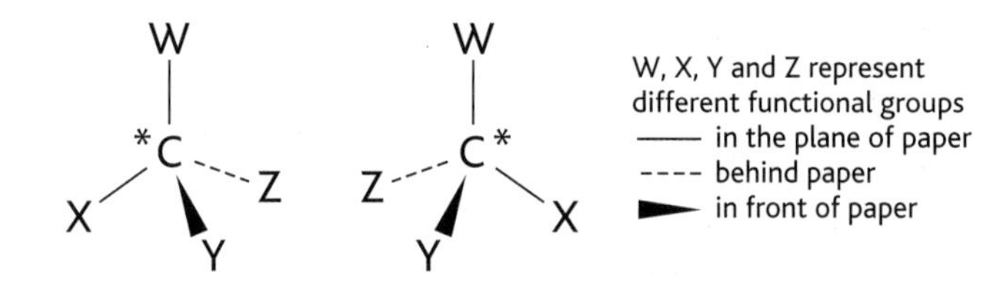

fig. 1.6.3 Representing three-dimensional molecules in two dimensions.

The 2,3-dihydroxypropanal molecule (structural formula $CH_2OHCHOHCHO$) is useful to consider because it is the standard for comparison to all other potentially chiral molecules. The non-systematic name for this is glyceraldehyde.

The three-dimensional arrangement of the bonds means there are two different forms. They are mirror images of each other, given the labels D- (dextro) and L- (laevo). However hard you try, you cannot **superimpose** the image of D-glyceraldehyde onto L-glyceraldehyde (**fig. 1.6.4**).

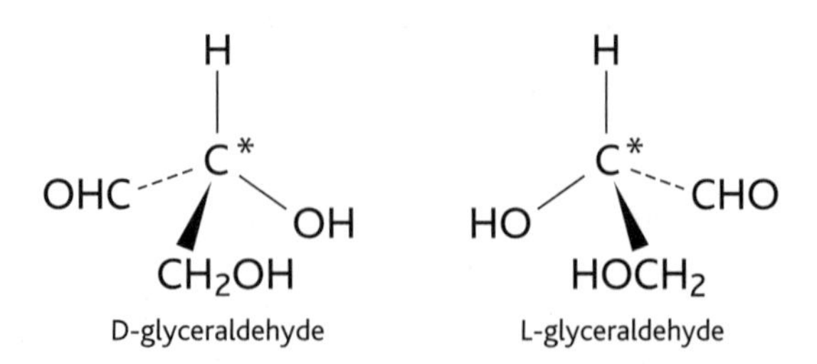

fig. 1.6.4 The two optical isomers of glyceraldehyde. They are mirror images.

Isomers that are mirror images of each other are called **enantiomers**. They are remarkably similar in both their physical and chemical properties. Unlike structural and geometric isomers, they cannot be told apart by differences in their physical properties or chemical reactivities. However, they do have one difference which allows us to identify them – they have different effects on **plane-polarised light** (see over). Compounds such as these are said to be **optically active** and their isomers are known as optical isomers. Making ball-and-stick models can help understanding of molecular structure.

Summarise the different types of isomerism in a flow diagram, including definitions of the types of isomers.

Optical isomerism in biochemical molecules

Biological systems are sensitive to the difference between stereoisomers. For example, only the D-forms of sugars and only the L-forms of amino acids are found in living materials.

Propoxyphene is an organic molecule with the molecular formula $C_{22}H_{29}NO_2$. It can exist as four enantiomers but only α-d-propoxyphene (known as D-propoxyphene) and α-l-propoxyphene (known as L-propoxyphene) are useful medically.

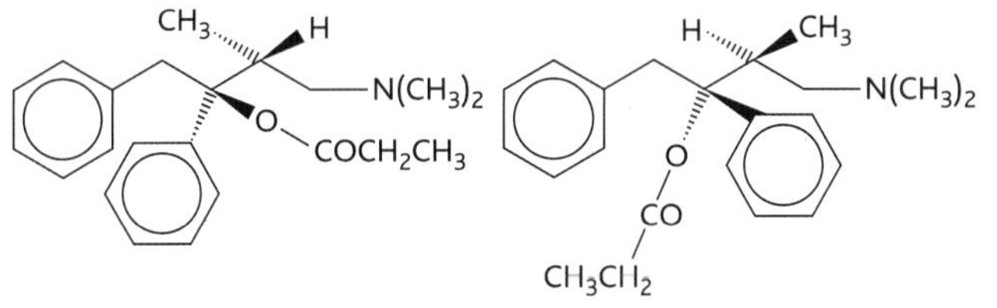

CH_3, H, $N(CH_3)_2$, O, $COCH_2CH_3$ — D-propoxyphene

H, CH_3, $N(CH_3)_2$, O, CO, CH_3CH_2 — L-propoxyphene

fig. 1.6.5 The optical isomers of propoxyphene.

D-propoxyphene is used as an analgesic (pain killer). The sensitive cells in the human body respond to the shape of this molecule, but do not respond to the mirror image form because of its different shape. However, L-propoxyphene may be no good as a pain killer but it is used in cough mixtures.

Look further into the controversy over the use of AAR (see HSW box) and compare it with the radiocarbon dating method.

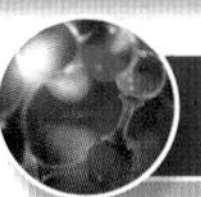

HSW Amino acid dating

After death, there is a slow conversion of L-amino acids to the D- enantiomers until equilibrium is established. There has been considerable argument in the scientific community about the use of this conversion process (known as amino acid racemisation or AAR) as a kind of 'clock' for dating ancient specimens of living organisms, including human remains. Unfortunately, the rate of the conversion is affected by a wide range of factors, from temperature and pH to the presence of aldehydes, and even the original arrangement of amino acids in the protein. Even preparing the specimen for measuring can affect the mixture. Most scientists now feel that, although interesting, it is of little absolute value in dating specimens – although some still support it strongly.

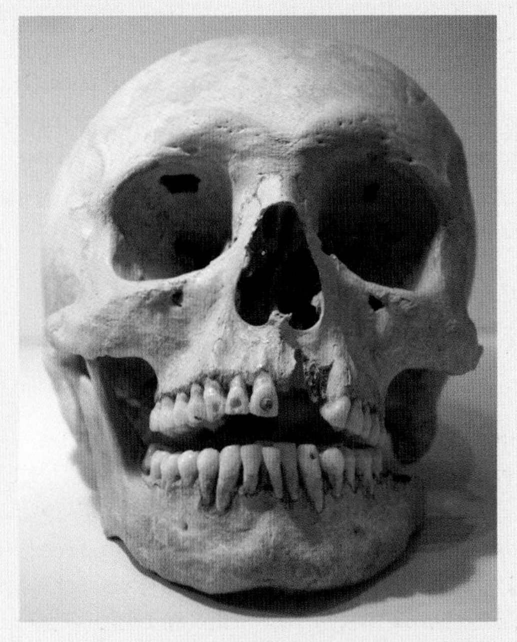

fig. 1.6.6 Teeth from different parts of the same mouth have been shown to have different ages using AAR dating – yet another reason why many scientists cast doubt on the method.

Questions

1 An alkane has a molecular formula C_7H_{16}. One of the structural isomers with this formula forms two optical isomers. Draw the displayed formula of this compound.

2 Two amino acids are glycine, NH_2CH_2COOH, and serine, $NH_2CH(CH_2OH)COOH$. Explain why one is optically active, but the other is not.

Optical activity of chiral molecules [4.8.1c]

We saw on the previous page that organic compounds containing chiral carbon atoms – those attached to four different groups – have optical isomers. Such enantiomers affect the plane of plane-polarised light by rotating it clockwise or anticlockwise.

Chirality

The subtle difference between molecules of optical isomers that we call chirality can be described in a less technical way as 'handedness'. While similar in appearance, your left hand and your right hand are not the same – after all, a left-handed glove simply does not fit the right hand. However, if you hold your left hand up to a mirror, it looks exactly like your right hand. We describe them as mirror images and the crucial difference is that the *image* of one cannot be superimposed on the other. It is only when two objects pass a test of superimposability that they can be described as truly identical.

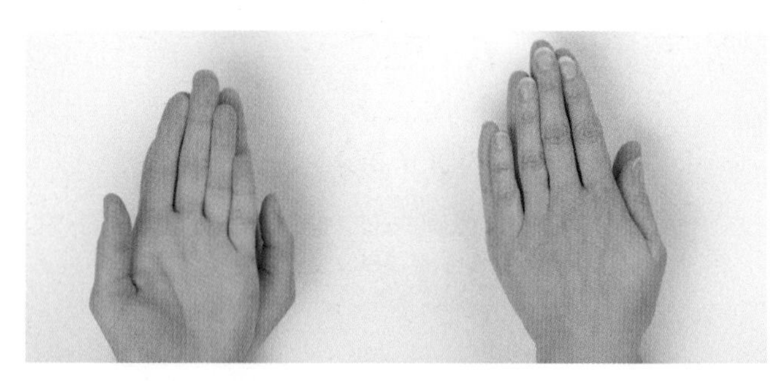

fig. 1.6.7 However hard you try, it is impossible to superimpose the image of one hand onto the other so that they match exactly.

Figure 1.6.8 shows two mirror images of 1-bromo-1-chloroethane. The central carbon atom is a chiral centre attached to four different groups. If you use a ball and stick model you will find it is possible to form two different models that are mirror images and are not superimposable.

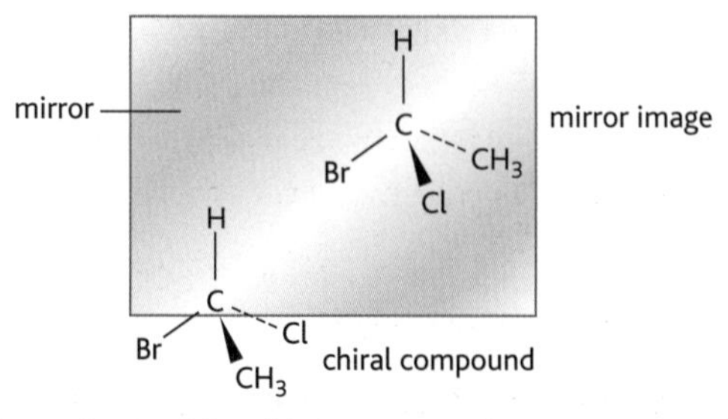

fig. 1.6.8 A molecule with a chiral centre can form two isomers that are mirror images.

If an organic compound does have D- and L-enantiomers then one of these will rotate the plane of plane-polarised light clockwise, and the other will rotate it anticlockwise. It is not possible to say which will be which, but the enantiomer that causes a clockwise rotation, whether it is the D- or the L-form, is called the '+' form, and the one that causes the anticlockwise rotation is the '–' form.

What is plane-polarised light?

The wave model envisages light as a form of electromagnetic radiation travelling as a transverse wave. This means that the direction in which disturbance takes place is at right angles to the direction in which the wave travels.

Figure 1.6.9 shows some of the planes in which the oscillations of a transverse wave may occur. A wave in which the oscillations take place in many planes is described as **unpolarised** and represents the situation in 'normal' light. A wave in which the oscillations occur in one plane only is said to be **plane polarised** in that direction.

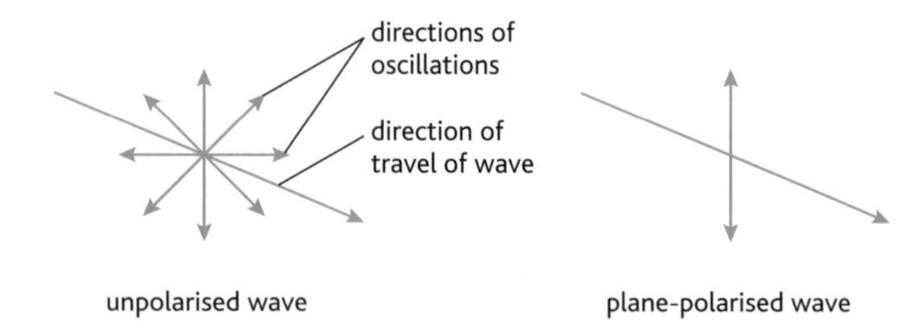

fig. 1.6.9 In an unpolarised wave, oscillations may occur in any plane, while in a plane-polarised wave they occur in only one plane.

A light beam that passes through a piece of Polaroid, the material used to make the lenses of some sunglasses, leaves with all the waves vibrating in the same plane. If a second piece of Polaroid is placed at right angles to the first, no light will pass through (**fig. 1.6.10**).

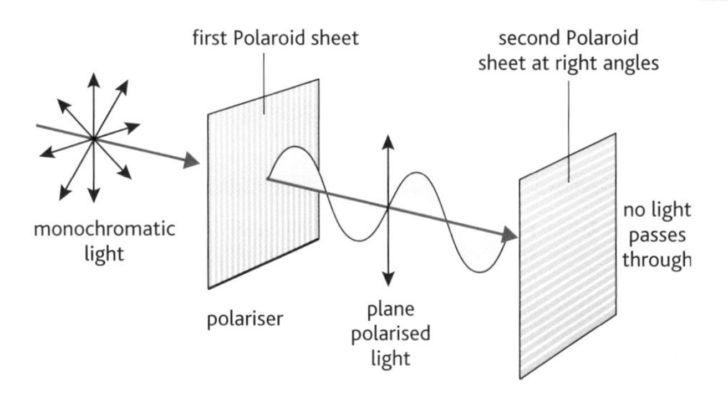

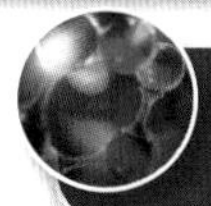

fig. 1.6.10 Polarisation of light.

You can see in **fig.1.6.10** that the plane of the polarised light (between the Polaroid sheets) is vertical. This means it passed through the first piece of Polaroid where the 'lines' are vertical but did not pass through where the 'lines' are horizontal. Now, if a transparent sample containing a single enantiomer is placed between the Polaroids, then it will rotate the plane of the polarised light because it is optically active. You would then have to rotate the second piece of Polaroid through an angle, called the angle of rotation, to return to the situation where no light passes through. The rotation may be clockwise (+) or anticlockwise (–). This principle is used in the polarimeter. This instrument measures the angle of rotation caused by a sample. The source of light is monochromatic, that is light of a single wavelength.

The optical activity of a substance is measured in terms of number of degrees by which plane-polarised light is rotated as it passes through a solution of the substance.

It is worth noting that some solutions of optically active substances have no effect on plane-polarised light. This is because it is a 1:1 mixture of the two enantiomers – called a **racemic mixture**. The clockwise rotation caused by one enantiomer is balanced by the anticlockwise rotation caused by the other.

SC Suggest how bacteria can cause a sample of 2-hydroxypropanoic acid that contains equal amounts of both of its optical isomers to cause a rotation of plane-polarised light.

HSW Planes of symmetry, Pasteur and tartaric acid

Molecules that are optically active do not have planes of symmetry. So the optical isomers of such a compound will not show a plane of symmetry. Conversely, compounds which are not optical active will have a plane of symmetry.

The optical isomers of tartaric acid were discovered by Louis Pasteur. Tartaric acid from living sources was found to rotate the plane of polarised light. Artificially synthesised tartaric acid did not. Pasteur noticed that the synthesised tartaric acid was made up of two types of crystals which were mirror images of each other. He sorted them by hand – and found that in solution they rotated light in opposite directions, although all their chemical properties were identical. Mixed together, they cancelled each other out and had no effect on light. It was the first clear demonstration of chiral molecules, (–)-tartaric acid and (+)-tartaric acid. Later, a third isomer of tartaric acid, called mesotartaric acid, was found. It is not optically active – and has a clear plane of symmetry!

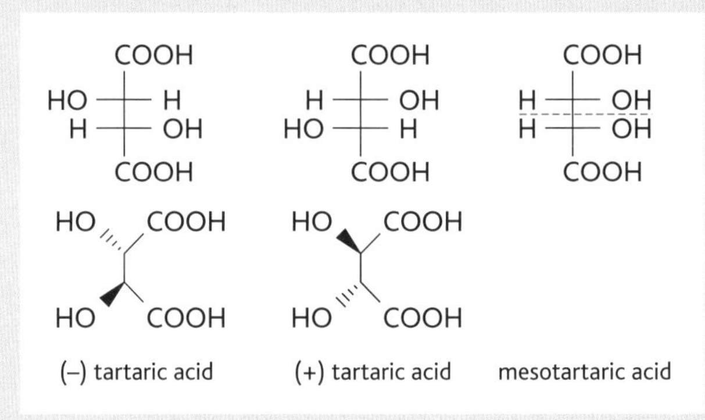

fig. 1.6.11 The three isomers of tartaric acid – note the plane of symmetry in mesotartaric acid.

Questions

1 One enantiomer of compound X rotates plane-polarised light in a polarimeter by +12°.

  a What rotation would you expect for the same amount of the other enantiomer of X?

  b What rotation would you expect for a mixture of equal amounts of the two enantiomers of X?

  c What name is given to a mixture of equal amounts of the two enantiomers of X?

2 What is the function of the polariser in a polarimeter?

Evidence for reaction mechanisms from optical activity [4.8.1d]

You will know from AS Chemistry that halogenoalkanes react with **nucleophiles**, such as hydroxide ions and cyanide ions, by **nucleophilic substitution**.

Earlier in this book, in **chapter 1.1**, you saw that using kinetic studies we could study this reaction more closely and identify two distinct routes for this reaction – S_N1 or S_N2. We can throw further light on the mechanism of this reaction using ideas from stereochemistry. Similar ideas can be used when considering the addition reactions of carbonyl compounds later in this chapter.

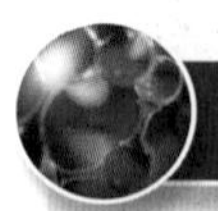

HSW Nucleophilic substitution

The S_N2 mechanism

We are going to consider the nucleophilic substitution reaction of an enantiomer that rotates the plane of plane-polarised light clockwise (+). This is summarised in **fig. 1.6.12**.

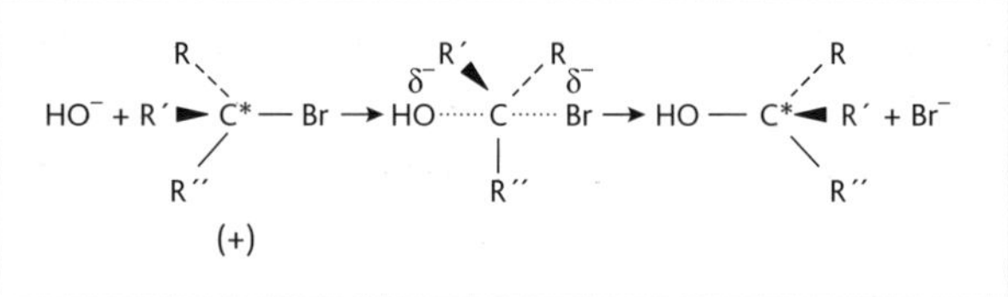

fig. 1.6.12 Nucleophilic substitution by the S_N2 mechanism.

The arrangement of the three groups R, R' and R'' attached to the chiral carbon atom is effectively turned inside out during the reaction. This is called **inversion** of the configuration. If the product could be a bromide rather than an alcohol, we know that rotation of plane-polarised light would be in the opposite direction (i.e. –). Since the product is an alcohol, we cannot tell merely by observing the direction of optical rotation whether it has the same or the opposite configuration to the bromoalkane from which it is derived. Other than mirror images, compounds having opposite configurations do not necessarily cause opposite directions of optical rotation, any more than do compounds of the same configuration necessarily cause the same direction of optical rotation.

Further studies have shown that inversion of the configuration occurs when an S_N2 reaction occurs.

The S_N1 mechanism

We are going to consider again the case where the reactant molecule is an enantiomer which rotates plane-polarised light clockwise (+).

You may remember that in this mechanism the first step, the slow rate-determining step, involves the loss of the halogen as a halide ion and the formation of a carbocation. The carbocation is planar.

We might expect that the attacking nucleophile, e.g. OH^-, can attack the carbocation from either the front or the back, and so the product is likely to contain a 1:1 mixture of the two enantiomers of the alcohol product. Hence, we would expect the product to be a racemic mixture and the product to be optically inactive. This is shown in **fig. 1.6.13**.

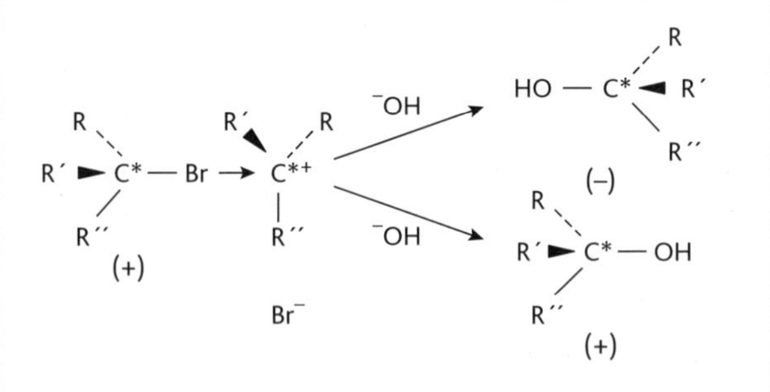

fig. 1.6.13 Nucleophilic substitution by the S_N1 mechanism.

In practice, the product often shows complete, or virtually complete, formation of the racemic mixture but it depends how rapidly the nucleophile attacks the central carbon atom of the carbocation. If the solvent can also act as a nucleophile (e.g. H_2O) then attack is likely to be more rapid even before the carbocation can achieve its planar form. The result will be more inversion.

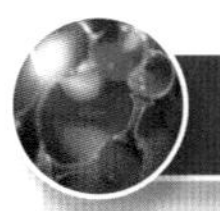

HSW Thalidomide – hope out of tragedy?

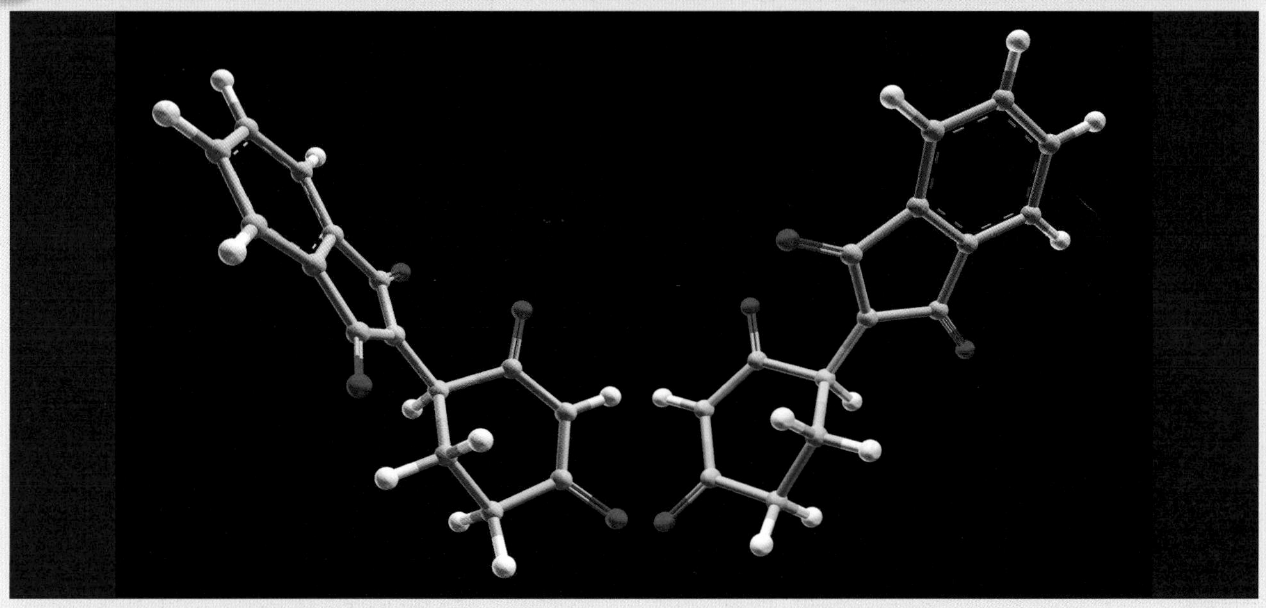

fig. 1.6.14 It is hard to understand how such a small change in a molecule could result in such devastating effects on a developing foetus.

Many medicines are mixtures of enantiomers. In the 1950s a new drug called thalidomide was introduced very successfully as a sedative. Doctors noticed that when it was prescribed to pregnant women, it eased the symptoms of morning sickness, and so they used it in early pregnancy. However, by the late 1950s and through the 1960s, 12 000 babies worldwide were born with unusual birth defects – their limbs simply did not grow properly. Eventually it was realised that the cause of the problem was thalidomide – the drug was affecting the foetuses of the pregnant women.

When the drug was synthesised on a trial basis, the process involved an S_N2 mechanism and only the L-isomer of the drug was produced – this has a sedative effect but no impact on a developing foetus. However, in the large-scale production process a planar intermediate was introduced. This led to an S_N1 mechanism, which produced a racemic mixture of the L- and D- enantiomers. The D-isomer is teratogenic – in other words, it affects the limb development of a foetus. So when women took the drug, they unwittingly took a compound that was going to cause terrible damage. However, even if they had only taken L-thalidomide, there would still have been an element of risk because there is some conversion between the two forms in the body.

The use of thalidomide was stopped completely, but scientists maintained an interest because – apart from its tragic effect on unborn children – it is an exceptionally safe drug. In recent years studies have shown that it is very useful in the treatment of conditions as diverse as leprosy and myeloma, a form of bone marrow cancer. It is now being used again, with many constraints when used in young fertile women. The biggest problem for scientists now is to reassure the public and convince people that, now the risks of the drug are known, the benefits it can bring to very sick people far outweigh the carefully managed risk.

Questions

1 An optically active halogenoalkane reacts with cyanide ions, CN^-, in a nucleophilic substitution reaction. The product is a racemic mixture of two enantiomers. What does this suggest about the mechanism of this reaction? Explain your answer.

2 Explain how different reaction mechanisms could result in the two different isomers of thalidomide.

Introduction to aldehydes and ketones [4.8.2a, b]

Aldehydes and **ketones** are important chemicals in many ways. They play a major part in many of the aromas and flavours associated with food – and also with wine. The two photographs illustrate some uses of aldehydes and ketones. **Figure 1.6.15** shows a brain bank in the USA where over 3000 healthy and diseased human brains are stored in methanal (formaldehyde) and used for medical research. Artists use methanal too – the artist Damian Hirst is well-known for artwork featuring various animals, including whole sheep and cattle, preserved in methanal.

Many ketones have distinctive smells. Patients with diabetes lack the hormone insulin, which breaks down glucose in the blood. The body utilises fats as an alternative energy source, leading to a build-up of ketones in the blood and urine that may eventually cause convulsions and diabetic coma. The distinctive smell of ketones on the breath can be a giveaway, but by testing the level of ketones in the urine, diabetes can be controlled and diabetic coma can be avoided in a much more reliable way. **Figure 1.6.16** shows a test being carried out on the urine of a diabetic patient. The gloved hands hold a test stick dipped in urine against a reference chart.

fig. 1.6.15 Human brain sections stored in methanal at the Harvard Brain and Tissue Resource Center, USA.

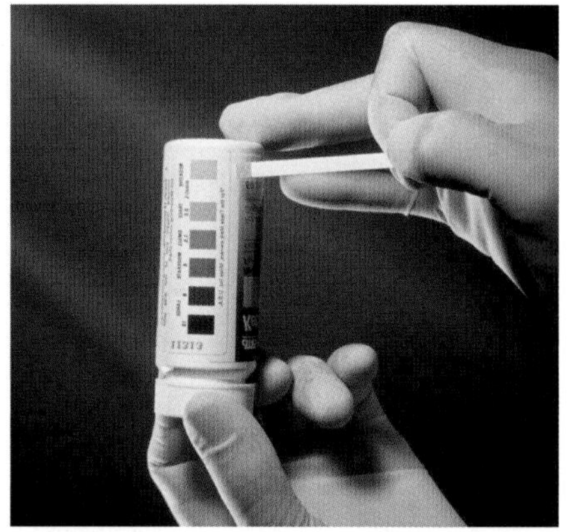

fig. 1.6.16 Test for diabetes – a negative result for ketones is shown here.

The nature of carbonyl compounds

You were introduced to alkenes in chapter 1.7 of *Edexcel AS Chemistry* – they are unsaturated hydrocarbons containing a C=C double bond. Aldehydes and ketones contain a C=O group called a **carbonyl** group.

Figure 1.6.17 shows that the bond between the two carbon atoms in an alkene consists of a σ bond (sigma bond) and a π bond (pi bond), each containing two electrons. Both carbon atoms are of identical electronegativity, so the electron density is equally distributed and the double bond is not polar.

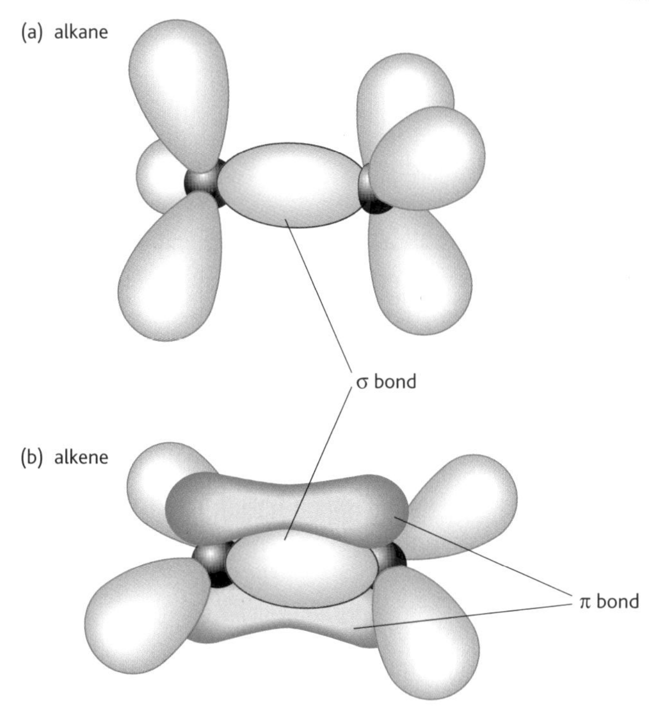

fig. 1.6.17 The nature of the double bond in alkanes and alkenes.

The situation is different in a carbonyl group, as shown in **fig. 1.6.18**. Again, there is a σ bond and a π bond, each containing two electrons. However, oxygen has a much higher electronegativity than carbon, so the electrons in the double bond are attracted more towards the oxygen atom – this is especially true of the π bond. The bond is permanently polarised and the result is a δ+ charge on the carbon atom and a δ– charge on the oxygen atom. As you will see, this affects the physical and chemical properties of aldehydes and ketones.

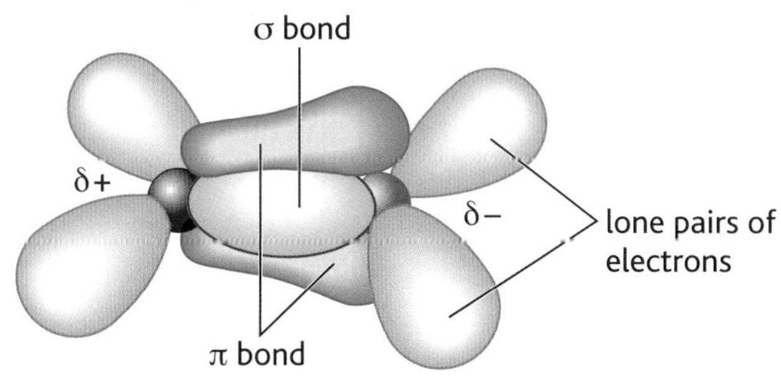

fig. 1.6.18 The pair of electrons in a carbonyl group spend more time around the more electronegative oxygen atom, shown here in red.

Aldehydes and ketones

In aldehydes, the carbonyl carbon atom has a hydrogen atom attached to it, together with either a second hydrogen atom or, much more commonly, a hydrocarbon group, which might be an alkyl group or one containing a benzene ring. Some simple aldehydes are shown in **fig. 1.6.19**. The general formula of aldehydes is shown in **fig. 1.6.20**.

Aldehydes can be produced by the oxidation of a primary alcohols.

In ketones, the carbonyl group has two hydrocarbon groups attached (and so has no hydrogen atoms). Again, these can be either alkyl groups or groups containing benzene rings. Some simple ketones are shown in **fig. 1.6.19**. The general formula of ketones is shown in **fig. 1.6.20**.

Ketones can be produced by the oxidation of secondary alcohols.

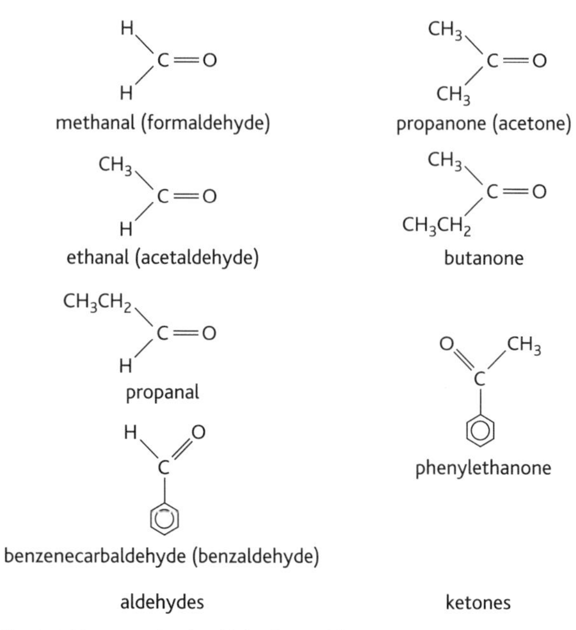

fig. 1.6.19 Some simple aldehydes and ketones.

$$\text{H}(\text{R})\text{C}{=}\text{O} \quad \text{aldehyde} \qquad \text{R}'(\text{R})\text{C}{=}\text{O} \quad \text{ketone}$$

fig. 1.6.20 The general formulae of aldehydes and ketones.

Physical properties of the aldehydes and ketones

The physical properties of the aldehydes and the ketones are influenced by the presence of the carbonyl group. This influence is particularly marked in compounds with molecules of relatively short-length carbon chains – as the chains get longer they have an increasing effect on the chemistry of the compounds. Aldehydes and ketones occupy territory part way between the alkanes (very volatile, with only **van der Waals** weak intermolecular forces) and the alcohols (low volatility, with van der Waals forces and much stronger hydrogen bonds between the molecules). The carbonyl group cannot take part in hydrogen bonding, but there are permanent dipole–dipole interactions between aldehyde molecules and between ketone molecules. This gives them boiling temperatures which are higher than those of the corresponding alkanes, but lower than those of the corresponding alcohols (see **fig.1.6.21**).

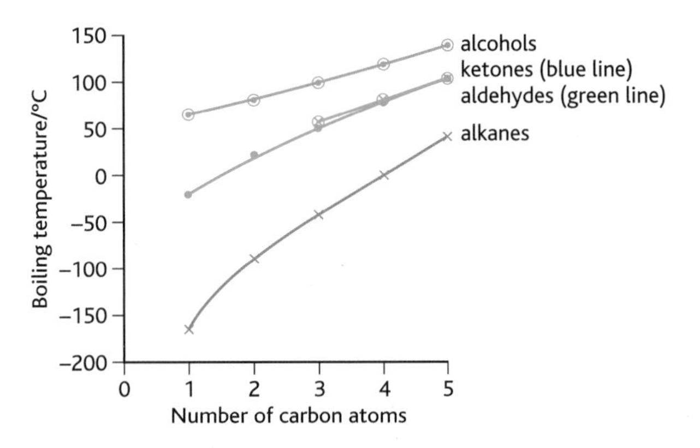

fig. 1.6.21 The boiling temperatures of some alkanes, aldehydes, ketones and alcohols. The effect of the intermolecular forces on the boiling temperatures can be seen clearly, with dipole–dipole attractions, which result from the carbonyl groups of the aldehydes and ketones, causing higher boiling temperatures than the equivalent alkanes but lower than the alcohols with their hydrogen bonding.

A further point about the boiling temperatures of the aldehydes and ketones is that molecules of straight-chain isomers can pack closer together than those of branched-chain molecules. This means that the forces between molecules of straight-chain isomers are stronger than those between molecules of branched-chain isomers – so the boiling temperatures of straight-chain isomers are higher. For example, butanal boils at 72 °C, while 2-methylpropanal boils at 64 °C; pentan-2-one has a boiling point of 102 °C compared to 94 °C for 3-methylbutan-2-one. **Figure 1.6.21** also shows that aldehydes are slightly more volatile than ketones.

Both the families of carbonyl compounds have members with strong odours, but the aldehydes are particularly responsible for imparting characteristic smells and flavours to a variety of foods and are also one of the groups of substances responsible for the bouquet of good wines.

HSW Should wine be bottled in plastic bottles rather than glass?

Wine has traditionally been bottled in glass. Some people have suggested that PET (polyethylene terephthalate) could be used as an alternative – it is cheaper to produce and requires less energy too. Like glass, it can be recycled.

Ethanal is a volatile component of wine – it is produced by partial oxidation of some of the ethanol during the maturing phase. In fact, it makes up about 90% of the aldehyde content of wine. At the low levels at which it is present, it enhances the flavour of the wine. At high levels the aroma is said to resemble rotting apples!

Studies using gas chromatography and mass spectrometry have shown that around 19 substances migrate from PET containers into the contents – to a greater extent at higher temperatures. Among these are appreciable amounts of ethanal. If this happens in PET wine bottles, the ethanal content of the wine will be increased, which might alter its aroma and flavour.

fig. 1.6.22 Plastic – or good old glass?

Solubility of aldehydes and ketones in water

Aldehydes and ketones composed of small molecules – such as methanal, ethanal and propanone – are miscible with water in all proportions, but this decreases with increasing chain length.

The reason for the solubility is that they can form hydrogen bonds with water molecules even though they cannot form hydrogen bonds with themselves.

With its δ+ charge, one of the water molecule's hydrogen atoms can be sufficiently attracted to one of the lone pairs on the oxygen atom of a carbonyl group for a hydrogen bond to be formed – as shown in **fig. 1.6.23**.

H–O–H$^{\delta+}$----O$^{\delta-}$=C(H)–CH$_3$

fig. 1.6.23 Hydrogen bonding between water molecules and ethanal molecules is stronger than the intermolecular bonds in the pure aldehyde.

Dispersion forces and **dipole–dipole** attractions between the aldehyde or ketone and the water molecules will also exist. Forming these attractions releases energy, which helps to supply the energy needed to separate the water molecules and aldehyde or ketone molecules from each other so that they can mix together.

As chain length increases, the hydrocarbon portions of the molecules prevent such attractions occurring.

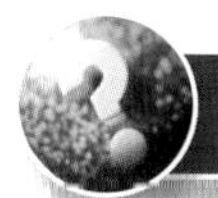

Questions

1 Draw displayed formulae and write the names of all the aldehydes and ketones with molecular formula C_4H_8O.

2 The compounds A–D (right) are either aldehydes or ketones. Which are aldehydes and which are ketones? Explain your answers.

A: $CH_3CH_2C(=O)CH_3$

B: CH_3CH_2CHO

C: $C_6H_5C(=O)C_6H_5$

D: $C_6H_5C(=O)H$

Testing and identifying carbonyl compounds [4.8.2c(iv)]

All aldehydes and ketones contain the carbonyl group, C=O. In this section we are going to consider a chemical test that can be used to identify a particular aldehyde or ketone.

Brady's reagent

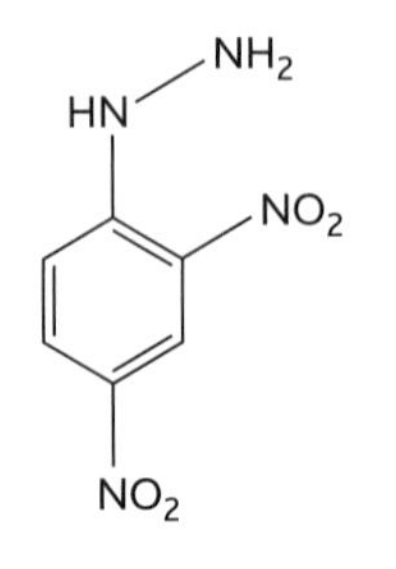

fig. 1.6.24 2,4-dinitrophenylhydrazine.

Brady's reagent (2,4-dinitrophenylhydrazine) is a red–orange solid, usually supplied wet to reduce the risk of explosion. **Figure 1.6.24** shows the displayed formula of this molecule.

Aldehydes and ketones react with **Brady's reagent** to form yellow/orange/red crystalline solids. These solids are called 2,4-dinitrophenylhydrazine **derivatives**. They have no practical use other than in identifying aldehydes and ketones. **Figure 1.6.25** shows the equation for the reaction between 2,4-dinitrophenylhydrazine and ethanal, and **fig. 1.6.26** shows the product.

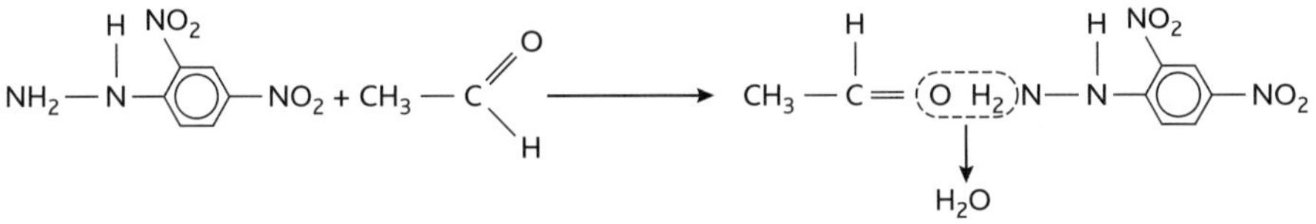

$$NH_2-NH-C_6H_3(NO_2)-NO_2 + CH_3-CHO \longrightarrow CH_3-CH{=}N-NH-C_6H_3(NO_2)-NO_2 + H_2O$$

2,4-dinitrophenylhydrazine ethanal ethanal 2,4-dinitrophenylhydrazone

fig. 1.6.25

The reaction that takes place is a **condensation reaction** – this involves addition followed by the elimination of water. The derivative can be removed by filtration and purified by recrystallisation. The process is summarised in **fig. 1.6.27**.

These compounds have characteristic melting temperatures which can be compared with values given in data books. **Table 1.6.1** shows the melting temperatures for the derivatives of some carbonyl compounds.

Carbonyl compound	Melting temperature of 2,4-dinitrophenylhydrazone/°C
Methanal	166
Ethanal	168
Propanal	155
Butanal	126
Benzaldehyde	237
Propanone	126
Butanone	115
Pentan-3-one	156

table 1.6.1 Melting temperatures for some 2,4-dinitrophenylhydrazones.

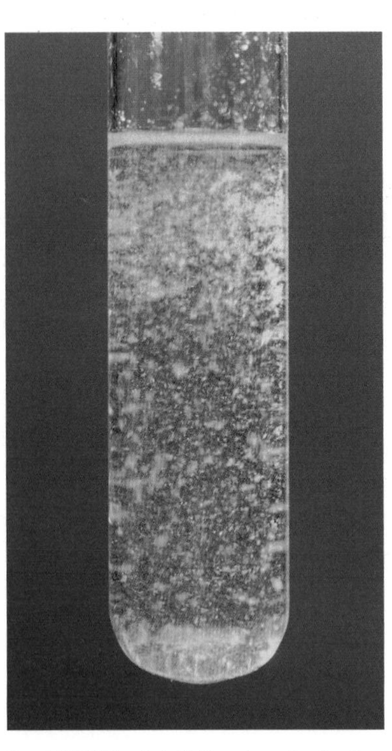

fig. 1.6.26 The 2,4-dinitrophenylhydrazine derivative of ethanal.

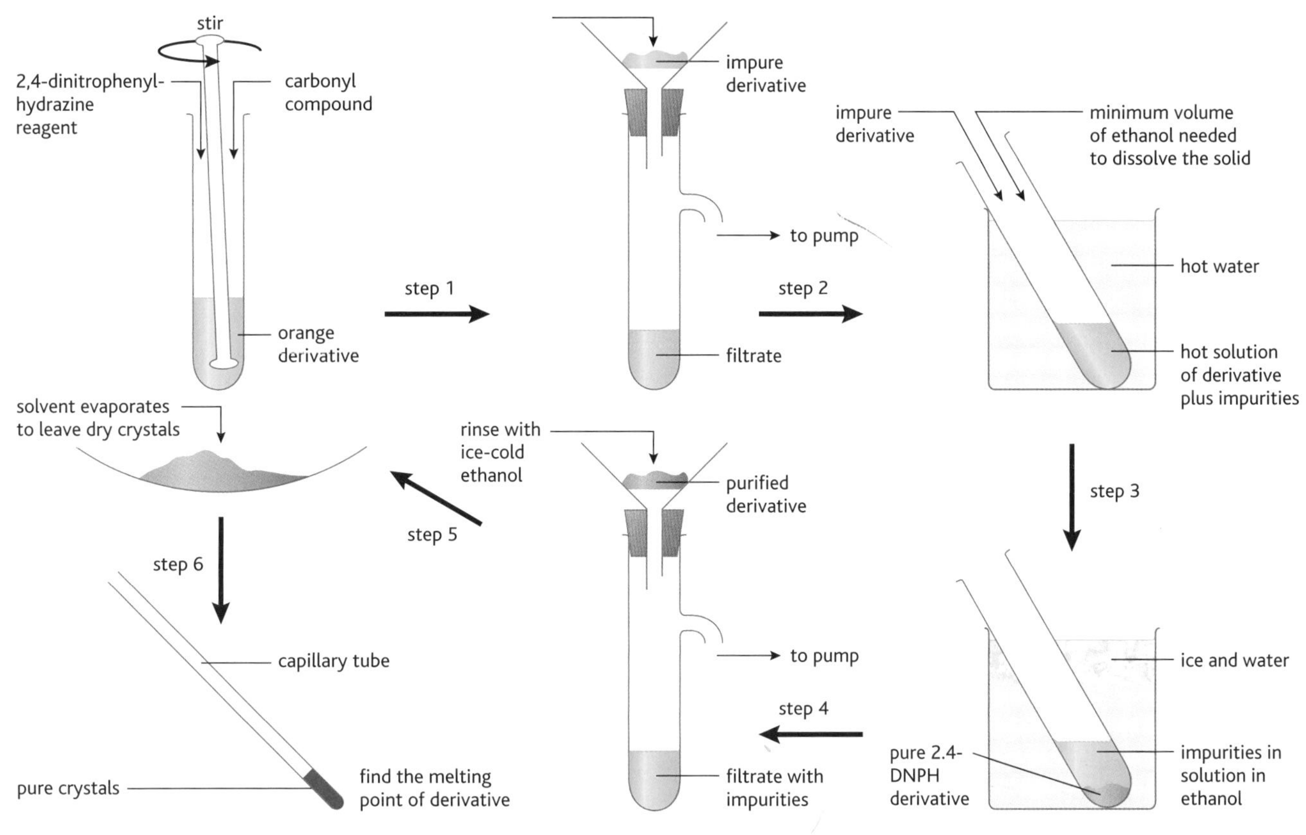

fig. 1.6.27 Small-scale preparation and characterisation of the 2,4-dinitrophenylhydrazine derivative of a carbonyl compound.

Why use Brady's reagent?

Other compounds could be used to form derivatives – hydroxylamine, hydrazine or dinitrohydrazine, for example. 2,4-dinitrophenylhydrazine derivatives are less soluble and therefore crystallise out more easily, so it is more useful.

However, more sophisticated techniques such as infrared spectroscopy have evolved, and have largely replaced the practically demanding derivative method.

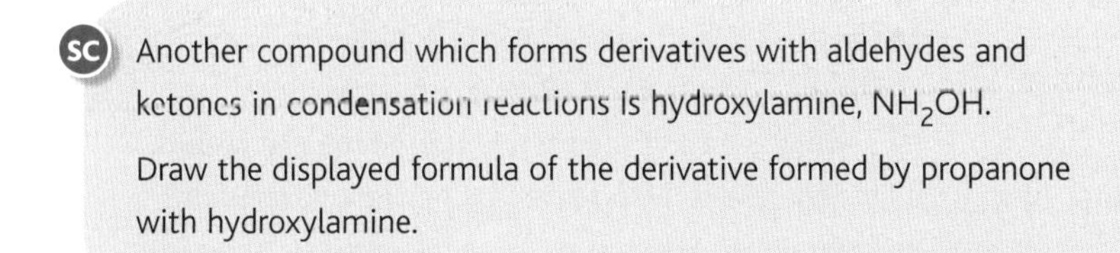

SC Another compound which forms derivatives with aldehydes and ketones in condensation reactions is hydroxylamine, NH_2OH.

Draw the displayed formula of the derivative formed by propanone with hydroxylamine.

Questions

1. An aldehyde forms a derivative with 2,4-dinitrophenylhydrazine which, after purification, is found to have a melting temperature of 156 °C. Suggest which aldehyde this is.
2. A student suggested that an aldehyde could be identified by finding its boiling temperature. Suggest why this is not as good as measuring the melting temperature of a derivative.
3. A ketone forms a 2,4-dinitrophenylhydrazine derivative that melts between 118 and 124 °C. Suggest what this could mean.

Reactions of carbonyl compounds [4.8.2c(i–iii, v)]

Oxidation reactions of aldehydes

There are many similarities in the reactions of aldehydes and ketones with a wide variety of reagents. However, they can be readily distinguished by the way in which they react with oxidising agents. Aldehydes have a hydrogen atom that forms part of the functional group. This hydrogen is activated by the carbonyl group and, as a result, aldehydes are readily oxidised to carboxylic acids by even quite mild oxidising agents. In contrast, ketones have no such hydrogen atom and so they are very resistant to oxidation – they only react after prolonged treatment with a very strong oxidising agent.

The oxidation of aldehydes can be carried out by warming with a solution of potassium dichromate(VI) acidified with sulfuric acid. For example:

$$\underset{\text{ethanal}}{3CH_3CHO(l)} + Cr_2O_7^{2-}(aq) + 8H^+(aq) \rightarrow \underset{\text{ethanoic acid}}{3CH_3COOH(aq)} + 4H_2O(l) + 2Cr^{3+}(aq)$$

Testing carbonyl compounds

There are three reactions that are regularly used as qualitative tests to determine if an unknown compound is an aldehyde or a ketone. These are Fehling's solution, Tollens' reagent and acidified dichromate(VI) ions.

The most common tests to identify aldehydes and ketones all involve oxidation reactions.

Fehling's or Benedict's solution

This contains an aqueous alkaline solution containing Cu^{2+} ions. Normally these would give a precipitate of copper(II) hydroxide under these conditions, but in **Fehling's solution** the copper(II) ions are prevented from precipitating by complexing them with 2,3-dihydroxybutanedioate ions.

If an aldehyde is warmed with Fehling's solution then the aldehyde is oxidised to a carboxylic acid, and the Fehling's solution is reduced, giving a red-brown precipitate of copper(I) oxide (**fig. 1.6.28**). The half-equations are:

$$2Cu^{2+}(aq) + OH^-(aq) + 2e^- \rightarrow Cu_2O(s) + H^+(aq)$$

$$RCHO(aq) + H_2O(l) \rightarrow RCOOH(aq) + 2H^+(aq) + 2e^-$$

so overall:

$$2Cu^{2+}(aq) + RCHO(aq) + OH^-(aq) + H_2O(l) \rightarrow Cu_2O(s) + RCOOH(aq) + 3H^+(aq)$$

If a ketone is warmed with Fehling's solution in the same way then there is no reaction – so the two can be distinguished by this method.

fig. 1.6.28 Fehling's solution (blue, left) – the reagent has a blue colour because it contains copper(II) ions. The test tube on the right contains the products of the reaction with an aldehyde – it has a suspension of red copper(I) oxide.

HSW Fehling's solution or Benedict's solution?

You are probably familiar with Benedict's solution from biology. Both this and Fehling's solution give a positive test with reducing sugars and some aldehydes. Both consist of a solution of copper(II) ions in alkaline solution but they use different chemicals to prevent the copper(II) hydroxide precipitating. In the case of Benedict's solution, citrate ions are used. In both cases, a red-brown precipitate of copper(I) oxide is formed. Benedict's solution can be kept in a bottle ready for use and is safer, so it is used in biology lessons for identifying reducing sugars. Fehling's solution is less stable and has to be prepared by mixing two solutions just before use, but it gives more reliable results with aldehydes.

Tollens' reagent

This is produced by dissolving silver nitrate in water and adding aqueous ammonia. Silver(I) oxide is formed as a precipitate initially but this dissolves as more ammonia is added forming a complex – the diamminesilver(I) ion, $[Ag(NH_3)_2]^+$.

When **Tollens' reagent** is warmed gently with an aldehyde, the aldehyde is oxidised and the silver(I) ions in the complex are reduced to form silver metal. With careful preparation, this coats the inside of the test tube as a silver mirror (**fig. 1.6.29**) – or less successfully as a black precipitate. For this reason the reaction is known as the 'silver mirror test'.

fig. 1.6.29 The silver mirror produced when an aldehyde reacts with Tollens' reagent.

The half reactions are:

$$2Ag^+(aq) + 2e^- \rightarrow 2Ag(s)$$

$$RCHO(aq) + H_2O(l) \rightarrow RCOOH(aq) + 2H^+(aq) + 2e^-$$

so overall:

$$2Ag^+(aq) + RCHO(aq) + H_2O(l) \rightarrow 2Ag(s) + RCOOH(aq) + 2H^+(aq)$$

If a ketone is warmed with the same reagent, there is no reaction.

Reduction reactions of aldehydes and ketones

You may recall that aldehydes and ketones are conveniently prepared in the laboratory by the oxidation of alcohols. Partial oxidation of a primary alcohol produces an aldehyde, and oxidation of a secondary alcohol produces a ketone (see *Edexcel AS Chemistry* page 210).

These reactions can be reversed by reduction – so aldehydes are readily reduced to primary alcohols, and ketones to secondary alcohols. Suitable reducing agents include lithium tetrahydridoaluminate ($LiAlH_4$) dissolved in dry ether (ethoxyethane), or sodium tetrahydridoborate ($NaBH_4$) dissolved in ethanol. **Figure 1.6.30** shows equations for the reduction of propanal and propanone.

$$CH_3CH_2CHO + [H] \xrightarrow{LiAlH_4} CH_3CH_2CH_2OH$$

propanal → propan-1-ol

$$CH_3COCH_3 + [H] \xrightarrow{LiAlH_4} CH_3CH(OH)CH_3$$

propanone → propan-2-ol

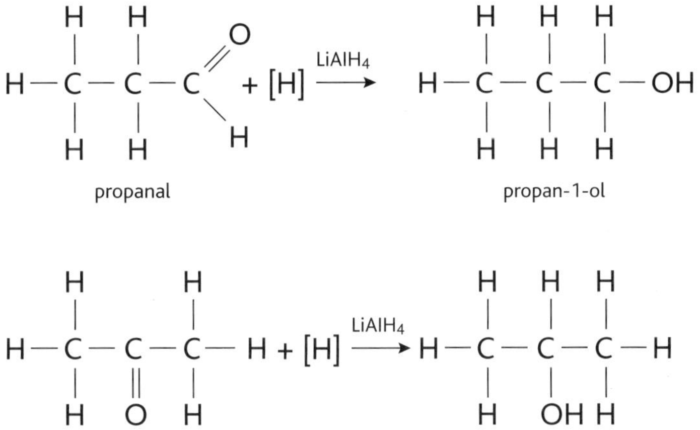

fig. 1.6.30 The reduction of propanal and propanone – [H] is used as an abbreviation for the reducing agent.

HSW Manufacture of fexofenadine

Fexofenadine hydrochloride is an antihistamine drug used in the treatment of hayfever and other allergy conditions. The manufacturing process involves the reduction reaction shown in **fig. 1.6.31**.

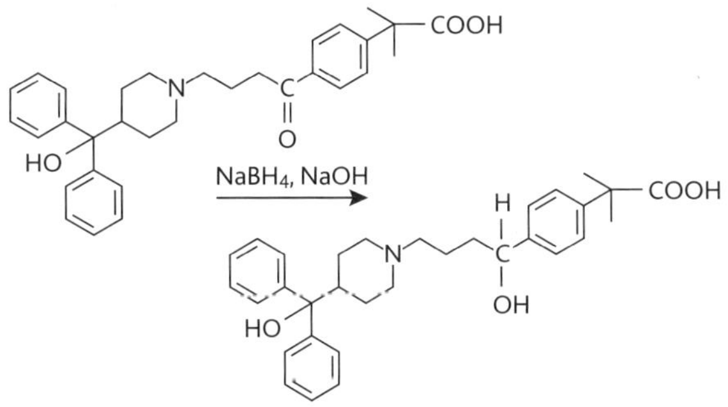

$NaBH_4$, $NaOH$

fig. 1.6.31 Reduction of a carbonyl group in a large molecule.

The conditions involve using the reducing agent sodium tetrahydridoborate and sodium hydroxide dissolved in a mixture of water and ethanol. It is kept at pH values between 7 and 8 at room temperature for 3 hours.

Why is sodium tetrahydridoborate used rather than the more powerful reducing agent lithium tetrahydridoaluminate (lithium aluminium hydride)? The latter also reduces esters, amides and carboxylic acids. In this reaction it would reduce the –COOH group to $-CH_2OH$.

Addition reactions with hydrogen cyanide

Aldehydes and ketones undergo an addition reaction with hydrogen cyanide to produce hydroxynitriles (or cyanohydrins). The HCN molecule adds across the C=O bond. Examples are shown in **fig. 1.6.32**.

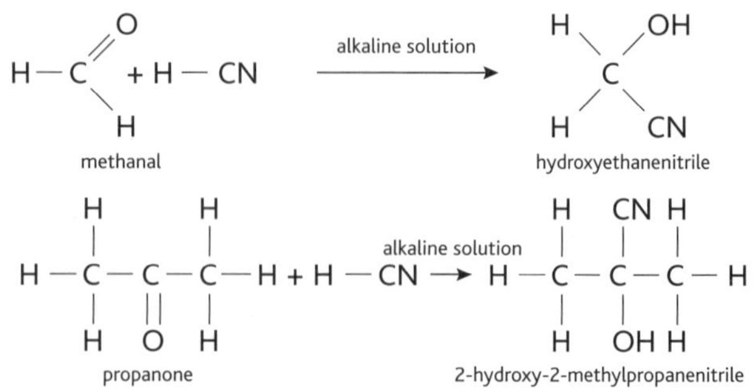

$$HCHO + H{-}CN \xrightarrow{\text{alkaline solution}} H_2C(OH)CN$$

methanal → hydroxyethanenitrile

$$CH_3COCH_3 + H{-}CN \xrightarrow{\text{alkaline solution}} CH_3C(OH)(CN)CH_3$$

propanone → 2-hydroxy-2-methylpropanenitrile

fig. 1.6.32 Examples of addition reactions with hydrogen cyanide.

This happens in the presence of potassium cyanide in cold, alkaline solution. In aqueous solution, the acid–base equilibrium:

$$K^+(aq) + CN^-(aq) + H_2O(l) \rightleftharpoons K^+(aq) + HCN(aq) + OH^-(aq)$$

is established. On the left-hand side of this equilibrium is the species CN^-, which is a weak Brønsted base (page 75). On the right-hand side is the OH^- ion, which is a much stronger Brønsted base than CN^-. The net result is that the hydroxide ion competes more effectively for the proton than the cyanide ion, and the reaction is pushed to the left, forming CN^-. This is the nucleophile needed for attacking the carbon atom of the carbonyl group.

The mechanism of the reaction is shown in **fig. 1.6.33**. In the first step, the incoming nucleophile, CN^-, has a lone pair of electrons. It attacks the central carbon atom, displacing a pair of electrons in the double bond and pushing them onto the oxygen atom. The oxygen then has a negative charge.

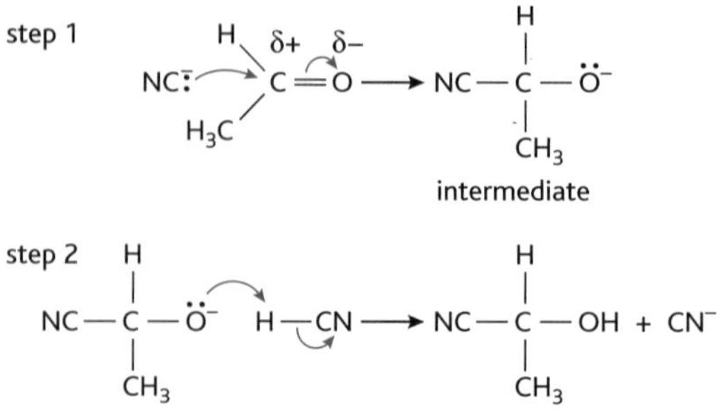

step 1

$$NC^{:-} + (H)(H_3C)C^{\delta+}{=}O^{\delta-} \longrightarrow NC{-}C(H)(CH_3){-}O^-$$

intermediate

step 2

$$NC{-}C(H)(CH_3){-}O^- + H{-}CN \longrightarrow NC{-}C(H)(CH_3){-}OH + CN^-$$

fig. 1.6.33 The mechanism of the nucleophilic addition reaction of carbonyl compounds.

Finally, in the second, fast, step the negatively charged oxygen acts as a base and gains a proton to make the final product.

This reaction was the first to be explained in terms of mechanisms for organic reactions (1902).

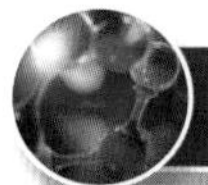

HSW Stereochemistry of the reaction

When hydrogen cyanide reacts with ethanal to form a hydroxynitrile, the molecule formed contains an asymmetrical carbon atom, and is therefore optically active. Despite this, the products of the reaction appear to show no optical activity. **Figure 1.6.34** shows why this is.

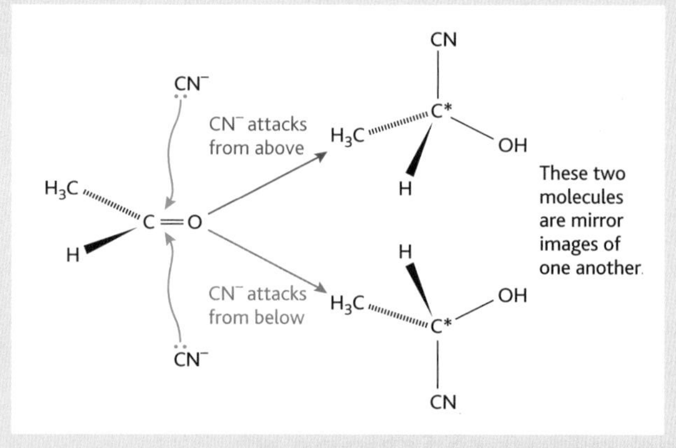

fig. 1.6.34 The reaction of ethanal with hydrogen cyanide, showing the two resultant products as mirror images of one another.

The reaction produces equal quantities of the two optical isomers of the hydroxynitrile, which rotate the plane of plane-polarised light in opposite directions. Each isomer cancels out the effect of the other, so no overall rotation of the plane of polarisation is seen.

The triiodomethane reaction

When ethanal is warmed with iodine in alkaline solution, a two-stage reaction takes place (**fig. 1.6.35**).

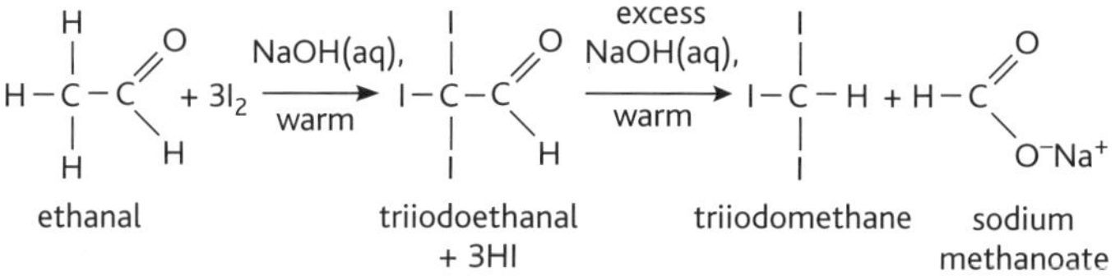

CH_3CHO (ethanal) $+ 3I_2 \xrightarrow[\text{warm}]{\text{NaOH(aq)}}$ CI_3CHO (triiodoethanal) $+ 3HI \xrightarrow[\text{warm}]{\text{excess NaOH(aq)}}$ CHI_3 (triiodomethane) $+ HCOO^-Na^+$ (sodium methanoate)

fig. 1.6.35 The reaction between ethanal and iodine in alkaline solution.

First, the aldehyde reacts with the iodine to form triiodoethanal; and then this reacts with the alkali to form triiodomethane. Triiodomethane (or iodoform, to give it its old non-systematic name) is an insoluble yellow solid with a characteristic smell. Because it is insoluble, it forms as a yellow, crystalline precipitate.

This reaction used to be known as the **iodoform reaction**. It is not specific to ethanal, but it is specific to organic compounds such as those in **fig. 1.6.36**, that contain a methyl group next to a carbonyl group or can react to give a methyl group next to a carbonyl group.

These functional groups give a positive iodoform reaction.

CH_3-CO-R — R may be hydrogen or a hydrocarbon chain

$CH_3-CH(OH)-R$ — R may be hydrogen or a hydrocarbon chain

ethanol: $CH_3CH_2OH \rightarrow CH_3CHO$ — oxidised by I_2 to give ethanal; gives the positive iodoform reaction

propan-2-ol: $CH_3CH(OH)CH_3 \rightarrow CH_3COCH_3$ — oxidised by I_2 to give propanone; gives the positive iodoform reaction

butanone: $CH_3COCH_2CH_3$ — gives the positive iodoform reaction

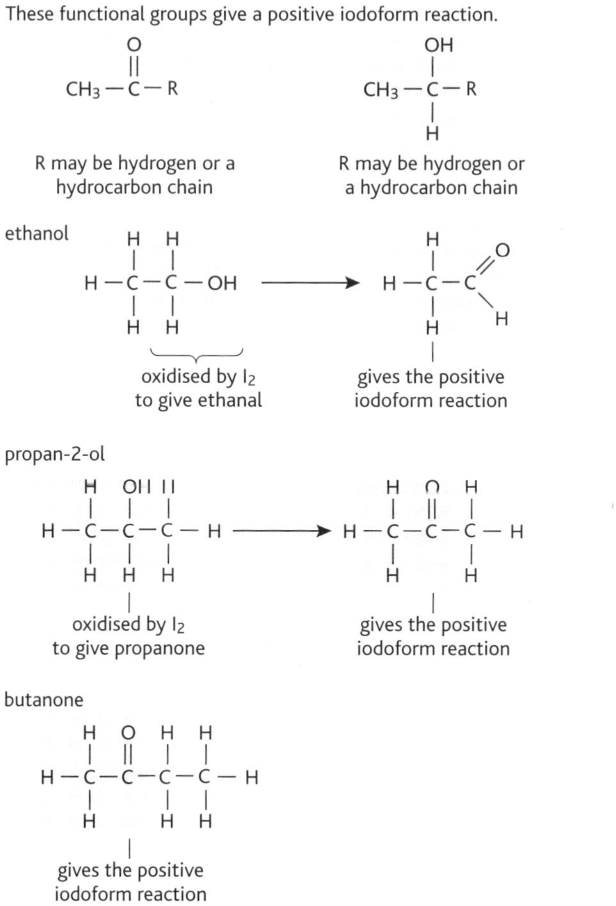

fig. 1.6.36 The molecular arrangements that will give the triiodomethane reaction.

Uses of aldehydes and ketones

Propanone (acetone) is a common fingernail-polish remover and is a widely used solvent. 2-butanone (methyl ethyl ketone) is used as a solvent and paint stripper. Benzaldehyde is an almond extract. Carvone gives an example of how the different optical isomers have different uses: (–)-carvone is used as spearmint flavouring, and (+)-carvone is used as caraway seed flavouring. Vanillin is the vanilla flavouring.

Questions

1 Suggest a test that could be used to distinguish between:
   a propanal and propanone
   b ethanal and propanal.

2 A compound with a molecular formula $C_5H_{10}O$ gives a negative test with Fehling's solution but a positive triiodomethane (iodoform) test. Identify a compound that would do this.

3 A compound A with a molecular formula C_3H_8O is oxidised with acidified potassium dichromate(VI) to form a liquid B. B reacts with hydrogen cyanide to form a compound C that contains three carbon atoms. Draw the structures of compounds A, B and C.

4 An alkene D with a molecular formula C_3H_6 can be converted into an alcohol E. This alcohol is oxidised to form a compound F, which does not give a silver mirror with Tollens' reagent. Draw the structures of compounds D, E and F.

An introduction to carboxylic acids: examples, physical properties and preparation [4.8.3a, b, c]

In the previous section you saw that complete oxidation of either primary alcohols or aldehydes produces carboxylic acids. Carboxylic acids contain the carboxyl group. This is a complex functional group which is made up of a hydroxyl group –OH (like the alcohols) and a carbonyl group C=O (like the aldehydes and ketones). Now, these two functional groups (**fig. 1.6.37**) are so close together they affect each other. Because of this, the reactions of the carboxylic acids tend to differ from those of the corresponding alcohols, aldehydes and ketones.

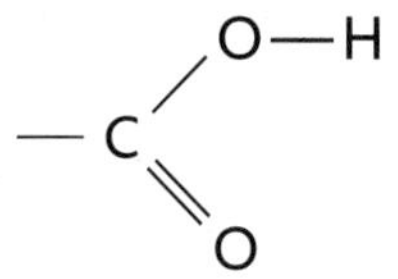

fig. 1.6.37 The carboxyl functional group.

The general formula of the carboxylic acids is RCOOH, where R can be a hydrogen atom, an alkyl group or an aryl group. For the straight-chain monocarboxylic acids, this results in a general formula of $C_nH_{2n+1}COOH$.

Carboxylic acids are named from the corresponding hydrocarbon. The stem (e.g. methan-, propan-, octan-) indicates the number of carbon atoms *including* the carbon of the carboxyl group. This is followed by the suffix -oic acid. Examples (see **fig. 1.6.38**) include methanoic acid, ethanoic acid and propanoic acid. Some acids contain two carboxylic groups and these are known as the dicarboxylic acids, e.g. ethanedioic acid.

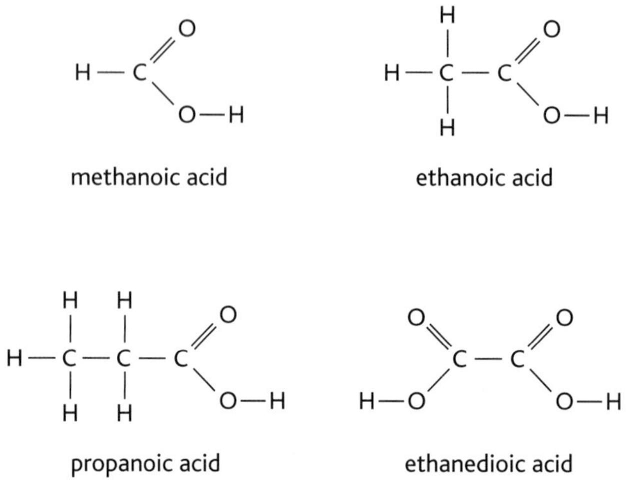

methanoic acid

ethanoic acid

propanoic acid

ethanedioic acid

fig. 1.6.38 Simple carboxylic acids.

The properties of the carboxyl group

The properties of the carboxylic acids, both physical and chemical, are largely governed by two factors:

- the polarity of the carboxyl group
- the length of the carbon chain.

As the carbon chain length increases, the influence of the functional group decreases. We have seen this in other homologous series – such as the alcohols and aldehydes.

Physical properties of carboxylic acids

The presence of two polarised groups (C=O and O–H) means that carboxylic acid molecules experience relatively strong intermolecular forces. So they have relatively high melting and boiling temperatures. Furthermore, the polarity and structure of the molecules is such that they can exist as hydrogen-bonded dimers in the vapour, liquid or solid state – this is shown in **fig. 1.6.39**.

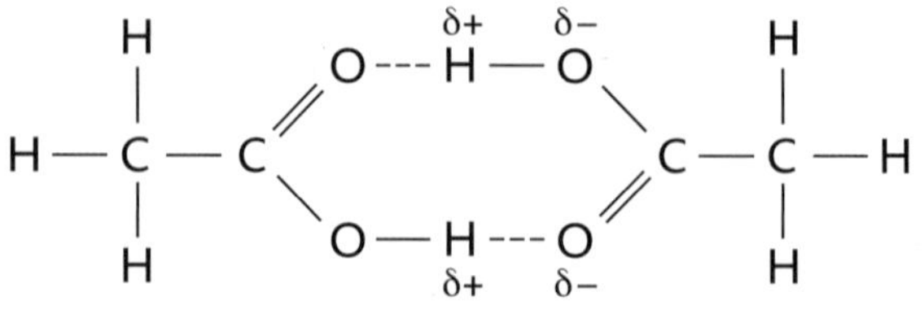

fig. 1.6.39 Hydrogen bonding in carboxylic acids is very efficient.

Hydrogen bonding is also responsible for the high solubility of the smaller carboxylic acid molecules in water and other polar solvents (**fig. 1.6.40**). Longer-chained acids become less soluble because the organic nature of the carbon chain becomes dominant.

fig. 1.6.40 The solubility of carboxylic acids in polar solvents.

Pure ethanoic acid is a liquid at normal room temperature. However, in an unheated storeroom in winter it is likely to freeze – its melting temperature is 17 °C. Solid ethanoic acid looks remarkably like ice, and is sometimes referred to as glacial ethanoic acid. It owes this relatively high melting temperature to hydrogen bonding.

The effect of this dimer formation can be seen by comparing the melting temperatures of different compounds each containing two carbon atoms (**fig. 1.6.41**).

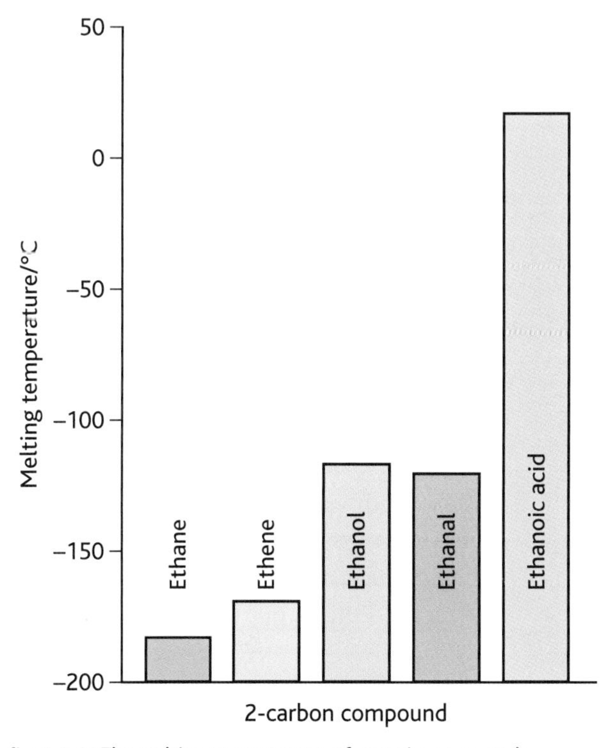

fig. 1.6.41 The melting temperatures of organic compounds containing two carbon atoms.

Carboxylic acids with more than 8 carbon atoms are solids at room temperature. They are usually only very slightly soluble in cold water, but more soluble in hot water.

Preparation of carboxylic acids

Carboxylic acids can be prepared by the following methods.

- Prolonged oxidation of primary alcohols or aldehydes:

  $CH_3CH_2OH(l) + 2[O] \rightarrow CH_3COOH(aq) + H_2O(l)$

  ethanol → ethanoic acid

One way of doing this is heating the alcohol with acidified potassium dichromate(VI) under reflux.

The orange solution turns green as the chromium(VI) compound is reduced to chromium(III).

- Hydrolysis of a nitrile by boiling with a dilute acid:

  $CH_3CH_2CN(l) + 2H_2O(l) + H^+(aq) \rightarrow CH_3CH_2COOH(aq) + NH_4^+(aq)$

  propanenitrile → propanoic acid

SC Look up the K_a or pK_a values of a range of carboxylic acids and try to explain any trends you can see. Include acids of increasing chain length and those with different atoms or groups such as Cl, Br, I and OH substituted at different points in the chain. Remember – the more easily the O–H bond is broken, the stronger the acid.

Questions

1 Benzenecarboxylic acid, C_6H_5COOH, is a carboxylic acid.

   a Why is this compound almost insoluble in water?

   b Draw the dimer structure that exists in a non-polar solvent.

2 Both ethanal (boiling temperature 240 K) and ethanoic acid (boiling temperature 391 K) are produced by heating ethanol with acidified potassium dichromate(VI). Suggest how you would prepare good yields of **a** ethanal; **b** ethanoic acid.

Reactions of carboxylic acids [4.8.3d(i–iii)]

In this section we are going to consider some of the general reactions of carboxylic acids – **reduction**, neutralisation and **substitution** of a halogen. We will consider reactions that form esters in the next section.

Reduction

Carboxylic acids are themselves the products of the oxidation of alcohols or aldehydes by strong oxidising agents. They are not easy to reduce and reduction is only possible back to the alcohol and not the aldehyde. A suitable reducing agent is lithium tetrahydridoaluminate (lithium aluminium hydride) suspended in dry ether (ethoxyethane) at room temperature.

$$CH_3CH_2COOH(l) + 2[H] \rightarrow CH_3CH_2CH_2OH(l) + H_2O(l)$$

Addition of water at the end of the reaction will destroy any excess lithium tetrahydridoaluminate.

Acidic properties of carboxylic acids

When a carboxylic acid is dissolved in water, it dissociates (**fig. 1.6.42**), forming hydrogen ions and carboxylate ions, so behaving as an acid.

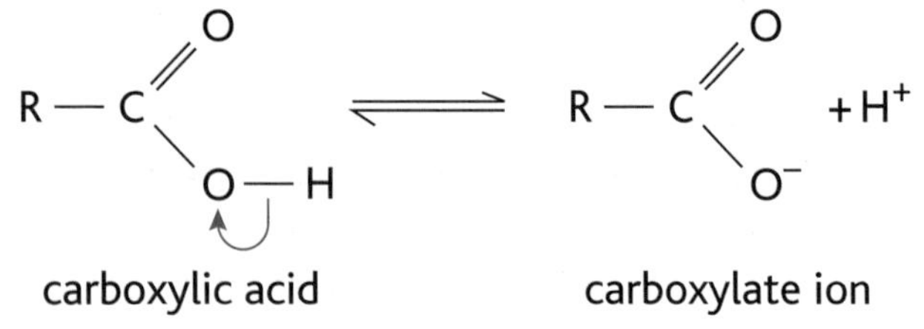

R—C(=O)—O—H ⇌ R—C(=O)—O⁻ + H⁺

carboxylic acid   carboxylate ion

fig. 1.6.42 The dissociation of a carboxylic acid.

Although the carboxylate ion is often drawn with one double carbon–oxygen bond and one single carbon–oxygen bond, X-ray diffraction analysis and observations of the ways in which it reacts suggest that the two carbon–oxygen bonds are actually identical – see **fig. 1.6.43**.

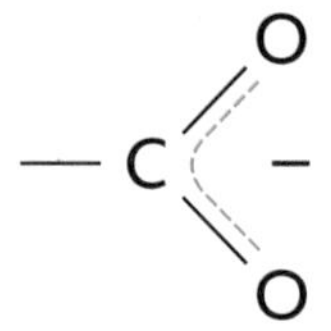

fig. 1.6.43 The carboxylate ion with the charge and double bond character evenly spread between the two oxygen atoms.

This implies that the negative charge is **delocalised** over the whole of the carboxylate group, which means that it will not attract positively charged particles as strongly as might be expected. As a result, the equilibrium for the reaction shown in **fig. 1.6.42** is further over to the right than would be expected. So, the carboxylic acids are relatively strong acids, and the carboxylate ion (the conjugate base) is a relatively weak base. This said, they are classified as weak acids, but stronger than alcohols and phenols.

Neutralisation

Like any acid, carboxylic acids will react with a base to form a salt and water:

$$CH_3COOH(aq) + NaOH(aq) \rightleftharpoons CH_3COO^-(aq) + Na^+(aq) + H_2O(l)$$

The salt which is formed by the reaction between a carboxylic acid and an alkali is very different in character from its organic acid parent. Such salts are ionic rather than molecular, and so have crystalline structures and are readily soluble in water.

Though it is not a neutralisation reaction, carboxylic acid salts can also be produced by the reaction of the acid with reactive metals, for example zinc:

$$Zn(s) + 2CH_3COOH(aq) \rightarrow (CH_3COO)_2Zn(aq) + H_2(g)$$

Salts with short carbon chains show mainly ionic character, and those with very long chains show mainly organic character. The salts of the carboxylic acids of medium chain length have a combination of organic and ionic character, and this is put to use in **soaps** and **detergents**.

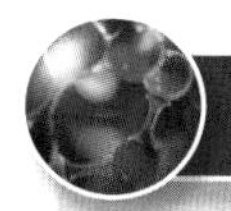

HSW Soaps and detergents

Washing our clothes and ourselves has been part of human civilisation for a long time. Human dirt includes oils and fats from our skin and food, carboxylic acids and salt from sweat, mud, dead skin cells and a wide variety of other substances – see **fig. 1.6.44**.

fig. 1.6.44 Scanning electron micrograph showing dirt trapped on a piece of fabric.

$$\begin{array}{l} CH_2-O-C(=O)-C_{17}H_{35} \\ | \\ CH-O-C(=O)-C_{17}H_{35} + 3NaOH \\ | \\ CH_2-O-C(=O)-C_{17}H_{35} \end{array}$$

a triglyceride – an ester of long-chain carboxylic acids and a triol

$$\rightarrow \begin{array}{l} CH_2OH \\ | \\ CHOH \\ | \\ CH_2OH \end{array} + 3C_{17}H_{35}C(=O)O^-Na^+$$

propane-1,2,3-triol (glycerol)

sodium octadecanoate (sodium stearate)

fig. 1.6.45 The hydrolysis of an ester in soap-making.

Water has long been our washing medium and, though very effective at removing water-soluble dirt, it is poor at removing material that does not dissolve in water. A detergent is a substance that acts as a cleaning agent, improving the ability of water to wash things clean.

The earliest, and for a long time the only, detergent known to the human race was soap. Soaps are produced by the action of sodium hydroxide on natural fats and oils that contain natural esters – see **fig. 1.6.45**. This hydrolysis reaction is sometimes called saponification.

Soapless detergents are produced by the action of concentrated sulfuric acid on hydrocarbon residues from fractional distillation of crude oil. A typical soapless detergent is shown in **fig. 1.6.46**.

$$C_{12}H_{25}-C_6H_4-S(=O)_2-O^-\ Na^+$$

fig. 1.6.46 A molecule of a soapless detergent.

How soaps work

Soap is a mixture of salts of medium chain length carboxylic acids. As we have seen, these are compounds with a combination of ionic and organic characteristics – polar 'heads' and non-polar 'tails'. The polar part of the molecule is the carboxylate group, which is **hydrophilic** – it is attracted to water. The non-polar part is the carbon chain, which is **hydrophobic** – it is repelled from water. Similar descriptions can be applied to molecules of soapless detergents.

When soap, or a soapless detergent, is in water, the non-polar parts arrange themselves together as far from water as possible, with polar ends pointing outwards – this is shown in **fig. 1.6.47** overleaf. The anions form clusters known as **micelles**.

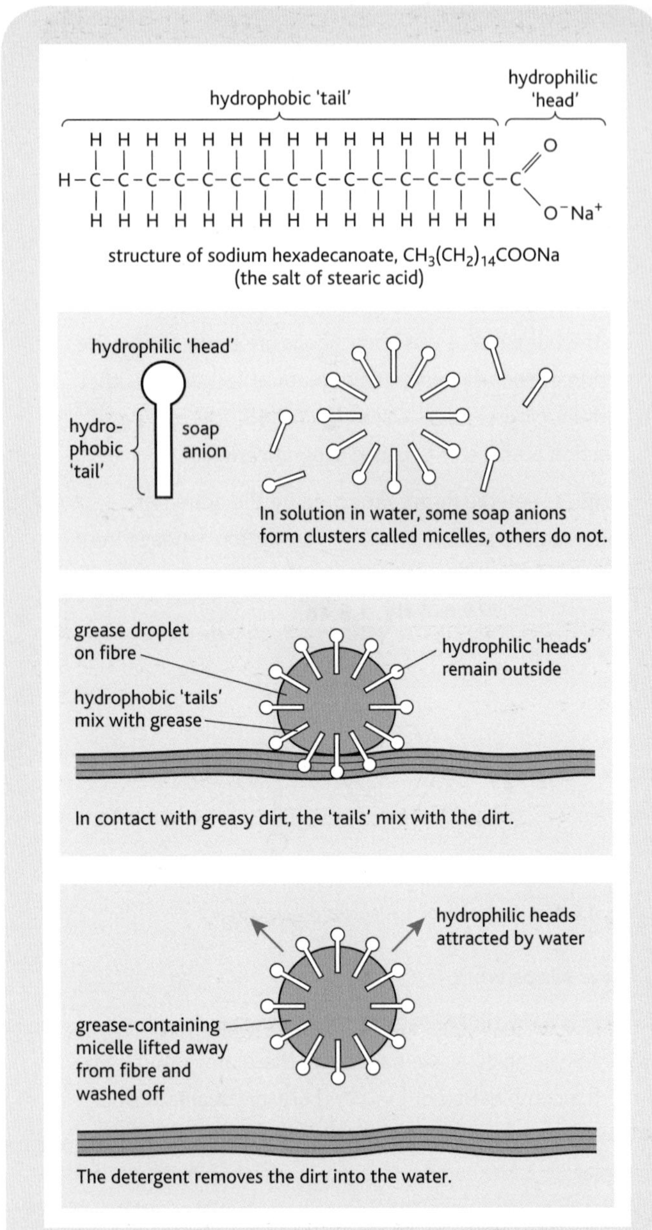

fig. 1.6.47 The cleaning action of soap.

When soap is in contact with fat or greasy dirt, the non-polar tails mix with non-polar dirt, so that the dirt becomes surrounded by polar 'heads' – these negatively charged heads repel each other and keep the grease separated. These loaded micelles are then carried away by the water and the material or skin left clean.

Halogenation reactions

Carboxylic acids react with solid phosphorus(V) chloride to form acyl chlorides. **Figure 1.6.48** shows the formation of ethanoyl chloride from ethanoic acid.

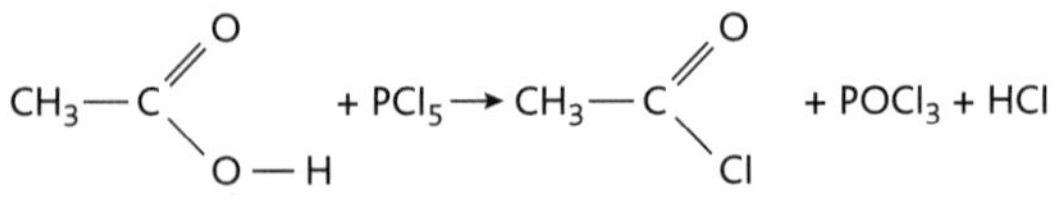

$$CH_3{-}C({=}O){-}O{-}H + PCl_5 \rightarrow CH_3{-}C({=}O){-}Cl + POCl_3 + HCl$$

fig. 1.6.48 Formation of an acyl chloride.

The reaction is accompanied by the white fumes of hydrogen chloride gas. This is commonly used as a test for –OH groups in an unknown organic compound. This happens because acyl chlorides react readily with water vapour (**fig. 1.6.49**).

fig. 1.6.49 Take care when working with acyl chlorides.

Acyl chlorides are useful materials for organic synthesis.

Questions

1 Write an equation for the reaction of propanoic acid with calcium carbonate.

2 Reduction of the compound shown below with lithium tetrahydridoaluminate produces a mixture of two isomers. Explain why.

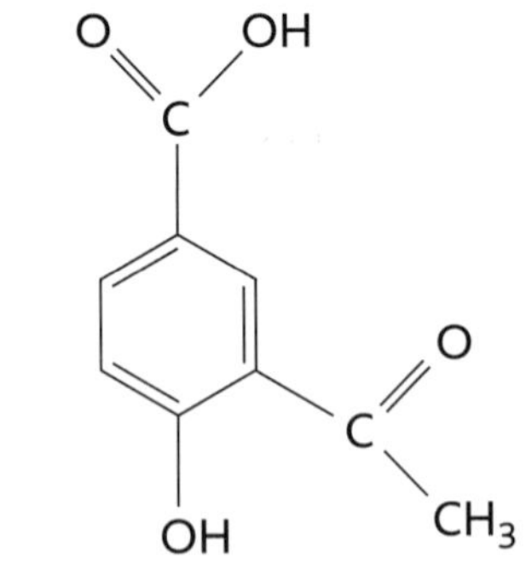

Synthesis of esters [4.8.3d(iv)]

Esters are one of several families of organic compounds that are direct derivatives of the carboxylic acids. When a carboxylic acid is mixed with an alcohol in the presence of a strong acid catalyst, such as sulfuric acid or hydrochloric acid, an ester is formed – **fig. 1.6.50** shows the general reaction, and **fig. 1.6.51** shows the formation of two specific esters.

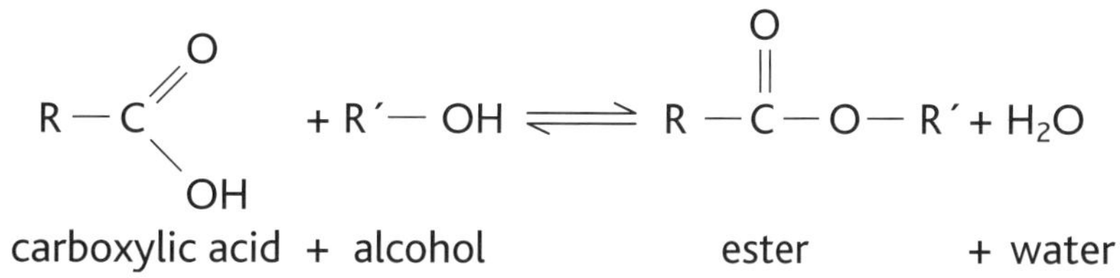

$$R{-}C(=O){-}OH + R'{-}OH \rightleftharpoons R{-}C(=O){-}O{-}R' + H_2O$$

carboxylic acid + alcohol ⇌ ester + water

fig. 1.6.50 The general equation for the esterification reaction.

The reaction is called **esterification**. It can be considered as a condensation reaction – one between two molecules to form a larger molecule with the loss of a small molecule, in this case water.

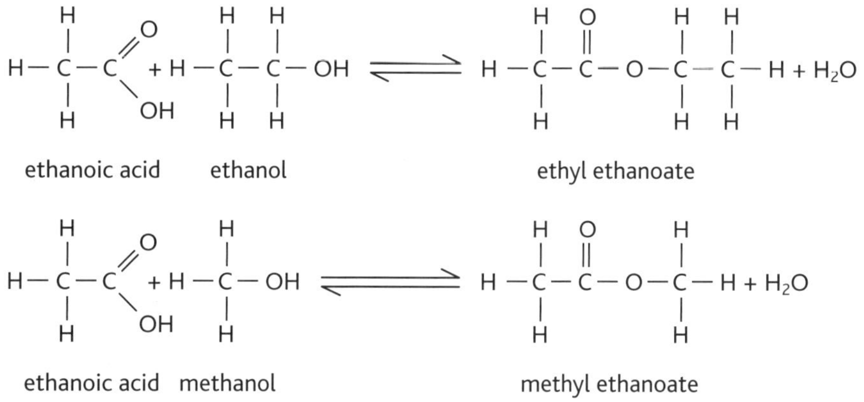

$$CH_3COOH + CH_3CH_2OH \rightleftharpoons CH_3COOCH_2CH_3 + H_2O$$

ethanoic acid + ethanol ⇌ ethyl ethanoate

$$CH_3COOH + CH_3OH \rightleftharpoons CH_3COOCH_3 + H_2O$$

ethanoic acid + methanol ⇌ methyl ethanoate

fig. 1.6.51

Esterification is a relatively slow reaction and is subject to equilibrium considerations. It can be more convenient to prepare esters using the more reactive acyl chloride or acid anhydride. Examples are shown in **fig. 1.6.52**.

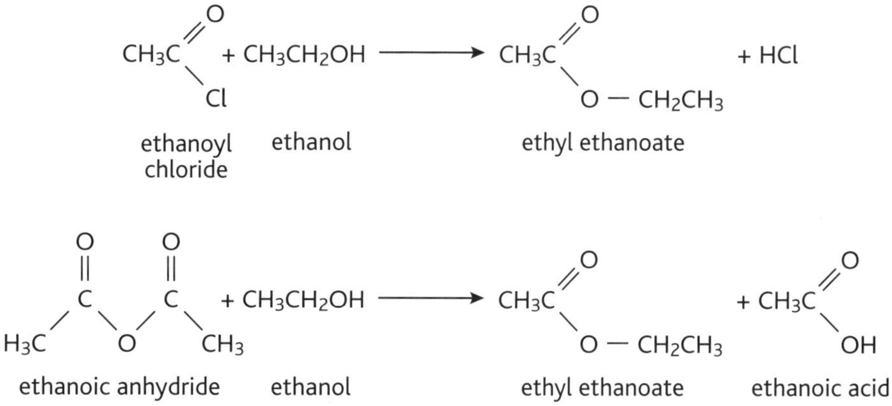

$$CH_3COCl + CH_3CH_2OH \longrightarrow CH_3COO{-}CH_2CH_3 + HCl$$

ethanoyl chloride + ethanol → ethyl ethanoate

$$(CH_3CO)_2O + CH_3CH_2OH \longrightarrow CH_3COO{-}CH_2CH_3 + CH_3COOH$$

ethanoic anhydride + ethanol → ethyl ethanoate + ethanoic acid

fig. 1.6.52 Other ways of making esters.

Naming esters

Esters have complicated-looking molecules but naming them is relatively simple. The name starts with an 'alkyl word' derived from the alcohol used in the reaction – methyl from methanol, ethyl from ethanol, and so on. This is followed by the name of the acid used in the reaction, replacing the -oic ending with -oate to show that it is an ester – so a methanoate ester is made from methanoic acid, an ethanoate from ethanoic acid, etc.:

propanoic acid + methanol →
methyl propanoate (+ water)

Preparation of an ester

You may carry out an experiment to prepare a sample of an ester from a carboxylic acid and an alcohol in the laboratory.

You may also prepare an ester using microscale techniques.

Properties of esters

Esters are volatile compounds. They do not form hydrogen bonds and so their melting and boiling temperatures are relatively low and they are not very soluble in water or other polar solvents. One typical characteristic of esters is a pleasant smell. They are responsible for many of the scents of flowers and fruits – for example, the smell of pineapple is due to methyl butanoate.

fig. 1.6.53 The flavours and odours of fruits like these are based on esters – but they are much more complex than the artificial esters you can make in the lab!

Using esters

Once scientists unravelled the chemistry of flavours and scents and discovered that esters were largely responsible, they could then be synthesised and used in manufacturing foods and cosmetics. **Table 1.6.2** shows the ingredients of a number of esters that could be made in a school laboratory and the smells they produce.

Smell	Reactants needed
Banana or pear	propan-1-ol and ethanoic acid
Pineapple	1-pentanol (amyl alcohol) and ethanoic acid
Peach	phenylmethanol (benzyl alcohol) and ethanoic acid
Raspberry	2-methylpropan-1-ol and methanoic acid

table 1.6.2 Esters and smells.

Synthetic ester flavours are made industrially for the enhancement of some foods. However, they are only rough approximations to the real thing. For example, the sweets known as 'pineapple chunks' do not taste much like fresh pineapple because the odour and flavour of the fresh fruit is the sum of lots of different esters, and other compounds, interacting.

Many natural scents contain combinations of esters. For example, the fragrance of lavender flowers is made up of linalyl, terpenyl, lavandulyl and geranyl ethanoates among others.

fig. 1.6.54 The esters that make up the fragrance of lavender flowers are widely used in a number of ways.

These esters can be extracted in a variety of ways, ranging from simple squashing to using specialist solvents, as well as a number of distilling methods. The essential oils extracted are widely used in perfumes and cosmetics, in aromatherapy and in healing. The best techniques extract the esters without damaging them or changing them chemically in any way.

Some esters are used as solvents – for example, ethyl ethanoate is used extensively in the manufacture of adhesives. They are also used in the manufacture of soaps and detergents.

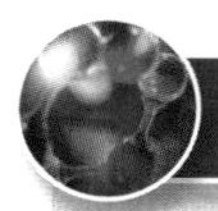

HSW Where does the bridging oxygen come from?

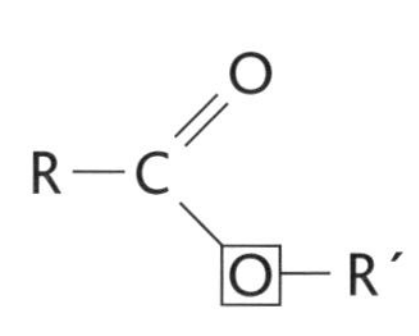

A distinctive feature of an ester molecule is the oxygen atom which forms a bridge between the two parts. This 'oxygen bridge' is formed (**fig. 1.6.55**) as the carboxylic acid and alcohol molecules join.

fig. 1.6.55 The 'bridging' oxygen atom in an ester.

Where does this oxygen atom come from – the alcohol or the carboxylic acid? The answer to this question was first provided by two American chemists, Roberts and Urey. They used a radioactive isotope of oxygen, ^{18}O, to solve the problem. Using a technique similar to that outlined in **fig. 1.6.56** it was shown that the bridging oxygen atom in the ester comes from the alcohol molecule and not the carboxylic acid.

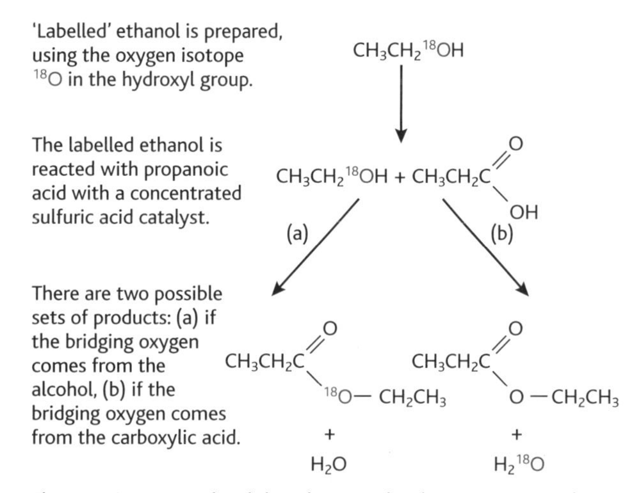

fig. 1.6.56 Chemical detective work on the esterification reaction.

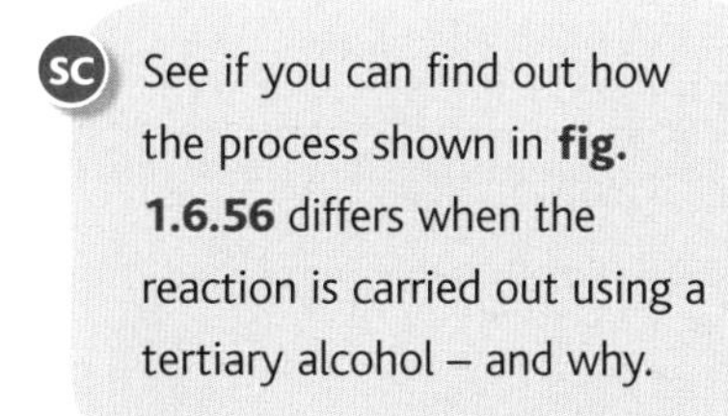

SC See if you can find out how the process shown in **fig. 1.6.56** differs when the reaction is carried out using a tertiary alcohol – and why.

Questions

1 Draw the displayed formula and give the name of the ester produced from methanol and ethanoic acid.

2 Name the carboxylic acid and the alcohol that react to form ethyl hexanoate.

3 Explain the similarities between natural and artificial flavours, and say why they differ.

Reactions of esters [4.8.4a, c]

Hydrolysis

Earlier in this chapter, it was stated that the esterification reaction between a carboxylic acid and an alcohol is subject to equilibrium considerations. The reverse reaction, in which an ester splits up to reform the original acid and alcohol, is called **hydrolysis** – so called because water causes the compound to split up:

$$CH_3COOC_2H_5 + H_2O \rightarrow CH_3COOH + C_2H_5OH$$

Acid hydrolysis

Esters can be split up by boiling with water and, much more effectively, by using a dilute acid – **fig. 1.6.57** shows the role of the water in the reaction.

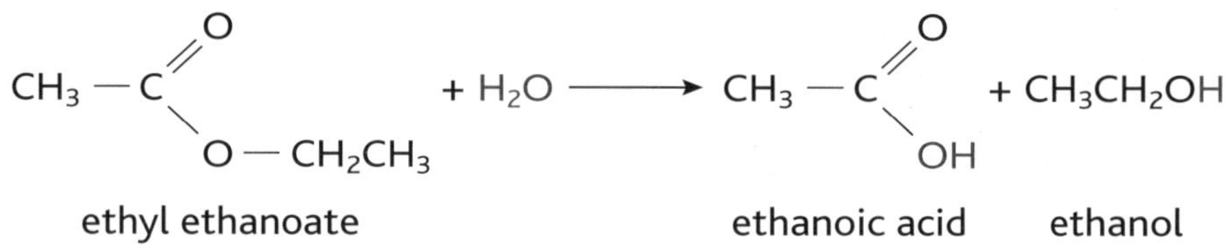

$CH_3-C(=O)-O-CH_2CH_3$ + H_2O ⟶ $CH_3-C(=O)-OH$ + CH_3CH_2OH

ethyl ethanoate ethanoic acid ethanol

fig. 1.6.57 The role of water in the hydrolysis of esters.

The dilute acid provides hydrogen ions, which catalyse the reaction.

Base hydrolysis

An ester can also be hydrolysed by boiling it with potassium hydroxide solution – see **fig. 1.6.58**. This is a more efficient process because when the acid is released it is neutralised by alkali to form the potassium salt – this does not react with the alcohol. This process is sometimes called saponification – we met this earlier in soap-making.

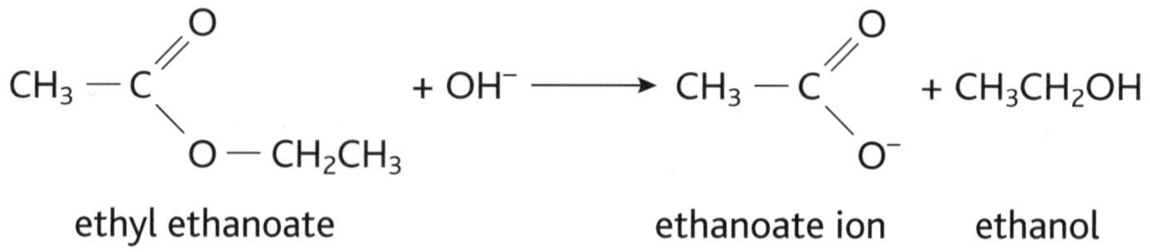

$CH_3-C(=O)-O-CH_2CH_3$ + OH^- ⟶ $CH_3-C(=O)-O^-$ + CH_3CH_2OH

ethyl ethanoate ethanoate ion ethanol

fig. 1.6.58

HSW Biofuels

In *Edexcel AS Chemistry* (page 242) you will have read that alternative fuels for vehicles can be made from biomass. Biodiesel can be made from vegetable oil, including waste vegetable oils from fast-food stores – **fig. 1.6.59** shows a bus that uses biodiesel. Not only does this reduce the use of crude oil, it also reduces exhaust emissions.

fig. 1.6.59 Biofuels will become increasingly important.

Making such fuels from waste vegetable oil starts with filtration to remove dirt, charred food and other non-oil materials. Water is removed using anhydrous magnesium sulfate. If a batch is left to stand, the **triglycerides** may hydrolyse to give salts of the carboxylic acids, instead of undergoing **transesterification** to give biodiesel. The drying agent can be separated by decanting or by filtration.

Transesterification involves the reaction of a triglyceride (a fat or an oil) with an alcohol to form esters and glycerol. The esters are then used as the fuel we call biodiesel.

Different fatty acids produce different esters, which in turn affect the nature of the biodiesel produced.

The most commonly used method for producing biodiesel involves base-catalysed transesterification of vegetable oils, that is, the reaction of vegetable oils with alcohols – usually methanol or ethanol – in the presence of a strong alkali such as sodium hydroxide, which acts as a catalyst. This forms monoalkylesters and glycerol (see **figs 1.6.60** and **1.6.61**). This transesterification process is very efficient, with a conversion yield of about 98%. It is also economical, as it uses low temperatures (around 50 °C) and atmospheric pressure as well as a cheap and readily available catalyst.

$$R'OH + R''O{-}C({=}O){-}R \longrightarrow R''OH + R'O{-}C({=}O){-}R$$

fig. 1.6.60 Transesterification.

$$3CH_3OH + \begin{matrix} CH_2OOCR' \\ | \\ CHOOCR'' \\ | \\ CH_2OOCR''' \end{matrix} \longrightarrow \begin{matrix} CH_2OH \\ | \\ CHOH \\ | \\ CH_2OH \end{matrix} + R'COOCH_3 + R''COOCH_3 + R'''COOCH_3$$

fig. 1.6.61 Using transesterification to produce methyl esters for use as biodiesel.

The lower layer of the batch is composed primarily of glycerine and other waste products. The glycerol can be purified for use in the cosmetic or pharmaceutical industries, or it can be sold as it is. It is certainly not wasted! The top layer, a mixture of biodiesel and excess alcohol, is decanted and the alcohol is distilled off. The fuel produced can be used in diesel engines or mixed with diesel from petroleum.

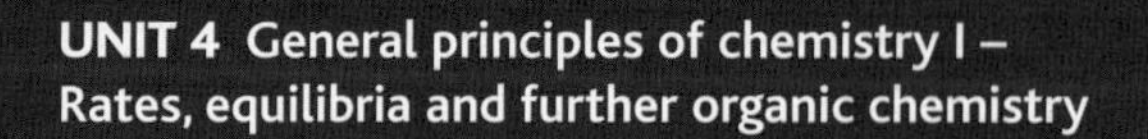

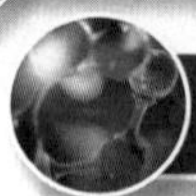

HSW Margarines and low-fat spreads

In 1869, Emperor Louis Napoleon III of France offered a prize to anyone who could make a satisfactory substitute for butter. This was for use by the army and the lower classes. French chemist Hippolyte Mège-Mouriés invented a substance he called **oleo-margarine** – a name which became shortened to the trade name, 'margarine'.

Manufacturers produced margarine by taking clarified vegetable fat, extracting the liquid portion under pressure, and then allowing it to solidify. By 1900 artificial butters were available throughout the world.

fig. 1.6.62 Different types of spread.

Vegetable oils are perceived to be healthier than animal fats such as butter. During the manufacture of margarine the vegetable oil is 'hardened' by reaction with hydrogen in the presence of a nickel catalyst.

Several types of butter-substitute spreads are common:

- hard, usually uncoloured, margarine – for cooking or baking
- 'traditional' margarines for such uses as spreading on toast – they contain saturated fats and are mostly made from vegetable oils
- margarines high in mono- or polyunsaturated fats – made from sunflower, soybean or olive oil
- low-fat spreads where more water is incorporated in the emulsion to lower the fat content and so the calorie count.

Many of today's popular spreads are blends of margarine and butter, and are designed to combine the lower cost and easy-spreading of artificial butter with the taste of the real thing.

Modern margarine can be made from a wide variety of animal or vegetable fats, and is often mixed with skimmed milk, salt and emulsifiers. In terms of microstructure, margarine is a water-in-oil emulsion, containing dispersed water droplets of typically 5–10 μm diameter. Low-fat spreads involve producing an emulsion which has a higher proportion of water and a lower proportion of oils and hydrogenated fats. These low-fat spreads are seen as healthy as they reduce the calorie intake of people eating them. The big problem is getting the water–oil emulsion to stay together and to form a soft solid at the temperatures it is used at. Chemical emulsifiers are an important part of the process.

However, it is the amount of crystallising fat in the oil/fat phase which determines the firmness of the product. In the relevant temperature range, saturated fats contribute most to the amount of crystalline fat, whereas mono- and polyunsaturated fats contribute relatively little to the amount of crystalline fat in the product.

Traditional hydrogenation using nickel catalysts can lead to partial hydrogenation of the fats, but can lead to the formation of **trans-fats** as well. In trans-fats hydrogen is added to trans positions in the fat molecule. Research has shown that these fats, first developed about 1910, can be harmful in the diet. In fact trans-fats are now regarded as being just as much of a health risk as the saturated fats they were designed to replace!

Transesterification provides an alternative way of transforming mono- and polyunsaturated fats and oils into chemicals which will be solid at room temperature. It has the advantage that it produces far fewer trans-fats. This transesterification can be brought about by inorganic catalysts, but scientists are increasingly using enzymes or genetically modified bacteria to produce the partially saturated fats needed to produce margarine and low-fat spreads.

Questions

1 Write equations for the acid hydrolysis and base hydrolysis of:

a ethyl methanoate b methyl ethanoate

Polyesters [4.8.4d]

Carboxylic acids react with alcohols to form an ester and water in a condensation reaction.

If a *di*carboxylic acid and a *di*ol are used it is possible to form a **condensation polymer,** as shown in **fig. 1.6.63**, in which the black squares and circles represent different hydrocarbon chains.

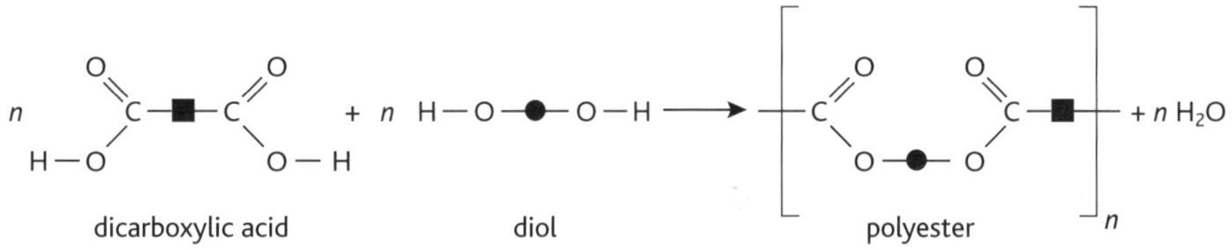

fig. 1.6.63 Forming a polyester – a condensation polymer.

An example, shown in **fig. 1.6.64**, is the polymer formed from benzene-1,4-dicarboxylic acid (a dicarboxylic acid sometimes called terephthalic acid) and ethane-1,2-diol (a dialcohol sometimes called glycol). The polymer is called polyethylene terephthalate.

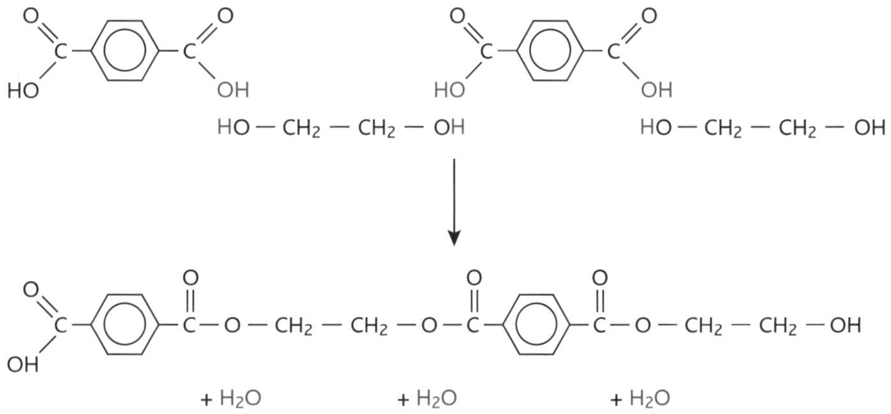

fig. 1.6.64 Making polyethylene terephthalate (PET).

Likewise, diacyl chlorides and diacid anhydrides can be used in place of dicarboxylic acids to make polyesters.

When made into fibres, polyesters have trade names such as Terylene and Dacron. Polyesters have a wide range of uses, including making plastic bottles for soft drinks.

HSW Polyesters everyday

Polyesters are widely used because of their versatility. In fact, 18% of all the polymers made in the world are polyesters. Polyester is sold under the trade names PET and PETE. It can range from semi-rigid to rigid, depending on its thickness, and is very lightweight. It is a good barrier to gas, alcohol and solvents, as well as being a fairly good barrier to moisture. It is strong and impact-resistant, naturally colourless with high transparency, but when it is semi-crystalline it is opaque and white.

PET bottles are widely used for soft drinks; for certain special bottles, there is an additional polyvinyl alcohol layer between two layers of PET to further reduce its oxygen permeability. It is used as a thermal insulation layer on the outside of the International Space Station.

fig. 1.6.65 Chips of PET plastic ready for recycling.

While most thermoplastics can, in principle, be recycled, PET recycling is more practical than for many other plastics – see **figs 1.6.65** and **1.6.66**. There is a recycling code on PET bottles which makes them easy to identify and recycle.

fig. 1.6.66 The recycling symbol for PET.

One of the uses for recycled PET bottles is the manufacture of polar fleece material. It can also be made into fibres for making polyester products such as duvets. Many yacht sails are made of the polymer Dacron. Terylene and other polyester fibres are widely used for making clothing fabrics – it is hard-wearing and easy to wash and dry.

100% POLYESTER

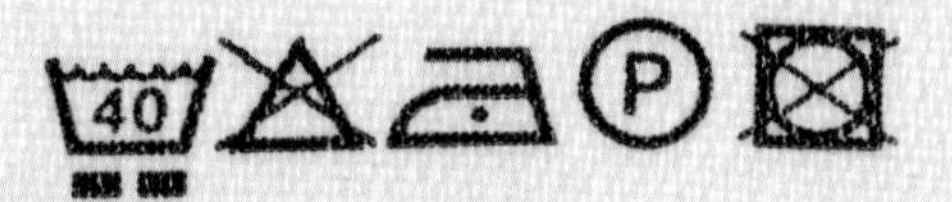

Wash deep colours together

fig. 1.6.67 Care instructions for polyester garments.

Using propane-1,2,3-triol in place of ethane-1,2-diol, it is possible to make a polyester resin with a three-dimensional structure.

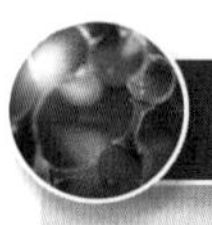

HSW Biological polyesters

In 2007, a new type of suture was introduced to doctors. It was a **biopolymer** – a polymer made by genetically engineered bacteria. The research team set out to produce a biopolymer on an industrial scale, and they focused on enzymes in a pathway that produces natural polyesters. The idea was for these polyesters to be capable of being broken down in the human body, causing no harm as they are absorbed. Genes were successfully manipulated and transferred into bacteria, which can then produce a strong, flexible fibre in large quantities in industrial fermenters. Stitches made from such a material are ideal for closing incisions in the abdomen and for stitching tendons and ligaments. They are 30% stronger than the sutures currently used and are very flexible, so they are easy for surgeons to use – and better for patients too.

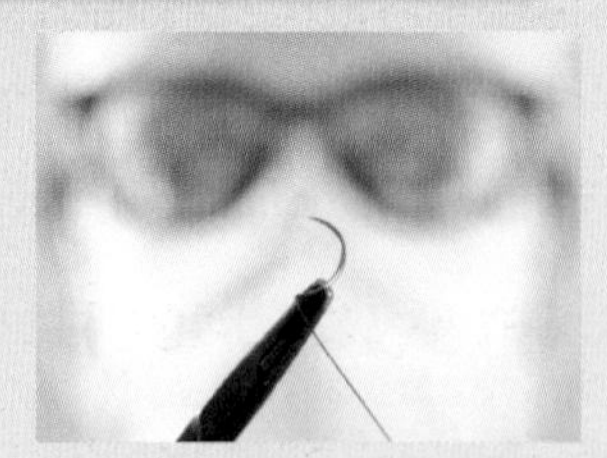

fig. 1.6.68 Biological polyester – using natural polymers to improve medical success.

Questions

1 A polyester can be made by polymerising the monomer shown here.

HO–C(=O)–C₆H₄–O–H

Draw the displayed formula of the repeating unit of the polymer.

Reactions of acyl chlorides [4.8.4a, b]

Acyl chlorides are reactive compounds produced when phosphorus(V) chloride (phosphorus pentachloride) reacts with a carboxylic acid – as shown in **fig. 1.6.69**.

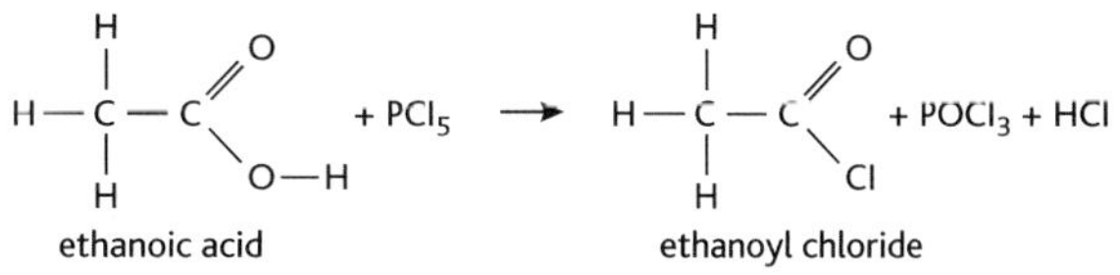

$$CH_3COOH + PCl_5 \rightarrow CH_3COCl + POCl_3 + HCl$$

ethanoic acid → ethanoyl chloride

fig. 1.6.69 The formation of a typical acyl chloride.

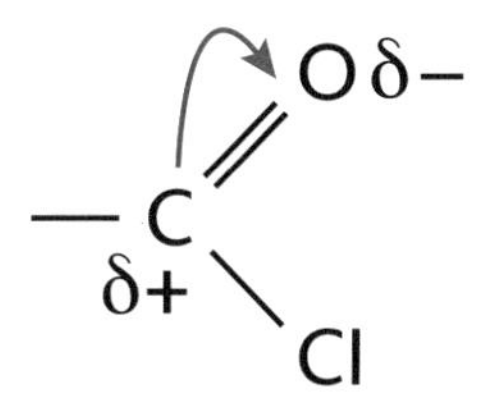

fig. 1.6.70 The polarity of the acyl group.

Acyl chlorides contain a carbonyl group with a chlorine atom attached to the carbon atom in the carbonyl group. The oxygen in the carbonyl group withdraws electrons from the double bond, giving the central carbon atom a δ+ charge (**fig. 1.6.70**) – making it susceptible to attack by nucleophiles.

The reactions of the acyl group involve addition followed by elimination. The addition/elimination reactions can also be called condensation reactions. It is worth noting that acyl chlorides react in exactly the same way as those of carboxylic acids – only much more readily.

Reaction with water

Acyl chlorides are hydrolysed rapidly with cold water (**fig. 1.6.71**):

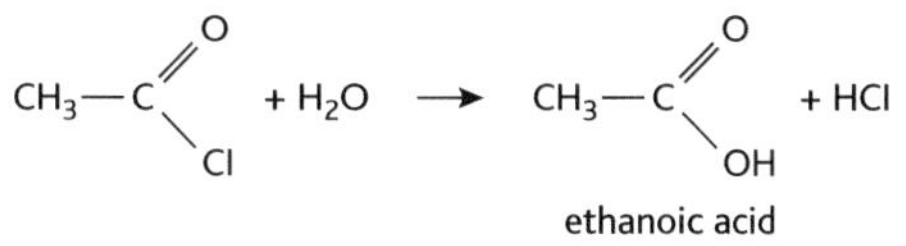

$$CH_3COCl + H_2O \rightarrow CH_3COOH + HCl$$

ethanoic acid

fig. 1.6.71 Hydrolysis of ethanoyl chloride.

fig. 1.6.72 Acyl chlorides are very reactive.

The vigour of the reaction can be gauged from the fact that if the stopper is removed from a bottle of ethanoyl chloride or it is in an open beaker (**fig. 1.6.72**), hydrogen chloride fumes will be seen as the organic compound reacts with the water vapour in the air.

Reaction with alcohols

Acyl chlorides react with alcohols to produce esters. These reactions take place on mixing the acyl chloride and the alcohol without heating (**fig. 1.6.73**):

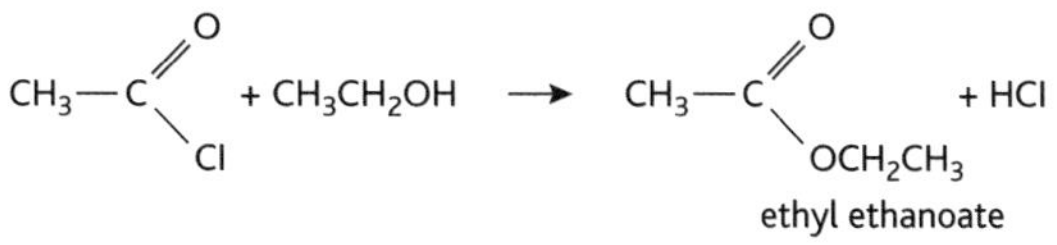

$$CH_3COCl + CH_3CH_2OH \rightarrow CH_3COOCH_2CH_3 + HCl$$

ethyl ethanoate

fig. 1.6.73 Reaction of acyl chlorides with alcohols.

The mechanism for this reaction is shown in **fig. 1.6.74**. It involves nucleophilic addition, gain and loss of protons, and elimination of water.

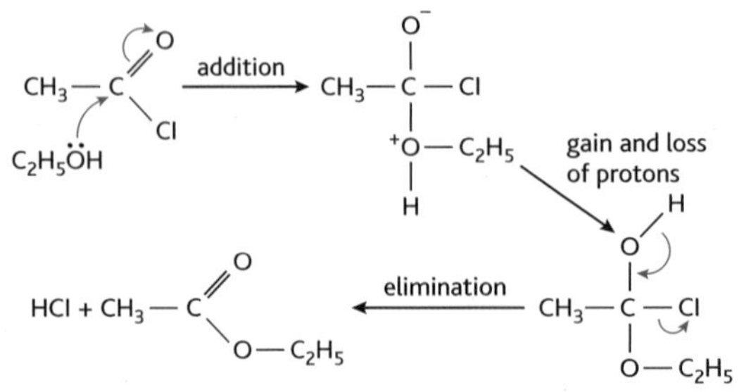

addition

gain and loss of protons

elimination

fig. 1.6.74 The mechanism for a nucleophilic addition/elimination reaction.

Reaction with concentrated ammonia

Acyl chlorides react with concentrated ammonia to produce amides. These reactions take place on mixing the reactants and without heating (**fig. 1.6.75**).

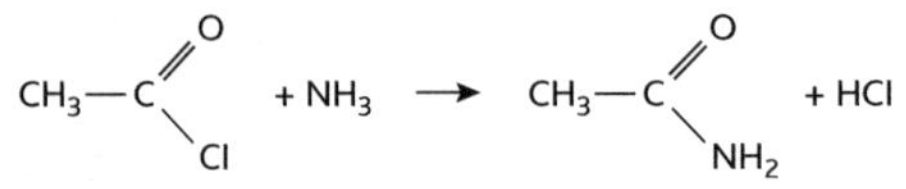

ethanoyl chloride ethanamide

fig. 1.6.75 Reaction of acyl chlorides with ammonia.

Because of the acid/base nature of the products, a further reaction takes place:

$$HCl(g) + NH_3(g) \rightarrow NH_4Cl(s)$$

You will find out more about amides in Unit 5.

Reaction with amines

A similar reaction occurs between acyl chlorides and amines to produce a substituted amide (**fig. 1.6.76**):

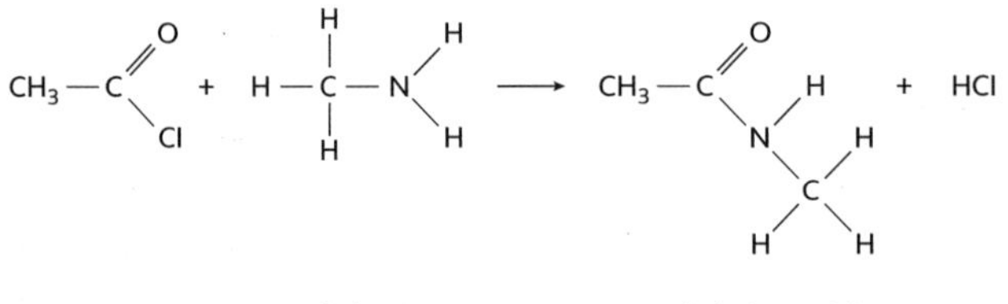

methylamine *N*-methylethanamide

fig. 1.6.76 Reaction of acyl chlorides with amines.

In naming compounds such as this, '*N*-methyl' means that a methyl group is attached to the nitrogen atom of the amide.

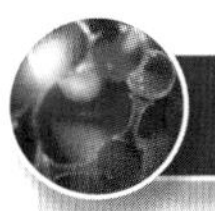

HSW Acylation/ethanoylation

In all of these reactions, a hydrogen atom attached to an atom of an electronegative element such as oxygen or nitrogen is replaced by an RCO– group, called an acyl group. The replaced hydrogen atom is called an 'active' hydrogen atom. When ethanoyl chloride is used the process is called ethanoylation and the other product is hydrogen chloride.

Ethanoyl chloride is a difficult substance to handle because it is very reactive. Ethanoic anhydride is often used in ethanoylation instead.

It reacts more slowly, and produces ethanoic acid as the other product instead of hydrogen chloride.

O
CH_3—C
O
CH_3—C
O

fig. 1.6.77 Ethanoic anhydride.

In chemical synthesis, chemists might want to protect an amine group from further reaction. Since an amine group is relatively reactive, it is liable to undergo further change. Ethanoylation is a means of protecting the amine group during later reaction steps. De-ethanoylation at the end releases the amine in the final product.

Making aspirin

In Unit 5 you may produce acetylsalicylic acid (aspirin) or paracetamol using ethanoic anhydride in an ethanoylation reaction.

SC Nylon is a condensation polymer produced using a reaction between molecules containing –COOH and –NH_2 groups. Find out what has to be special about the molecules for the polymer to be formed and find out the difference between nylon-6,6 and nylon-6,10.

HSW Nylon – a synthetic polymer

Nylon is a totally synthetic fibre. It is a condensation polymer made up of two monomers, which are often hexanedioic acid and 1,6-diaminohexane. However, in the laboratory it is easier to make it from hexanedioyl dichloride and 1,6-diaminohexane at room temperature. The monomer units in nylon are joined together in condensation reactions.

fig. 1.6.78

Nylon was invented by Wallace Carothers in 1934 and it went on the market in 1938. The fibres are used in a wide variety of ways. Nylon is very tough and is used in clothing, carpets, nylon ropes and bearings. However, nylon also produces environmental problems – it does not rot and biodegrade, and so disposal of this useful material is not easy.

Questions

1 Describe and explain the mechanism for the reaction of ethanoyl chloride with ammonia.

2 Methyl ethanoate can be prepared using the reaction of methanol with ethanoic anhydride. Write the equation for the reaction.

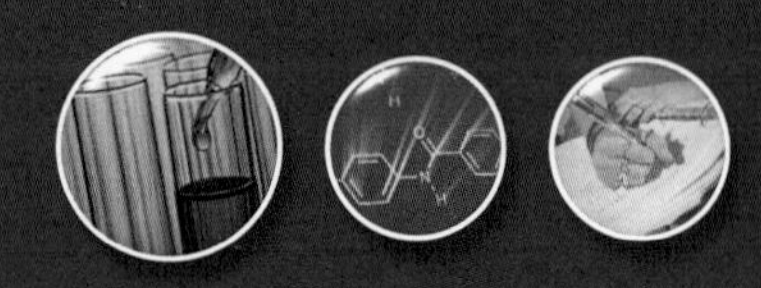

1.7 Spectroscopy and chromatography

How does radiation affect molecules?

[4.9a(i–iv), c]

In this chapter we are going to consider how the different types of radiation affect molecules and how this information can be used. First we need to revise some ideas about the **electromagnetic spectrum**. Many of the tools we use for analysis rely on equipment using different parts of the electromagnetic spectrum – e.g. infrared spectroscopy.

Electromagnetic spectrum

Electromagnetic waves are transverse waves produced by oscillating electric and magnetic fields at right angles to each another. Like water waves, they are oscillations; but unlike water waves, these oscillations involve not matter but electric and magnetic fields. **Figure 1.7.1** shows how an electromagnetic wave travels.

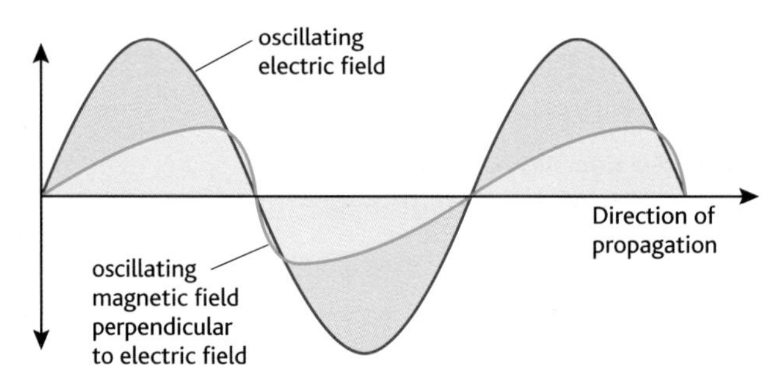

fig. 1.7.1 As an electromagnetic wave travels, the electric and magnetic fields oscillate together. In a vacuum, electromagnetic waves travel at the speed of light, $3 \times 10^8 \text{ m s}^{-1}$.

As with any wave, the speed of travel, c, of an electromagnetic wave is related to its wavelength, λ, and its frequency, f, by the relationship:

$$c = f \times \lambda$$

Using SI units, speed is measured in metres per second (m s^{-1}), wavelength in metres (m) and frequency in seconds^{-1} (s^{-1}). The unit of frequency is more commonly given as hertz (Hz).

Electromagnetic waves can cover a wide range of wavelengths or frequencies. These cover the electromagnetic spectrum, which can be split into seven overlapping regions – see **fig. 1.7.2** – within each of which the waves have different properties.

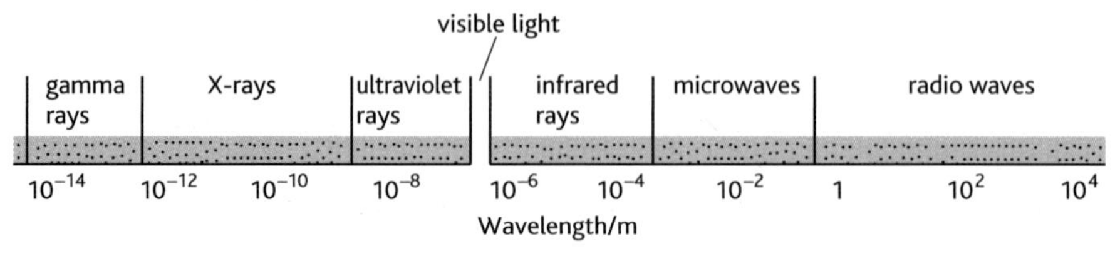

fig. 1.7.2 The electromagnetic spectrum.

We are now going to consider the effects and uses that some of these types of electromagnetic radiation have in chemistry.

Infrared radiation

We saw in chapter 2.10 of *Edexcel AS Chemistry* that molecules can absorb infrared radiation and that this causes the bonds in the molecules to vibrate by stretching and bending. Different bonds require different amounts of energy to make them vibrate. We also saw that not all bonds absorb infrared radiation – only molecules that change their polarity as they vibrate absorb infrared radiation. So H–H and Cl–Cl will not absorb infrared radiation, but H–Cl will. **Figure 1.7.3** shows the three peaks corresponding to the three types of vibration in the sulfur dioxide molecule.

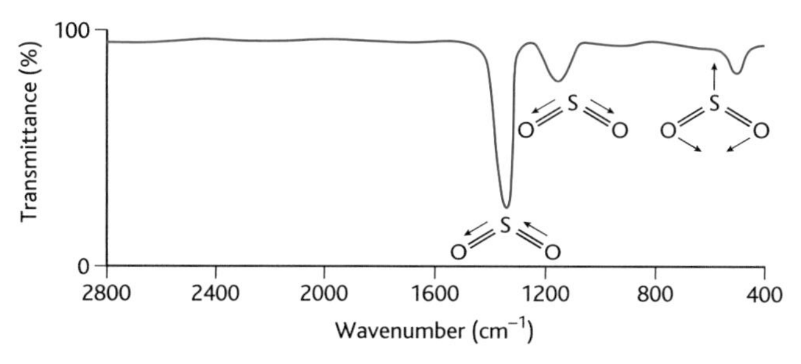

fig. 1.7.3 The IR spectrum for sulfur dioxide has three peaks corresponding to the three types of vibration of the molecule.

Table 1.7.1 shows the positions of infrared absorption peaks for bonds and functional groups of organic compounds.

Tracing the progress of a reaction using infrared spectroscopy

Figure 1.7.4 shows the infrared spectra for ethanol and ethanoic acid.

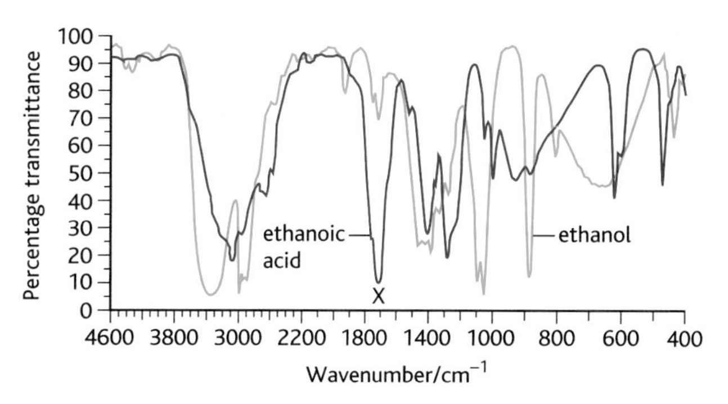

fig. 1.7.4 Infrared spectra for ethanol (blue) and ethanoic acid (red).

Wine contains up to about 15% ethanol. Oxidation of wine can occur through contact with oxygen – partial oxidation produces ethanal and this is an essential part of the maturing of the wine. However, further oxidation produces ethanoic acid, giving the wine an unpleasant taste and generating colours that alter its appearance. Wine makers must monitor the components of their wine as well as taste it!

In the spectrum of ethanoic acid there is a large peak labelled X at about 1700 cm^{-1} – this corresponds to the C=O bond stretching. (Ethanal also has this peak but no –OH group, so it can be identified.) There is no such bond in ethanol so chemists can check the extent of any oxidation by monitoring the growth of this peak.

Bond (stretching)	Intensity	Functional group	Wavenumber/cm^{-1}
C–H	medium–strong	alkanes	2850–3000
C–H	medium	alkenes	3095–3010
C–H	weak	aldehydes	2820–2900
C–O	variable	alcohols, esters, acids	1000–1300
O–H	variable (broad peaks)	alcohols	3200–3750
O–H	weak (broad peaks)	acids	2500–3300
N–H	medium	amines	3300–3500
C=O	strong	aldehydes, ketones, acids	1680–1740
C–X	strong	halogenoalkanes	500–1400

table 1.7.1 IR absorption peaks for bonds and functional groups of organic compounds.

Microwaves

We are familiar with microwave energy heating our food in microwave ovens. As we saw in Unit 2, microwave radiation creates an electric field which causes water molecules in the food to line up with the field. The field then switches to the opposite direction and the water molecules swivel round to line up again. This causes the energy in the microwaves to be converted into thermal energy and this heats the food. Of course, this also happens to other polar molecules that happen to be present – but not to non-polar molecules.

We saw in Unit 2 that using microwaves can speed up reactions significantly. Suppose a chemist is using water to extract an organic compound dissolved in tetrachloromethane (**fig. 1.7.5**). Also suppose that the organic compound becomes more soluble in water than in tetrachloromethane as the temperature increases.

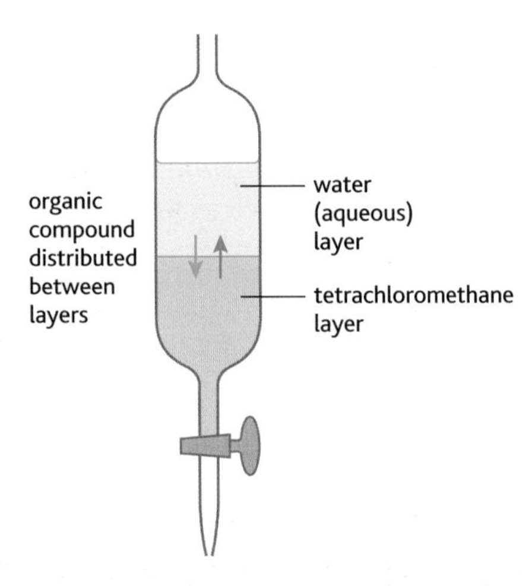

fig. 1.7.5 Separating funnel containing two solvents and a common solute.

If this system is subjected to microwave radiation, the temperature of the aqueous layer would rise (water is a polar solvent) but the temperature of the tetrachloromethane (non-polar) would not.

As a result, more of the organic compound would dissolve in the aqueous layer and the chemist would get a better separation.

Radio waves

Radio waves are used in the sophisticated technique called nuclear magnetic resonance spectroscopy. This will be considered next in this chapter.

Ultraviolet

In Unit 2 we saw that ultraviolet radiation can initiate reactions in the Earth's upper atmosphere by causing oxygen molecules to dissociate into reactive oxygen atoms by **homolytic fission**. These can react with other oxygen molecules to produce ozone.

The technique of flash **photolysis** (**fig. 1.7.6**) was developed about 50 years ago by George Porter and others. This involves subjecting a sample to intense ultraviolet radiation, or light from a laser, for a nanosecond. This excites the system and observations are made as the system relaxes.

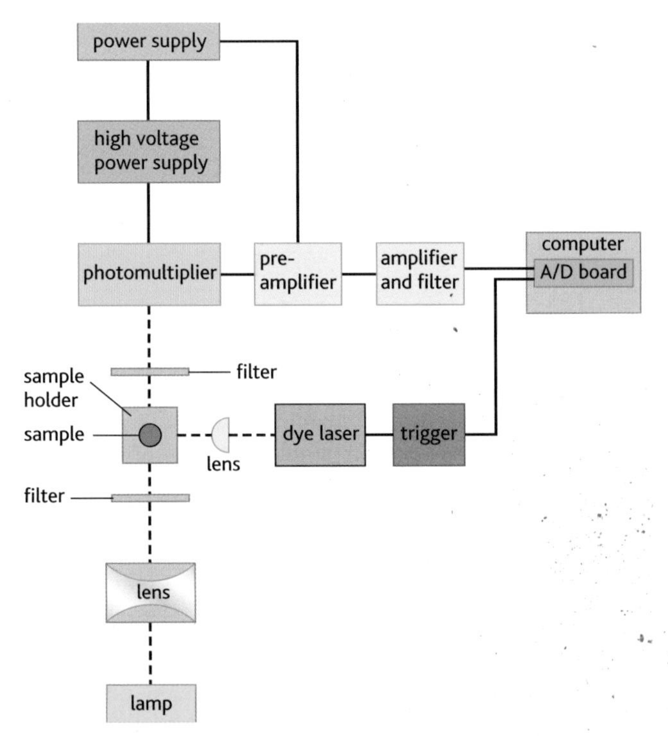

fig. 1.7.6 The flash photolysis technique.

Using this technique, scientists have studied organic molecules, polymers, **nanoparticles** and semiconductors. They have also been able to study the process of photosynthesis in great detail.

Questions

1 The techniques mentioned in this section have replaced traditional methods of chemical analysis. For example, in forensic science these techniques have largely replaced traditional qualitative and quantitative methods. Suggest reasons why.

High-resolution nuclear magnetic resonance spectroscopy [4.9b]

Like electrons, protons possess a property called spin. Because of this, the nuclei of certain atoms – including hydrogen – behave like tiny magnets. When these atoms are placed in a magnetic field, the nuclei align themselves with the field, just as a bar magnet aligns itself with the Earth's magnetic field – see **fig. 1.7.7**.

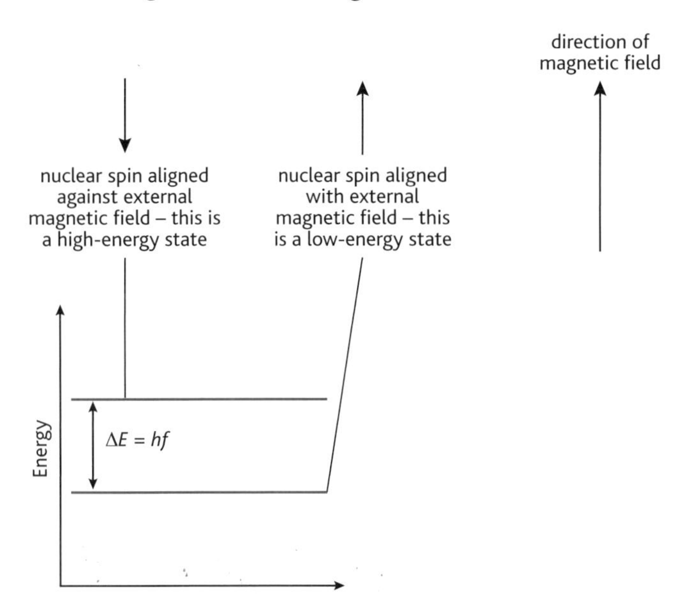

fig. 1.7.7 When a nuclear 'magnet' aligns against the direction of the applied field, it is less stable than when the two are aligned in the same direction. The energy difference between these two states corresponds to radio frequencies, according to the relationship $\Delta E = hf$.

Energy is needed to change this alignment, in the same way as energy is needed to change the alignment of a compass needle. In the case of the compass needle, this energy can be supplied by a push from your finger – the nuclei of atoms can be 'pushed' from one alignment to the other by the energy from radio waves.

Organic chemists use nuclear magnetic resonance (nmr) spectroscopy to find out about, particularly, the hydrogen atoms in molecules. When a hydrogen atom absorbs radio waves as it 'flips' between states, the frequency of radiation that it absorbs depends on its environment to some extent. This means that the spectrum of frequencies absorbed by a particular molecule provides a great deal of information about the position of the hydrogen atoms in it.

The hydrogen atoms in a molecule that is subjected to an externally applied magnetic field do not all experience the same magnetic field. This is because the proton in the nucleus of a hydrogen atom is shielded by electrons close to it, modifying the magnetic field it 'feels'. The amount of this modification depends on the electron density around the proton, which is influenced by the position of the hydrogen atom within the molecule, and on the other atoms surrounding it. Therefore, the frequency at which magnetic resonance occurs is an indication of the chemical environment of a proton within a molecule. The magnetic environment of a proton is quantified using its 'chemical shift' – this is on a scale often called the δ scale. The units of δ are parts per million (ppm).

The sample being investigated is dissolved in a solvent such as CCl_4, as this does not show the resonance effect because it contains no hydrogen atoms. A small amount of tetramethylsilane (TMS), $Si(CH_3)_4$, is also added to the solution to act as a reference substance for the δ scale. The twelve identical protons of TMS are arbitrarily assigned a chemical shift of zero.

TMS is chosen because:

- only a trace amount is needed
- it is a volatile liquid that can be evaporated off easily if recovery of the pure sample is required
- molecules of TMS do not readily interact with molecules of the sample. Such interactions would be undesirable, since they would modify the magnetic environment of the TMS protons and so affect their chemical shift
- protons in the majority of organic compounds show nuclear resonance at chemical shift values that are positive with respect to the protons of TMS, making the comparison of chemical shifts easy.

Low-resolution nmr

Figure 1.7.8 shows the nmr spectrum of ethanol.

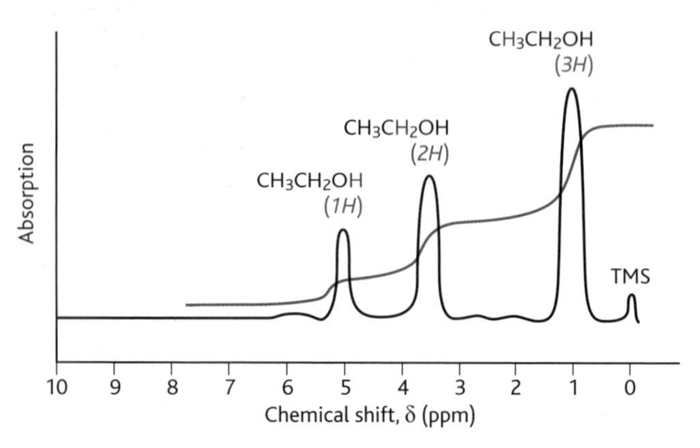

fig. 1.7.8 Low-resolution spectrum of ethanol, CH_3CH_2OH. Each set of peaks corresponds to one group of hydrogen atoms in the molecule.

The spectrum of ethanol shows a number of features. The black line in **fig. 1.7.8** shows the absorptions of the molecules. The blue line is generated by the spectrometer, which calculates the area below the black line – it is often called the 'integrated spectrum'. It corresponds to the number of hydrogen atoms represented by each set of peaks. Together, these lines show which set of peaks corresponds to which hydrogen atoms. This particular spectrum shows:

- a peak at δ = 5.0 ppm – due to the resonance of the proton of the single hydrogen atom of the –OH group
- a peak centred at δ = 3.7 ppm – due to the resonance of the two protons in the hydrogen atoms of the CH_2 group
- a peak centred at δ = 1.0 ppm – due to the resonance of the three protons of the CH_3 group.

Low-resolution nmr enables the identification of:

- the number of different types of proton – from the number of peaks
- the relative number of hydrogen atoms in each group – from the area under the peaks.

High-resolution nmr

Further resolution of the spectrum splits the signals into multiplets. This splitting occurs because of the interaction between protons on neighbouring atoms. This interaction is called **coupling** and causes the splitting into distinct lines. **Figure 1.7.9** shows the high-resolution spectrum for ethanol. The single peak is for the O–H. This is because of hydrogen bonding, which means that protons can exchange between the molecule and water.

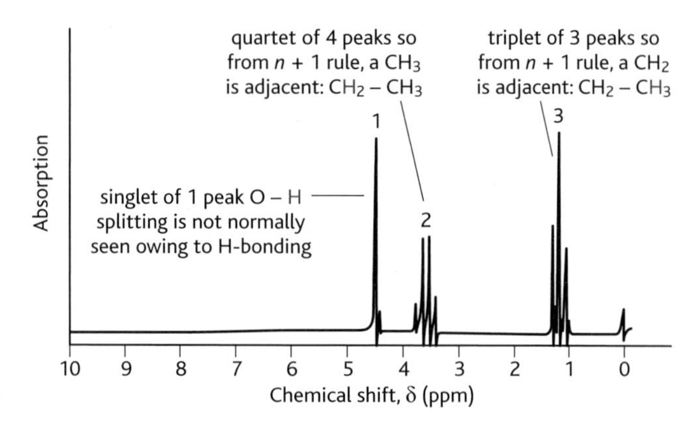

fig. 1.7.9 High-resolution spectrum of ethanol.

The other absorptions of the low-resolution nmr spectrum are split into separate lines in the high-resolution spectrum. When interpreting this, the $n + 1$ rule can be used – the number of lines in the split pattern for a peak is equal to one more than the number of hydrogen atoms in the adjacent group in the molecule. For example, this means that the peak formed by the protons in the CH_2 group in ethanol is split into 4 because this group is next to a group containing 3 hydrogen atoms (CH_3).

Table 1.7.2 summarises the number of lines formed in the spectrum for different numbers of protons in adjacent atoms and gives their intensity.

Number of protons on adjacent atom (n)	Number of lines (n + 1)	Relative intensity of lines
1	2	1 1
2	3	1 2 1
3	4	1 3 3 1
4	5	1 4 6 4 1

table 1.7.2 The n + 1 rule in high-resolution nmr spectroscopy.

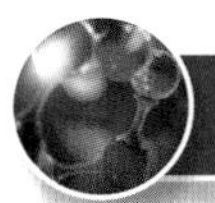

HSW Determining structures

Nuclear magnetic resonance and other methods of structural determination enable chemists to determine chemical structures very quickly – a process that used to take a very long time indeed. **Figure 1.7.10** shows the structure of γ-linolenic acid, an essential fatty acid present in oil of evening primrose.

fig. 1.7.10 γ-linolenic acid, the molecule responsible for the action of oil of evening primrose. Without modern instruments, determining the structure of a molecule like this would be much more difficult.

The oil has been used for hundreds of years for various conditions, including eczema, and recent research has revealed its active ingredient. Oil of evening primrose is now also being used to treat premenstrual tension.

Using nuclear magnetic resonance

We have seen that the nmr spectroscopy technique can be used to determine the molecular structure of organic compounds by studying the series of peaks in the spectrum. Nuclear magnetic resonance has been developed into other applications.

Medical imaging

X-rays are very good at showing doctors the state of bones inside the body. It is much harder to get clear pictures of soft tissues such as the brain, the heart and the gut. Using the concept of nuclear magnetic resonance, scientists and doctors have together developed a diagnostic technique called nmr imaging. This is an extremely rich source of information about what is going on inside patients. New techniques are continually being refined to increase the scope of the technique to show up crucial changes – for example, in the hours immediately after a stroke. A process known as diffusion weighted imaging produces an nmr image which allows very specific diagnosis of the area of damage after a stroke – see **fig. 1.7.11**.

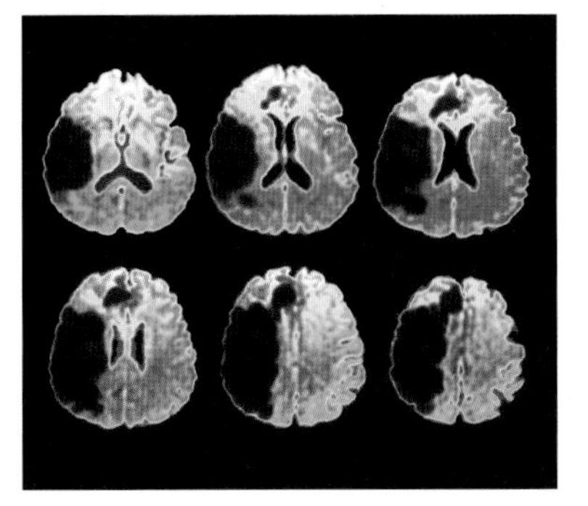

fig. 1.7.11 An imaging technique that enables scientists and doctors to pinpoint exactly where damage has occurred in the brain, in this case after a stroke, is bound to be a very useful tool indeed.

HSW Developing new techniques

Diagnosing the growth of a cancerous tumour in the brain at an early stage gives the patient a much better chance of being treated successfully. Because magnetic resonance imaging (MRI) does not require an incision into brain tissue, it is called a non-invasive procedure. For example, a glioma is a type of cancer that affects the brain and spinal cord – these can be very aggressive. The sooner they are diagnosed, the more effectively they can be treated, and the more likely it is that the patient will have a reasonable extension of life. One type of treatment is the drug temozolmide. In some patients this works well, in others it does not. Doctors need to know as soon as possible if the tumour is responding to the drug and shrinking. If it is not shrinking then other drugs, or even brain surgery, can be tried.

Inside the head, it is impossible to see what is happening and traditional imaging techniques are limited. Until recently, the only way to see if treatment was working was to wait for months and then monitor the size of the tumour using an MRI scanner. However, the use of nmr spectroscopy allows doctors to see what is happening much more rapidly.

The nmr equipment measures the amount of choline in the area of the tumour. A high choline level indicates that cells are dividing rapidly and, hence, that the tumour is growing. A low level means that cells are dividing slowly. Because this was a research project, the scientists also measured the size of the tumour using traditional MRI scans. There was good correlation between the results, suggesting that monitoring choline is a good way of getting an early indication of the effectiveness of the treatment.

As a result of these, and other, investigations, the use of nmr imaging in diagnosing and monitoring the treatment of brain disorders is steadily growing.

nmr in the pharmaceutical industry

Gas chromatography (see page 140) is a widely used method of testing the purity of chemicals. It is particularly important in the pharmaceutical industry.

Table 1.7.3 compares gas chromatography with analysis using nmr spectroscopy. You can see that nmr spectroscopy has a number of advantages.

	Gas chromatography	nmr spectroscopy
Size of sample needed	130 mg	20 mg
Sample compared to	Reference solution that has to be prepared	Built-in internal standard
Volume of solvent needed	25 cm^3	1 cm^3
Time required	112 minutes	32 minutes

table 1.7.3 Comparison of analysis requirements.

One example of the value of nmr in the pharmaceutical industry involves the production of nicotine. This is a chemical used as a pharmaceutical. It can be extracted from plants such as tobacco but can also be made synthetically.

If nicotine is made synthetically, it may contain a chemical called cotinine, which has a very similar structure to nicotine but has undesirable properties. It is an oxidation product of nicotine. Nmr spectroscopy can measure the purity of samples quickly and accurately, so that if cotinine appears in a batch of the drug it can be isolated and removed before the nicotine is used therapeutically.

Questions

1 The nmr spectrum of an organic compound shows a doublet at δ = 2.1 ppm and a quartet at δ = 9.8 ppm. It is known that the compound is either ethanal or propanone – which one is it?

2 The diagram shows the nmr spectrum of compound A, with a molecular formula C_3H_8O.

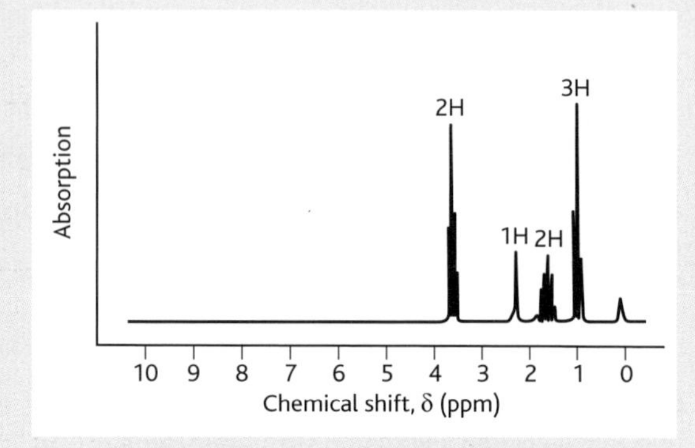

a Which substance is responsible for the peak at δ = 0 ppm? Why is this substance used in obtaining nmr spectra?

b Draw the displayed formulae of three isomers with a molecular formula C_3H_8O.

c Which of these is compound A? Explain your choice.

3 How does nmr spectroscopy differ from X-ray and MRI scans?

4 Outline the advantages of nmr spectroscopy over gas chromatography in the analysis of the purity of substances.

A review of mass spectroscopy [4.9d]

In *Edexcel AS Chemistry* (pages 52–3) you were introduced to the way a mass spectrometer is constructed. On pages 54–5 you saw how the mass spectrometer can be used to identify different isotopes in an element. In a later chapter, on pages 230–1, we considered how mass spectrometry could be used to identify some organic compounds. In this section, we will look at some further examples of how mass spectrometry is used in chemical analysis.

Finding the molar mass of naturally occurring magnesium

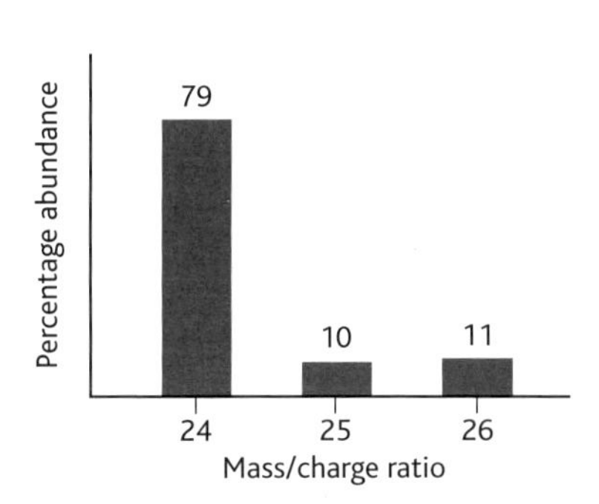

fig. 1.7.12 The mass spectrum of magnesium.

Figure 1.7.12 shows a typical mass spectrum for a pure element – this one is for naturally occurring magnesium.

The *x*-axis is labelled 'mass/charge ratio'. This scale is used because ions of the same mass but different charges give separate lines on the spectrum. For ions with a single charge, the mass/charge ratio is equal to the isotopic mass. **Table 1.7.4** shows the relative abundance of the different isotopes of magnesium, as obtained from the mass spectrum.

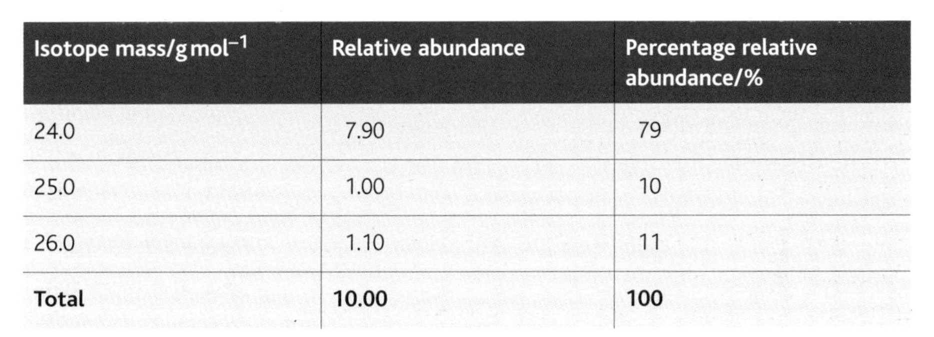

Isotope mass/$g\,mol^{-1}$	Relative abundance	Percentage relative abundance/%
24.0	7.90	79
25.0	1.00	10
26.0	1.10	11
Total	**10.00**	**100**

table 1.7.4

The data from the graph mean that in a random sample of 100 magnesium atoms, 79 will be magnesium-24, 10 will be magnesium-25 and 11 will be magnesium-26.

We find the mass of these 100 atoms, and then divide by 100 to get an accurate value for the molar mass of magnesium. This is done in **table 1.7.5**.

Isotope mass/$g\,mol^{-1}$	Number of atoms in 100	Relative mass in 100 atoms of mixture
24.0	79	1896
25.0	10	250
26.0	11	286
Total	**100**	**2432**

table 1.7.5

So, the average mass of 1 atom $= \dfrac{2432\,g\,mol^{-1}}{100} = 24.32\,g\,mol^{-1}$.

Interpreting a mass spectrum

Many organic molecules can be identified using a mass spectrometer. The machine can be connected to a computer, which can search a database of known mass spectra to identify the substance.

Alkanes

In *Edexcel AS Chemistry* (page 230) we examined the mass spectrum of butane C_4H_{10}. **Figure 1.7.13** shows the mass spectrum of a different alkane.

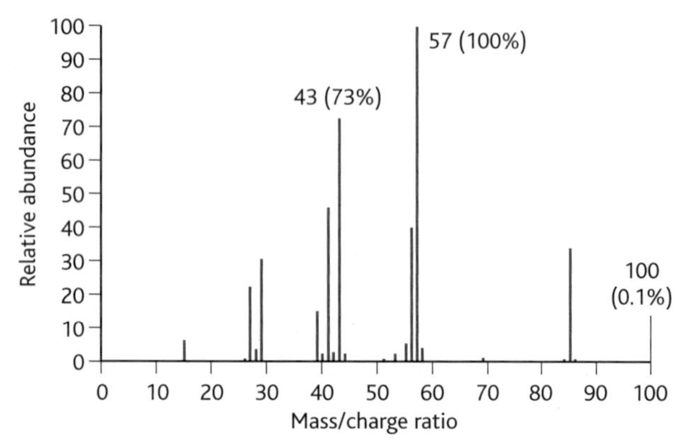

fig. 1.7.13 Mass spectrum of an unknown alkane.

The molecular ion peak has a value of 100. We know this compound is an alkane, so this suggests that the molecular formula is C_7H_{16}. The most abundant peak (the base peak) is at 57.

Now, 100 – 57 = 43, suggesting a $C_3H_7^+$

and 100 – 43 = 57, suggesting a $C_4H_9^+$ group.

The compound is likely to be 2,2-dimethylpentane, with the structure shown in **fig. 1.7.14**. Other peaks include CH_3^+ at 15, $CH_3CH_2^+$ at 29 and $C_6H_{13}^+$ at 85 – none of which contradict this conclusion.

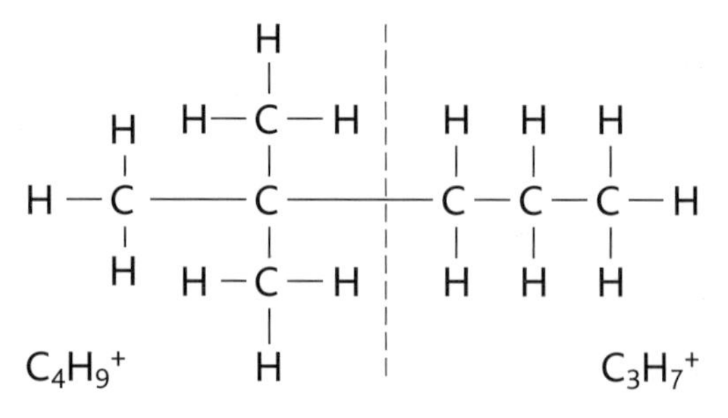

fig. 1.7.14 2,2-dimethylpentane.

Alcohols, aldehydes and ketones

In *Edexcel AS Chemistry* (page 231) you will find mass spectra for ethanol, propanal and propanone. In each case, the molecular ion peak has an even value for the mass/charge ratio – which is consistent with a compound containing an oxygen atom.

Carboxylic acids and esters

Figure 1.7.15 shows the mass spectrum of a carboxylic acid.

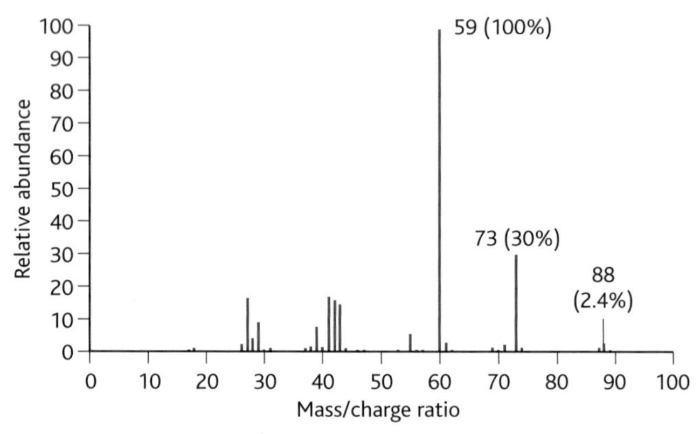

fig. 1.7.15 The mass spectrum of a carboxylic acid.

The molecular ion peak is at 88, so the relative formula mass is 88. The base peak at 59 indicates the loss of 29 units, which could be the loss of $C_2H_5^+$. The peak at 73 corresponds to a loss of CH_3^+. The compound is butanoic acid, $CH_3CH_2CH_2COOH$.

Figure 1.7.16 shows the mass spectrum of ethyl ethanoate. This also has a relative formula mass of 88 because the molecular ion peak is at 88. The losses of 45 units and 27 units are explained in **fig. 1.7.16**.

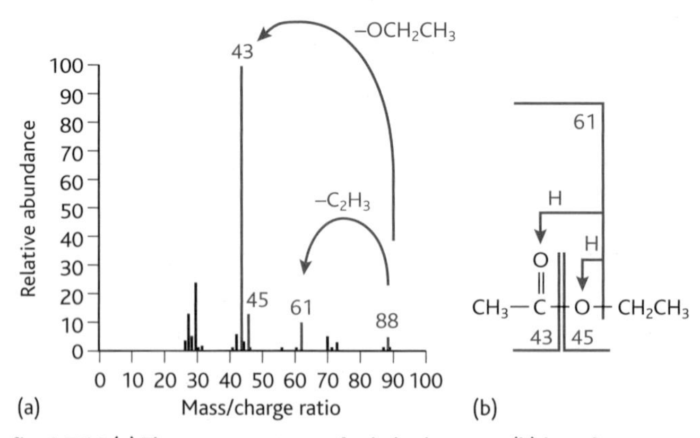

fig. 1.7.16 (a) The mass spectrum of ethyl ethanoate; (b) how fragments are formed.

Amines

Figure 1.7.17 shows the mass spectrum of an amine. Note that the odd number (73) for the relative formula mass is consistent with the presence of a nitrogen atom in the structure.

The cleavage of the C–C bond as shown in the diagram explains the loss of 43 units in the spectrum. The compound is 1-aminobutane.

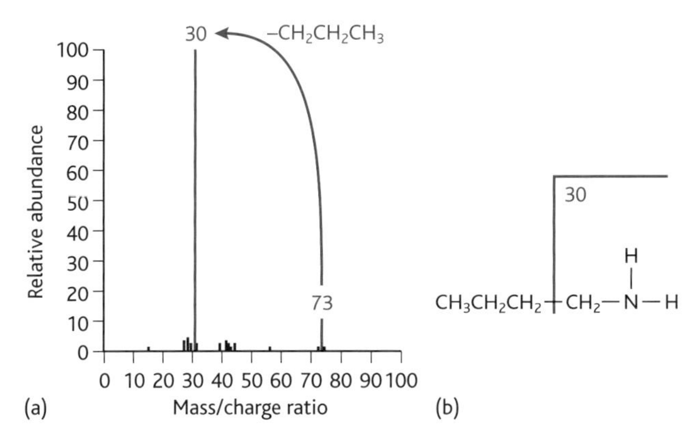

fig. 1.7.17 (a) The mass spectrum of 1-aminobutane; (b) how fragments are formed.

HSW Fermentation in antiquity

New findings suggest that Chinese villagers were brewing alcoholic beverages as long ago as 9000 years ago. This was before barley beer and grape wine were made in the Middle East. The results provide the earliest chemical evidence yet for fermented beverages.

Patrick McGovern and his colleagues evaluated shards of 16 pieces of pottery collected from a village in China's Henan province. The team used a variety of techniques – including mass spectrometry, chromatography and isotope analysis – to analyse sections from the bottom and sides of vessels that held liquid more than 9000 years ago. The team looked for tartaric acid, a key marker of fermentation, which is an organic acid present in grape wine. The tests revealed that 13 of the 16 remnants came from containers that had held the same liquid, a 'mixed fermented beverage of rice, honey and a fruit,' the authors report.

The researchers also analysed liquids that date back more than 3000 years ago, which were preserved inside sealed vessels. They determined that the 'wines' contained herb, flower and tree resins and are very similar to herbal drinks described in inscriptions from the Shang dynasty. 'The fragrant aroma of the liquids inside the tightly lidded jars and vats, when their lids were first removed after some 3000 years, suggests that they indeed represent Shang and Western Zhou fermented beverages,' McGovern says.

Questions

1 This mass spectrum is from a hydrocarbon with a molecular formula of C_4H_{10}. Explain how each peak is formed.

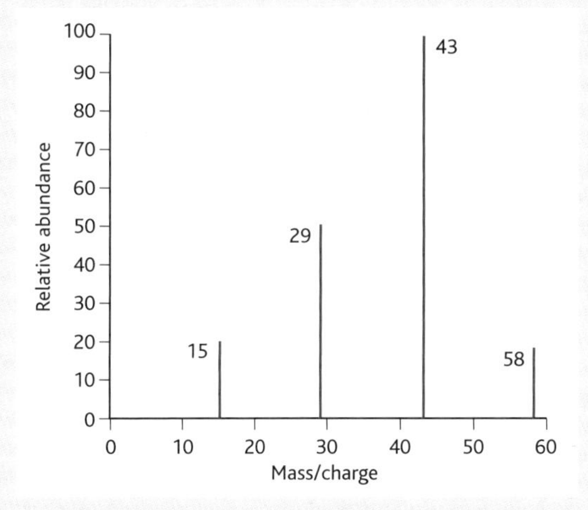

2 One of these mass spectra is for propanone; the other is for propanal. Which is which? Explain your reasoning.

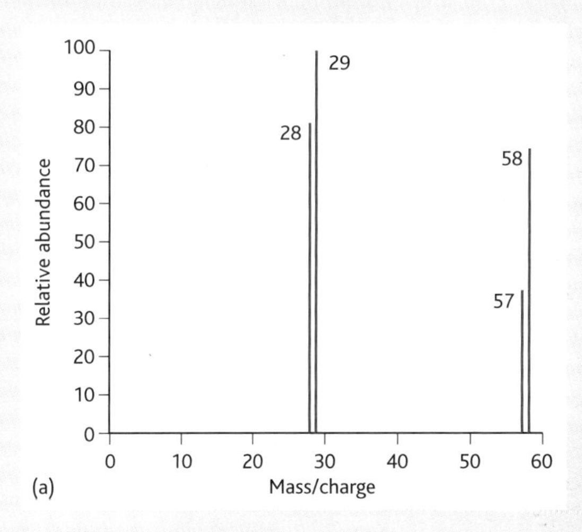

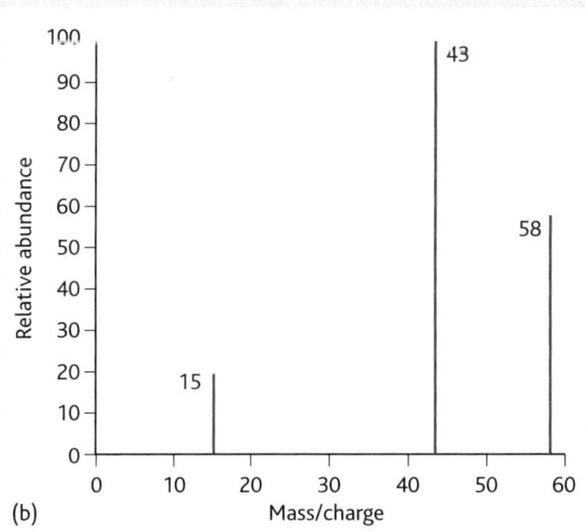

Gas chromatography and high-performance liquid chromatography [4.9e]

You probably have tried paper **chromatography**, and perhaps thin-layer chromatography. This section is about two other types of chromatography – gas chromatography and high-performance liquid chromatography, which is a development of column chromatography.

Gas chromatography

A schematic diagram of a simple machine is shown in **fig.1.7.18**.

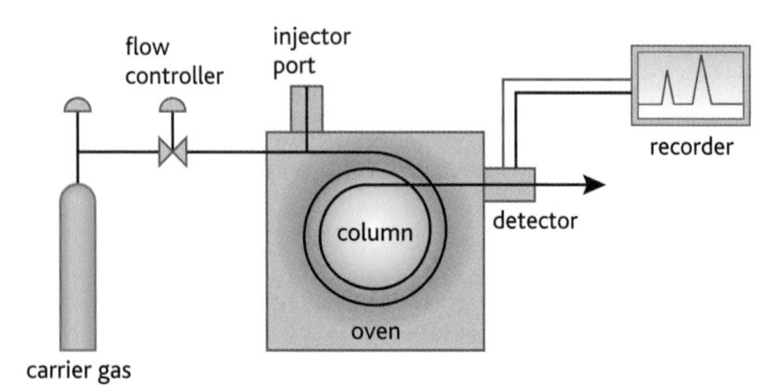

fig. 1.7.18 A gas chromatograph.

This type of chromatography, which is better called gas–liquid chromatography, involves the injection of a tiny amount of the sample into the head of a chromatography column. This is usually done with a microsyringe.

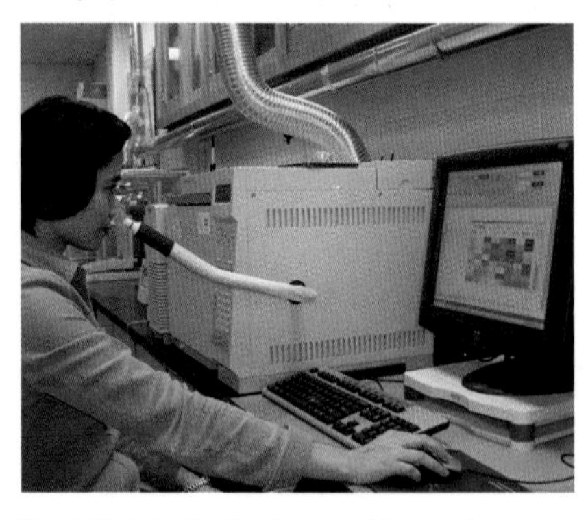

fig. 1.7.19 Perfecting beer – this gas chromatograph machine can separate out every single chemical in an essential oil extracted from high quality hops.

The sample is then vaporised – the temperature at the injection point is about 50 °C higher than the boiling temperature of the least volatile component of the sample. The gaseous sample is carried through the column by the flow of an inert gas, which acts as the mobile phase. Suitable gases are nitrogen, helium, argon and carbon dioxide – the type of detector used often determines which carrier gas is used.

The column itself contains a liquid, which is the stationary phase – this liquid is adsorbed onto the surface of an inert solid. There are different types of column but a common arrangement is the 'packed column' where the column itself contains a finely divided, inert, solid support material coated with liquid stationary phase. Most packed columns are 1–10 metres long and have an internal diameter of 2–4 mm. The temperature depends on the boiling point of the sample and must be carefully controlled.

As the sample passes through the column, the different components of the sample separate because they move at different rates. A simple analogy is to think of a swarm of bees and wasps moving over a field of flowers. The bees would tend to stop to pick up nectar from the flowers, but the wasps would not. Therefore, at the other side of the field the wasps would be detected first, and then the bees.

There are many types of detector that can be used but they all identify the separate substances leaving the column. Different detectors have different ranges of selectivity:

- a *non-selective* detector responds to all compounds except the carrier gas
- a *selective detector* responds to a range of compounds with a common physical or chemical property
- a *specific detector* responds to a single chemical compound.

One type of detector is the flame ionisation detector – see **fig.1.7.20**.

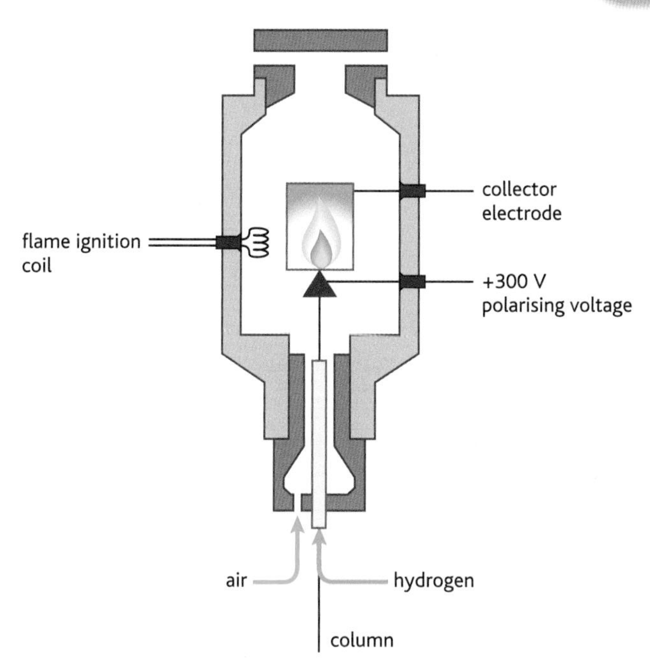

fig. 1.7.20 A flame ionisation detector.

In this type of detector, the gases leaving the column are mixed with hydrogen and air, and burnt. Organic compounds burn in the flame to produce ions and electrons, which can conduct electricity through the flame. A large electrical potential is applied at the burner tip, and a collector electrode is located above the flame. The current resulting from the burning of any organic compounds is measured.

This type of detector is suitable for detecting organic compounds – it is robust, relatively easy to use and can measure the mass of each substance identified. It does, however, destroy the sample. Other types of detector can identify the substances in a sample without destroying them and can also measure the concentrations of the components. The results from the detector are displayed on a monitor or printout.

High-performance liquid chromatography

Before trying to understand how this works, it is helpful to understand column chromatography, which was the first type of chromatography used by Michael Tswett in 1902. He was a Russian botanist who developed the process to separate coloured components in plant pigments.

The process involves having a column packed with a solid stationary phase, such as alumina or calcium carbonate. A solvent is added to the column to make the mobile phase. A solution of the plant pigment, for example, is added to the top of the column. Fresh solvent is then added to the column. The tap at the bottom of the column is open so liquid is flowing out constantly.

Figure 1.7.21 shows what happens as the process continues. The beaker should be changed as soon as the yellow pigment starts to drip out.

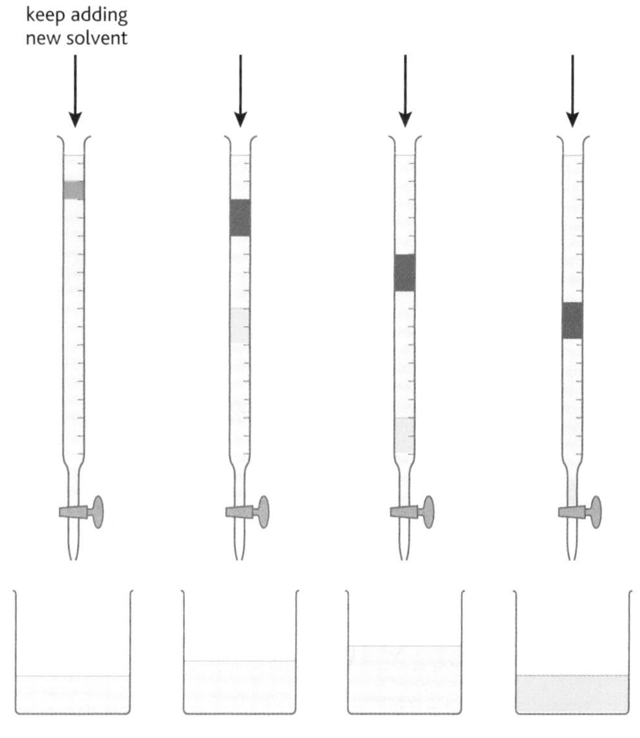

fig. 1.7.21 Separation of blue and yellow pigments in a column.

The mixture of green pigment is being separated. It is made up from a blue pigment and a yellow pigment. The blue pigment molecules are more polar than the yellow pigment molecules and, because they bond more strongly to the solid in the column the pigment does not pass through the column as quickly as the yellow pigment. The yellow pigment is therefore collected first.

The process can still be carried out if the compounds being separated are colourless, but you need to have some means of tracking them.

High-performance liquid chromatography (HPLC) is basically a highly improved form of column chromatography. Instead of a solvent being allowed to drip through a column under gravity, it is forced through under high pressure – up to 400 atmospheres. This makes the process much faster.

A very much smaller particle size for the column packing material can be used. This gives a much greater surface area for the interactions between the stationary phase and the molecules flowing past it. This allows a much better separation of the components of the mixture.

Another improvement with this method comes through using more sensitive detection equipment.

The method is summarised in the flow diagram in **fig. 1.7.22**.

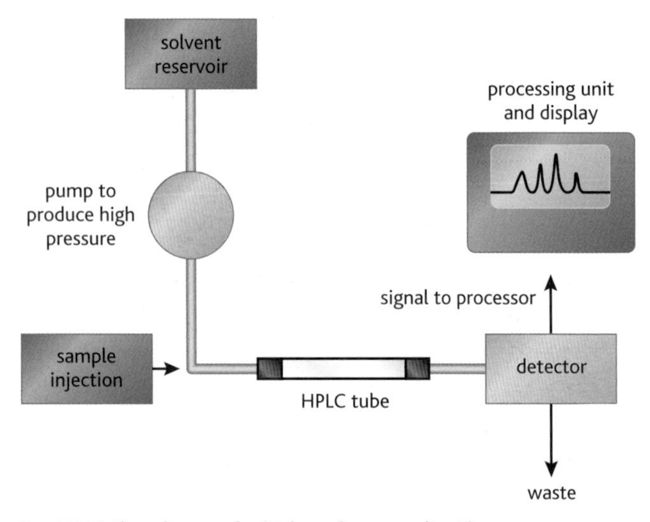

fig. 1.7.22 Flow diagram for high-performance liquid chromatography.

There are two types of HPLC:

- *Normal phase HPLC* – the column is filled with tiny silica particles, and the solvent is non-polar – hexane, for example. A typical column has an internal diameter of 4.6 mm (and may be less than that) and a length of 15–25 cm. Polar molecules in the test sample passing through the column will stick for longer to the polar silica than non-polar molecules. The non-polar ones will therefore pass through the column more quickly.
- *Reversed phase HPLC* – this is the more common type of HPLC. Basically this is similar, but the silica is modified to make it non-polar by attaching long hydrocarbon chains to its surface. These hydrocarbon chains have typically either 8 or 18 carbon atoms in them. A polar solvent (a mixture of water and methanol, for example) is used. There will be a strong attraction between the polar molecules of the solvent and the polar molecules in the test sample passing through the column. Attraction between the hydrocarbon chains attached to the silica (the stationary phase) and the polar molecules in the solution will be very weak. The polar molecules in the test sample will, therefore, spend most of their time moving with the solvent – so the polar molecules will travel through the column more quickly.

The **retention time** is the time taken for a particular compound to travel through the column and reach the detector. The time is measured from the point at which the sample is injected to the point at which the display shows a maximum peak height for that compound.

Different compounds have different retention times. Retention time depends on:

- the pressure used – because that affects the flow rate of the solvent
- the nature of the stationary phase – not only what material it is, but also the particle size
- the exact composition of the solvent
- the temperature of the column.

The consequence of this is that using retention times to identify compounds require careful control of conditions. Providing that conditions are kept constant, the retention time for a particular compound is also constant.

There are several ways of detecting when a substance has passed through the column – ultraviolet (UV) absorption is one such method. Many organic compounds absorb UV light of various wavelengths. If you have a beam of UV light shining through the stream of liquid coming out of the column, and a UV detector on the opposite side of the stream, you can get a direct reading of how much of the light is absorbed. The amount of light absorbed will depend on the amount of a particular compound that is passing through the beam at the time.

Solvents also absorb UV light – but in different parts of the UV spectrum. Methanol, for example, absorbs at wavelengths below 205 nm, and water below 190 nm. If you were using a methanol–water mixture as the solvent, you would therefore have to use a wavelength greater than 205 nm to avoid false readings from the solvent. The detector records a series of peaks, each peak representing a compound emerging from the column.

Suppose you are trying to identify a substance P in a sample. You could inject a small amount of pure P into the apparatus and find the time until a peak is detected – this is called the calibration peak. If your test sample, which might contain P, is then injected into the apparatus, a series of peaks will be formed. If a peak is formed at the same retention time as P, then it shows that P is present in the sample. Comparing the size of the calibration peak and the size of the peak at the same retention time enables the concentration of P to be calculated.

You cannot, however, compare the concentrations of different components in the sample by comparing the size of peaks because it depends on how much UV light is absorbed by the different components.

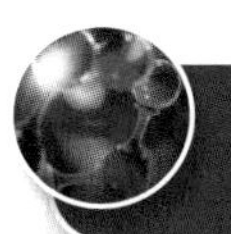

HSW Coupling HPLC to a mass spectrometer

We saw in *Edexcel AS Chemistry* (page 53) that using HPLC in conjunction with a mass spectrometer produces a very powerful tool for the analysis of chemicals such as pharmaceuticals.

The combination of HPLC and mass spectrometer can also be used for other purposes. The orange dye Sudan 1 (**fig. 1.7.23**) is added to dried chillies and curry powders. The photograph (**fig. 1.7.24**) shows chillies being dried in India.

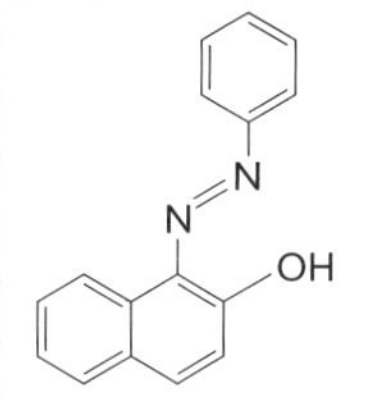

fig. 1.7.23 Structure of Sudan 1.

fig. 1.7.24 Group of women spreading out harvested chilli peppers to dry in the Sun.

However, this compound is carcinogenic. It is banned for use in foods in many countries but can be used in other products such as polishes. In 2005, some samples of Worcester sauce were found to be contaminated with this carcinogenic dye. The origin was traced to adulterated chilli powder. The sauce was used in hundreds of supermarket products, such as pizzas and ready-made meals, and the contamination led to over 400 products being taken off the shelves.

A combination of HLPC and mass spectrometry can be used to identify quickly whether Sudan 1 is present in samples of food. A sample of pure Sudan 1 is used to calibrate the equipment and then analysis of small samples of food will show its presence.

HSW Use of HPLC to detect olanzapine in pharmaceutical operations

Olanzapine (OLZ) is 2-methyl-4-(4-methyl-1-piperazinyl)-10*H*-thienol[2,3*b*][1,5]benzodiazepine (**fig. 1.7.25**).

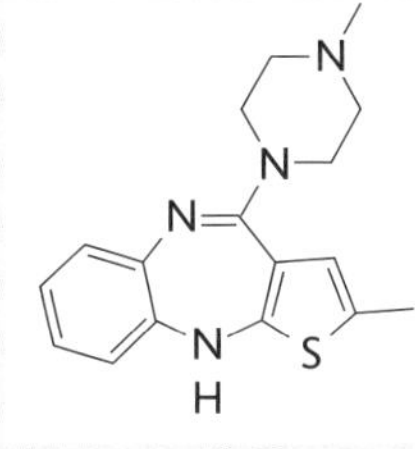

fig. 1.7.25 Olanzapine.

It is used to treat acute schizophrenia and other psychotic conditions. Various methods have been used to determine the OLZ in a sample including by non-aqueous titration and spectrophotometry. A new method has been developed using HPLC in reverse phase. The method was compared with alternative methods and it was found to have accuracy close to 100% – and there was no interference from other components. Unlike in titration, the sample is not destroyed so this method also could be used to separate OLZ from other components.

Questions

1 A sample containing two compounds, X and Y, is analysed using HPLC. X is more polar than Y.

- **a** Which compound is detected first, X or Y, if normal phase HPLC is used?
- **b** Which compound is detected first, X or Y, if reversed phase HPLC is used?

2 Why is it necessary to calibrate an HPLC machine every time?

Examzone

You are now ready to try the second Examzone test for Unit 4 (Examzone Unit 4 Test 2) on page 258, which tests you on what you have learnt in Chapters 6 and 7.

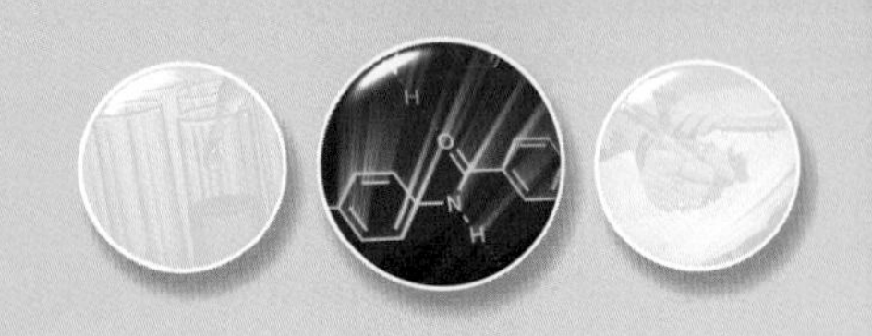

Unit 5 General principles of chemistry II – Transition metals and organic nitrogen chemistry

In Unit 5 you will learn about redox reactions and their vital role in chemistry. You will look at transition metals and develop an understanding of how their electronic structure affects their chemistry. You will learn more about organic chemistry, investigating how organic syntheses are planned. You will also consider the ethical dilemmas which face chemists as they attempt to improve our way of life.

Chemical ideas

In this unit you will bring together and apply your existing knowledge to develop an in-depth understanding of different aspects of chemistry.

You will extend your AS studies of redox reactions to find out how they can be used in titrations to find the concentration of an unknown solution. You will learn about standard electrode potentials and apply this knowledge to predict the thermodynamic feasibility and extent of a reaction. You will also develop your understanding of the transition metals and the importance of their variable oxidation numbers and the complex ions which they form.

In organic chemistry you will look at the arenes – organic chemicals based around the benzene ring. You will consider ideas about the structure of the benzene molecule and how its chemistry relates to the aliphatic straight-chain organic molecules you are already familiar with.

This unit also includes work on a number of polymers, including proteins and the amino-acid monomers from which they are formed. The way in which these complex molecules can be analysed will help you to understand how new designer polymers can be created. You will also look at how the synthesis of organic molecules is planned on both a laboratory and an industrial scale, including the importance of an awareness of optical activity in the development of potential drug molecules.

How chemists work

Throughout this unit you will hone your practical and analytical skills. You will look at the value of careful risk assessment, which plays a major part in the way chemists work. You will learn a wider range of analysis techniques and develop your titration skills. You will also look at how chemists develop their big ideas, including the roles of inspiration and sheer hard work.

Chemistry in action

In this unit you will meet many issues linked to the environment and drug safety. You will look at how chemists are developing new ways of storing energy by developing different types of batteries and hydrogen cells. The use of transition metal catalysts is one way in which chemists are reducing the environmental impact of many different industrial processes in organic chemistry. You will consider the role of society as a whole in making decisions in some areas, for example in the continued use of drugs such as thalidomide. You will find out about the input of chemistry into detecting drivers under the influence of alcohol – a clear example of the way chemistry can be used to make our lives safer.

In chapter 2.1 you will look at redox potentials and standard electrode potentials. One of the ways this type of chemistry is used is in the development of energy sources such as ethanol-based fuel cells based on biofuels.

Transition metals (see chapter 2.2) are used in many ways from catalysts to dyes. They involve some of the most complex and colourful chemistry you will encounter in your GCE course.

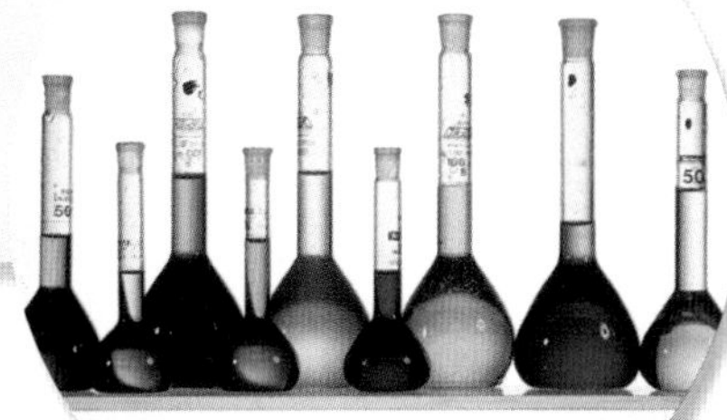

As you will learn in chapter 2.3, the first really useful model for the structure of benzene was developed by a chemist dozing by the fire! Derivatives of this molecule are immensely useful – for example lindane was formerly used to kill parasites on sheep – but some of these derivatives, including lindane, also damage the environment and people, raising a number of ethical issues.

As chemists have developed a better understanding of how proteins are built up from amino acids (see chapter 2.4), their ability to identify proteins in many situations has also grown. For example, the purple dye ninhydrin can be used to produce crime-scene fingerprints as a result of the reaction between it and the protein left behind in sweat from the skin.

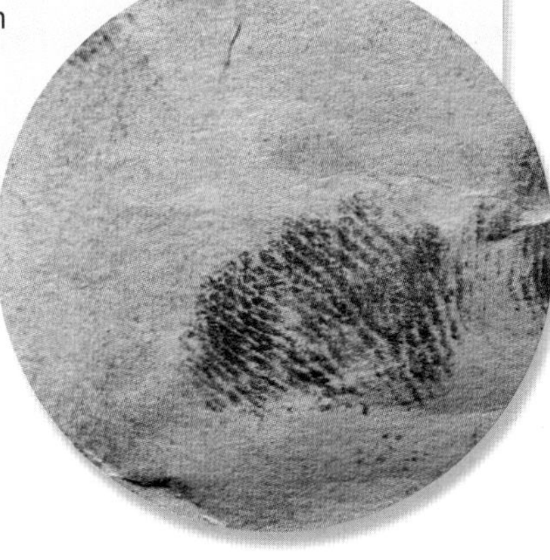

The synthesis of organic molecules usually involves many steps and a lot of time (see chapter 2.5). However, the development of combinatorial chemistry enables pharmaceutical chemists to develop thousands of closely related, but different, chemicals in a much shorter timeframe.

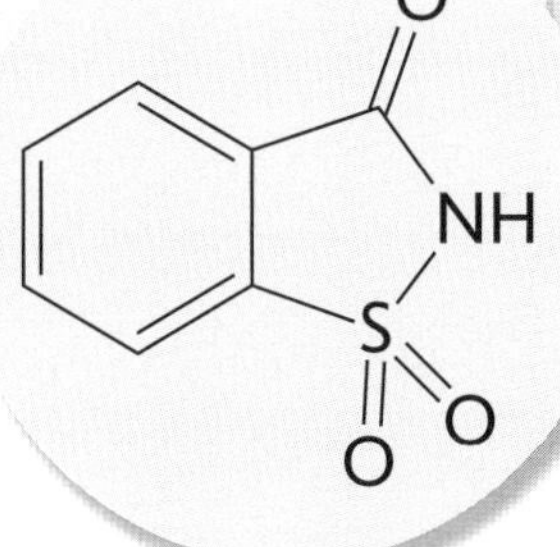

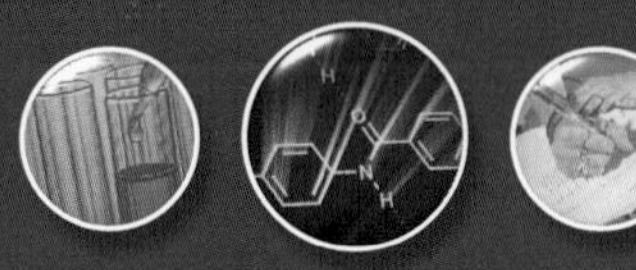

2.1 Redox chemistry

fig. 2.1.1 Redox reactions play a vital role in photosynthesis, the process by which plants make food, and in the respiration reaction by which living organisms obtain energy from their food.

Linking oxidation number and reaction stoichiometry [5.3.1a, b]

Many chemical systems involve reactions in which **reduction** and **oxidation** take place – some of these are included in the **redox reactions** that you met in *Edexcel AS Chemistry* (pages 170–175), along with the idea of oxidation numbers. The chemistry of redox reactions helps us to understand processes such as rusting, the manufacture of chlorine and the production of electricity from batteries, as well as enabling us to monitor the progress of chemical reactions by using electrical measurements. In this section, you will learn more about these important reactions and their importance in the chemistry of the transition metals.

Basic definitions

Oxidation number

The **oxidation numbers** of the elements in a compound are the charges they would have if the electrons in each bond of the molecule or ion belonged to the more electronegative element. You can calculate oxidation numbers by remembering that the oxidation number of an uncombined element is always 0, and that the oxidation numbers in any formula must always balance. For example, in a simple compound such as ammonia, NH_3:

- the oxidation number of H = +1 and there are 3 of these
- so, the oxidation number of N = –3

You can work out the oxidation numbers of individual elements in more complex molecules and ions by addition of the oxidation numbers that you know. For example, in a compound such as sulfuric acid, H_2SO_4, how do you work out the oxidation number of sulfur? The sum of the oxidation numbers of all the elements in the molecule is 0, so you use the oxidation numbers of the elements that you know about (H and O) to work out the oxidation number (o.n.) of the sulfur:

- the oxidation number of H = +1 and there are 2 of these
- the oxidation number of O = –2 and there are 4 of these
- $2 \times (+1) + (\text{o.n. of S}) + 4 \times (-2) = 0$
- $2 + (\text{o.n. of S}) - 8 = 0$
- so, in this compound the oxidation number of sulfur = +6.

The oxidation number of an ion is the number of electrons that have been removed (positive o.n.) or added (negative o.n.).

Redox

A redox reaction is one involving both reduction and oxidation. A process involving the loss of electrons is an oxidation reaction, and a process involving the gain of electrons is a reduction reaction (remember OIL RIG – Oxidation Is Loss, Reduction Is Gain).

The reaction of chlorine with potassium iodide solution is a redox reaction and the equation shown here is a **redox equation** for the reaction:

$$Cl_2(g) + 2I^-(aq) \rightarrow I_2(aq) + 2Cl^-(aq)$$

The chlorine atoms in the chlorine molecules are reduced (acting as an **oxidising agent**) and the iodide ions are oxidised (acting as a **reducing agent**).

Half-equations

An overall ionic equation such as the one above for the reaction of chlorine and iodide ions can be split up into two **half-equations**. These show electron transfer more clearly than the full equation:

$$Cl_2(g) + 2e^- \rightarrow 2Cl^-(aq)$$

$$2I^-(aq) \rightarrow I_2(s) + 2e^-$$

fig. 2.1.2 A redox reaction takes place when powdered zinc reacts with copper(II) sulfate solution. The copper ions are reduced to form reddish brown copper, and the zinc is oxidised to form zinc ions in solution.

Combining half-equations to form redox equations

In *Edexcel AS Chemistry* (page 177) you saw how half-equations could be combined to give an ionic equation that gives the **stoichiometry** of the reaction. For example, the reaction of acidified potassium manganate(VII) in acid solution with iron(II) ions:

$$MnO_4^-(aq) + 8H^+(aq) + 5e^- \rightarrow Mn^{2+}(aq) + 4H_2O(l)$$

$$Fe^{2+}(aq) \rightarrow Fe^{3+}(aq) + e^-$$

The electron numbers must balance in the overall redox equation – so we multiply the second equation by 5 and then add the two equations together:

$$MnO_4^-(aq) + 8H^+(aq) + 5Fe^{2+}(aq) \rightarrow Mn^{2+}(aq) + 4H_2O(l) + 5Fe^{3+}(aq)$$

This equation shows the stoichiometry of the reaction – that is, 1 mole of manganate(VII) ions react with 5 moles of iron(II) ions.

Another example is the reaction of thiosulfate ions in solution with iodine:

$$2S_2O_3^{2-}(aq) \rightarrow S_4O_6^{2-}(aq) + 2e^-$$

$$I_2(aq) + 2e^- \rightarrow 2I^-(aq)$$

The same number of electrons (2) are involved in each half-equation, so the equations can just be added together:

$$2S_2O_3^{2-}(aq) + I_2(aq) \rightarrow S_4O_6^{2-}(aq) + 2I^-(aq)$$

The equation shows the stoichiometry of the reaction – that is, 2 moles of thiosulfate ions react with 1 mole of iodine molecules.

SC Complex half-equations can be worked out from the charges on the ions involved. For example, vanadium has a number of oxidation states. The ions it forms include V^{2+}, V^{3+}, VO^{2+} and VO_2^+.

Work out the oxidation state of each of these ions. Then work out the half-equations and the overall redox equation for the reaction in which the V^{3+} ion becomes VO_2^+.

Questions

1 Give the half-equations and the full redox equation for the reaction shown in **fig. 2.1.2**.

2 Combine the half-equations for the reaction of hydrogen peroxide as a reducing agent with hydrogen peroxide as an oxidising agent. What is the significance of the overall equation?

Redox titrations with potassium manganate(VII) [5.3.1h (i)]

During your AS Chemistry studies you carried out acid–base titrations. You used an indicator to detect the end-point, when equivalent quantities of acid and base are present. There are other titrations that can be carried out. Many of these involve an oxidising agent and a reducing agent reacting together. Two examples involving potassium manganate(VII) acting as the oxidising agent are covered here.

Titration of potassium manganate(VII) with iron(II) ions

In this reaction, iron(II) ions are oxidised and manganate(VII) ions are reduced. The equations and the colours involved are:

- reduction: $MnO_4^-(aq) + 8H^+(aq) + 5e^- \rightarrow Mn^{2+}(aq) + 4H_2O(l)$
- oxidation: $Fe^{2+}(aq) \rightarrow Fe^{3+}(aq) + e^-$

$MnO_4^-(aq)$ (purple) $+ 8H^+(aq) + 5Fe^{2+}(aq)$ (colourless) $\rightarrow Mn^{2+}(aq)$ (colourless) $+ 5Fe^{3+}(aq)$ (colourless) $+ 4H_2O(l)$

Note that these colours apply only when the species are in dilute solution – at higher concentrations $Fe^{2+}(aq)$ is pale green, $Fe^{3+}(aq)$ is yellow and $Mn^{2+}(aq)$ is pale pink.

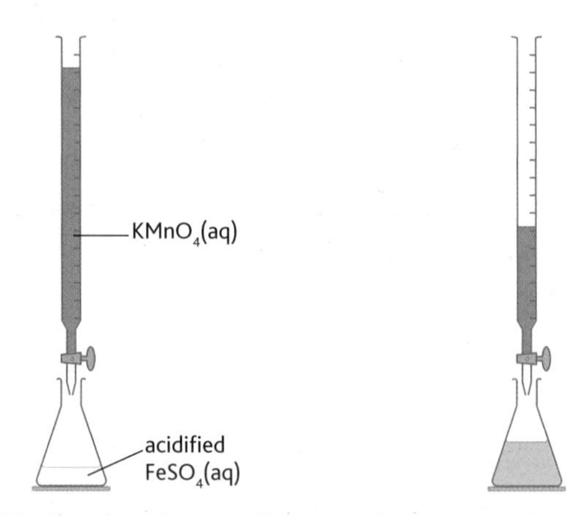

fig. 2.1.3 The colour change in the reaction between potassium manganate(VII) with iron(II) ions means it is self-indicating.

A measured volume of iron(II) solution is pipetted into a conical flask. This solution is acidified with a small amount of dilute sulfuric acid. Potassium manganate(VII) solution is then added slowly from a burette (**fig. 2.1.3**). After swirling to encourage mixing, the solution in the conical flask remains colourless until all of the iron(II) ions have been oxidised to iron(III) ions. Then, the addition of just one more drop of potassium manganate(VII) solution turns the solution pale pink. This colour change means that there is no need to use an additional indicator – the reaction is **self-indicating.**

Worked example

Titration calculation

25.0 cm^3 of iron(II) sulfate reacts with 24.3 cm^3 of 0.02 mol dm^{-3} potassium manganate(VII) solution. Calculate the concentration of the iron(II) sulfate solution.

Amount of potassium manganate(VII)

$= \frac{0.02 \text{ mol}}{1000 \text{ cm}^3} \times 24.3 \text{ cm}^3$

$= 4.86 \times 10^{-4}$ mol

From the redox equation:

$MnO_4^-(aq) + 8H^+(aq) + 5Fe^{2+}(aq) \rightarrow 5Fe^{3+}(aq) + 4H_2O(l) + Mn^{2+}(aq)$

1 mol MnO_4^- reacts with 5 mol of Fe^{3+}.

So, the amount of iron(II) sulfate = 4.86×10^{-4} mol $\times 5$

$= 2.43 \times 10^{-3}$ mol

Now this amount is present in 25.0 cm^3.

So the concentration of iron(II) sulfate

$= \frac{2.43 \times 10^{-3}}{25.0} \times 1000 \text{ mol dm}^{-3}$

= 0.097 mol dm^{-3}

Iron(II) sulfate is a common ingredient in the iron tablets given to pregnant women and anyone who is anaemic. It is very important that the correct dose of iron is present in each tablet. In the past, the iron content of tablets was checked by random sampling and titration, which gave a statistical measure of the variation in the tablets (see page 29 in *Edexcel AS Chemistry*). Nowadays titrations are rarely, if ever, used in this way – technology has taken over and machines carrying out high-performance liquid chromatography and different types of spectrometry are used with much higher convenience and reliability.

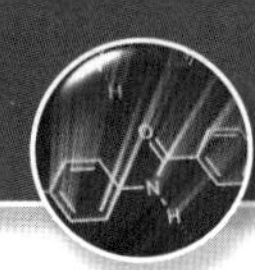

Iron in commercial iron tablets

You may measure the concentration of iron in some commercial iron tablets by titration with potassium manganate(VII) solution.

Titration of potassium manganate(VII) with ethanedioic acid

In this technique, potassium manganate(VII) is the oxidising agent and it is again reduced to manganese(II) in acidic solution. Ethanedioic acid ($H_2C_2O_4$), previously called oxalic acid, is oxidised to carbon dioxide.

fig. 2.1.4 Rhubarb stems are edible, but the leaves are toxic because they contain ethanedioic acid.

The equations and the colours involved are:

- reduction: $MnO_4^-(aq) + 8H^+(aq) + 5e^- \rightarrow Mn^{2+}(aq) + 4H_2O(l)$
- oxidation: $C_2O_4^{2-}(aq) \rightarrow 2CO_2(g) + 2e^-$

$$\underset{\text{purple}}{2MnO_4^-(aq)} + 16H^+(aq) + \underset{\text{colourless}}{5C_2O_4^{2-}(aq)} \rightarrow \underset{\text{colourless}}{2Mn^{2+}(aq)} + 8H_2O(l) + 10CO_2(g)$$

In the titration, ethanedioic acid (which is very poisonous) is pipetted into the conical flask and potassium manganate(VII) solution is added from the burette. However, the reaction is very slow and so the flask containing the ethanedioic acid is usually heated to about 60°C, to speed up the reaction before starting the titration.

The reaction is catalysed by manganese(II) ions. None are present at the beginning of the reaction, which is why it is slow and needs heating. But once some manganese(II) ions have been produced in the reaction, they catalyse it. This is an example of **autocatalysis** or **self-catalysis** – as the number of manganese(II) ions present increases, the reaction goes faster and faster.

SC Graph A is a typical reaction rate graph for a reaction. Graph B shows the rate of the reaction between potassium manganate(VII) and ethanedioic acid.

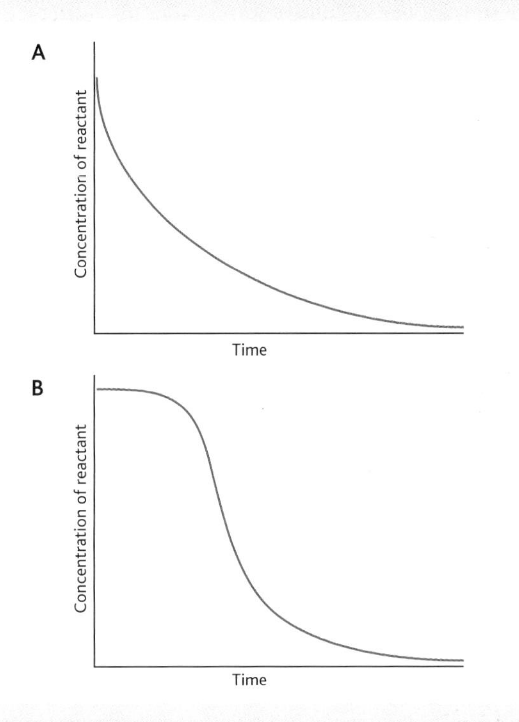

fig. 2.1.5

Explain the shape of both graphs, and how graph B demonstrates the principle of autocatalysis.

Questions

1. Explain why no external indicator is needed when titrating a solution containing ethanedioate ions with potassium manganate(VII).
2. Calculate the volume of 0.020 mol dm^{-3} potassium manganate(VII) solution needed to completely oxidise 25.0 cm^3 of:
   - a 0.010 mol dm^{-3} iron(II) sulfate solution
   - b 0.200 mol dm^{-3} hydrogen peroxide solution
   - c 0.100 mol dm^{-3} sodium ethanedioate solution.
3. A 1.340 g sample of iron ore was dissolved in dilute sulfuric acid to produce a solution of iron(II) sulfate. This was titrated with 0.020 mol dm^{-3} potassium manganate(VII) solution and it was found that 28.75 cm^3 was required. Calculate the percentage of iron in the iron ore.

Redox titrations with sodium thiosulfate [5.3.1h (ii)]

fig. 2.1.6 An iodometric titration.

Another example of a redox titration involves the titration of sodium thiosulfate solution and iodine solution. Titrations with iodine are sometimes called **iodometric titrations** (**fig. 2.1.6**).

Titration of sodium thiosulfate and iodine solutions

The equations and colours for this titration are:

$$2S_2O_3^{2-}(aq) \rightarrow S_4O_6^{2-}(aq) + 2e^-$$

$$I_2(aq) + 2e^- \rightarrow 2I^-(aq)$$

$$2S_2O_3^{2-}(aq) + I_2(aq) \rightarrow S_4O_6^{2-}(aq) + 2I^-(aq)$$

colourless brown colourless colourless

Iodine does not dissolve well in water so it is dissolved in potassium iodide solution. The iodine solution is pipetted into a conical flask and sodium thiosulfate solution added from the burette. The iodine solution is a pale yellow-brown colour (the depth of the colour depends on the concentration), and at the end-point this becomes colourless. In theory, you don't need an indicator but the iodine colour becomes very faint towards the end of the reaction, and this makes the accurate assessment of that final point extremely difficult. To overcome this, a small amount of starch is added as the end-point approaches (when the iodine solution is getting very pale). Starch reacts with free iodine to form a deep blue-black colour, which acts as an indicator. The colour disappears very suddenly when the last of the iodine is reduced, so the solution goes from blue-black to colourless. The starch is added only towards the end – if it is added too early the iodine is adsorbed strongly onto the starch, reducing the accuracy of the final reading.

Measuring the amount of chlorine in bleach

Household bleaches contain sodium chlorate(I) (NaClO) which is the active component. Chlorate(I) ions (hypochlorite ions) are reduced to chloride ions in acidic solution:

$$ClO^-(aq) + 2H^+(aq) + 2e^- \rightarrow Cl^-(aq) + H_2O(l)$$

The oxidation number of chlorine in the chlorate(I) ion is +1, and in the chloride ion it is –1. The decrease in oxidation number of chlorine during this process reveals that the chlorine atom acts as an oxidising agent, and in doing so becomes reduced. The two electrons on the left-hand side of the equation come from the substance that chlorate(I) ion is oxidising. The effectiveness of ClO^- as a bleaching agent depends on its ability to oxidise (remove electrons) from the coloured compounds in stains – a process that often eliminates their colours. The capacity of a given bleaching agent depends on the number of moles of electrons it can remove.

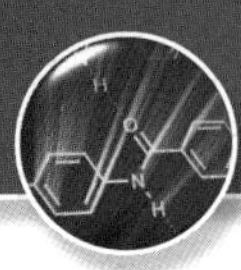

HSW Clean toilets – and NO chlorine!

There are many different products on the market specifically for cleaning toilets, but many people use simple household bleach.

fig. 2.1.7 Keeping toilets clean and hygienic is important both for preventing the spread of disease and for keeping the home smelling pleasant.

Both toilet cleaners and bleach are effective – so some people are tempted to use a bit of both. This can be a problem because many toilet cleaners contain an acid as part of their formula. When bleach mixes with these acidic cleaners, poisonous chlorine gas is produced. There are clear warnings on most cleaning products about not using them with bleach.

The strength of a bleach solution can be determined using a redox titration. This process involves three steps.

Step 1: The chlorate(I) ions in the bleach sample are reacted with excess iodide ions, I^-, in acidic solution:

$$ClO^-(aq) + 2H^+(aq) + 2I^-(aq) \rightarrow I_2(aq) + Cl^-(aq) + H_2O(l)$$

The molecular iodine produced is yellow-brown in solution.

Step 2: You determine the quantity of iodine produced in step 1 by **back-titrating** with a standardised solution of sodium thiosulfate, $Na_2S_2O_3$:

$$2S_2O_3^{2-}(aq) + I_2(aq) \rightarrow S_4O_6^{2-}(aq) + 2I^-(aq)$$

Step 3: You can now work backwards to the concentration of chlorate(I) ions in the bleach.

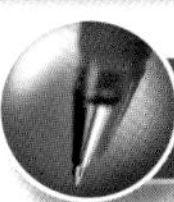

Worked example

The strength of bleach

10.0 g of a bleach is treated with 100 cm^3 of an acidified solution of potassium iodide, converting all the iodide ions to iodine molecules. 10.0 cm^3 of this solution requires 13.4 cm^3 of 0.20 mol dm^{-3} sodium thiosulfate solution to reduce all the iodine back to iodide ions. Calculate the percentage of chlorine in the bleach.

The amount of thiosulfate in 13.4 cm^3 of 0.20 mol dm^{-3} sodium thiosulfate solution

$= \frac{0.20}{1000} \times 13.4$ mol

$= 2.68 \times 10^{-3}$ mol

This reacts with the iodine according to:

$$2S_2O_3^{2-}(aq) + I_2(aq) \rightarrow S_4O_6^{2-}(aq) + 2I^-(aq)$$

So, the amount of iodine present is $\frac{2.68 \times 10^{-3}\text{ mol}}{2}$

$= 1.34 \times 10^{-3}$ mol

This is in 10.0 cm^3 of the initial reaction solution; so the 100 cm^3 reaction solution contains 1.34×10^{-2} mol of iodine.

This is produced in the reaction:

$$ClO^-(aq) + 2H^+(aq) + 2I^-(aq) \rightarrow I_2(aq) + Cl^-(aq) + H_2O(l)$$

So, there must be 1.34×10^{-2} mol of ClO^- in the original 10.0 g of the bleach.

This contains 1.34×10^{-2} mol of chlorine.

This has a mass of 1.34×10^{-2} mol $\times$ 35.5 g mol^{-1}, or 0.476 g.

So the percentage of chlorine in the bleach is

$\frac{0.476\text{ g}}{10.0\text{ g}} \times 100$,

or **4.76%**

Titration of copper(II) ions

The concentration of copper(II) ions in a sample can be found using an iodine/thiosulfate titration as part of a stepwise process.

Step 1: Add excess potassium iodide solution to a known volume of the solution containing copper(II) ions. Copper(I) iodide forms as a white precipitate, and iodine is formed and remains in solution:

$$2Cu^{2+}(aq) + 4I^-(aq) \rightarrow 2CuI(s) + I_2(aq)$$

Step 2: Titrate the iodine solution with standardised sodium thiosulfate solution as before to find the concentration of iodine in the solution.

Step 3: You can now work backwards to the concentration of copper(II) ions in the original sample.

Worked example

Percentage of copper in a substance

25.0 cm^3 of the solution made when 1.50 g of a copper alloy reacted with concentrated nitric acid is treated with excess potassium iodide solution. The iodine liberated is titrated against 0.100 mol dm^{-3} sodium thiosulfate solution. 20.5 cm^3 of sodium thiosulfate solution was needed. Calculate the percentage of copper in the alloy.

The amount of sodium thiosulfate

$= \frac{0.100}{1000} \times 20.5$ mol

$= 2.05 \times 10^{-3}$ mol

This reacts with the iodine according to:

$$2S_2O_3^{2-}(aq) + I_2(aq) \rightarrow S_4O_6^{2-}(aq) + 2I^-(aq)$$

So, the amount of iodine present is $\frac{2.05 \times 10^{-3}\text{ mol}}{2}$

$= 1.025 \times 10^{-3}$ mol

This is produced in the reaction:

$$2Cu^{2+}(aq) + 4I^-(aq) \rightarrow 2CuI(s) + I_2(aq)$$

So, the amount of Cu^{2+} ions

$= 1.025 \times 10^{-3}$ mol $\times 2$

$= 2.05 \times 10^{-3}$ mol

Mass of copper in alloy $= 2.05 \times 10^{-3}$ mol $\times 63.5$ g mol^{-1} = 0.130 g

So, the percentage of copper $= \frac{0.130\text{ g}}{1.50\text{ g}} \times 100$

= 8.68%

Percentage of copper in an alloy

You may be asked to find the percentage of copper in an alloy – and you will see that the worked example above is a simplification of the procedure.

fig. 2.1.8 Excess nitric acid oxidises iodide ions to iodine and produces fumes of harmful nitrogen dioxide gas.

First a weighed amount of the alloy is dissolved in concentrated nitric acid. However, the excess nitric acid oxidises iodide ions to iodine, which will affect the validity of the result. So the solution is neutralised with sodium carbonate solution. This causes some precipitation of the copper ions so a minimal amount of ethanoic acid is added to the mixture to keep the solution acidic enough for the copper ions to remain in solution, but not acidic enough for the iodide ions to be oxidised.

Excess potassium iodide is added and then the iodine liberated is back-titrated with standardised sodium thiosulfate solution. From this the mass of copper in the alloy can be found.

Questions

1 When finding the amount of copper in an alloy, suggest why it is necessary to neutralise the solution before titrating with sodium thiosulfate.

2 A sample of household bleach (2.00 cm^3) containing sodium chlorate(I) is added to excess potassium iodide solution. The iodine liberated is titrated with 0.050 mol dm^{-3} sodium thiosulfate solution. 32.4 cm^3 of this solution is required to discharge the colour of the solution using starch as an indicator. Calculate the strength of the bleach (ClO^-) in mol dm^{-3}.

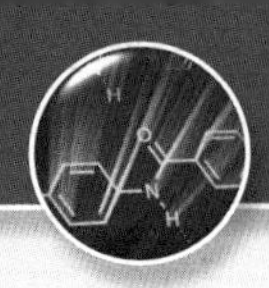

Measuring standard electrode potentials [5.3.1c]

HSW Ideas about electrochemistry

fig. 2.1.9 Luigi Galvani initially observed muscle contractions when he attached frogs to an iron railing by brass hooks.

In 1791, the Italian scientist Luigi Galvani observed that if he touched frogs' legs with two different types of metal then they twitched, even when they had been removed from the frog. He announced that he had discovered 'animal electricity'. Three years later, Alessandro Volta showed that this effect was nothing to do with living things, and that electricity can be produced whenever two metals are immersed in a conducting solution – which led to the development of the first battery. This development of ideas was the starting point for much of the chemistry you are going to be looking at in this chapter.

Competition for electrons

As you have seen, all redox reactions involve the transfer of electrons from one reactant to another. When metals react with non-metals, electrons are transferred from the metal to the non-metal every time, with the metal being oxidised and the non-metal being reduced. Non-metals always 'win' in the competition for electrons. However, the competition for electrons in redox reactions is not always this clear cut, particularly when two metals or complex ions are involved. So chemists have developed the idea of **electrode potentials**, which can be used to determine if a redox reaction will take place between two systems, and which direction it will go in.

Electricity from chemical change

If a piece of zinc metal is placed in a solution of copper(II) sulfate, the blue colour of the copper(II) sulfate fades slowly, the zinc dissolves and pink copper metal takes its place. The ionic equation for the overall reaction is:

$$Zn(s) + Cu^{2+}(aq) \rightarrow Zn^{2+}(aq) + Cu(s)$$

You can show that this is a redox reaction if you separate it into two half-equations:

$Zn(s) \rightarrow Zn^{2+}(aq) + 2e^-$ oxidation

$Cu^{2+}(aq) + 2e^- \rightarrow Cu(s)$ reduction

This reaction releases energy, which is lost as heat if the process is carried out in a single reaction vessel. However, by separating the reactions into two **half-cells**, we can harness the energy through the flow of electrons that takes place from one half-cell to the other. This combination of half-cells is called an **electrochemical cell**.

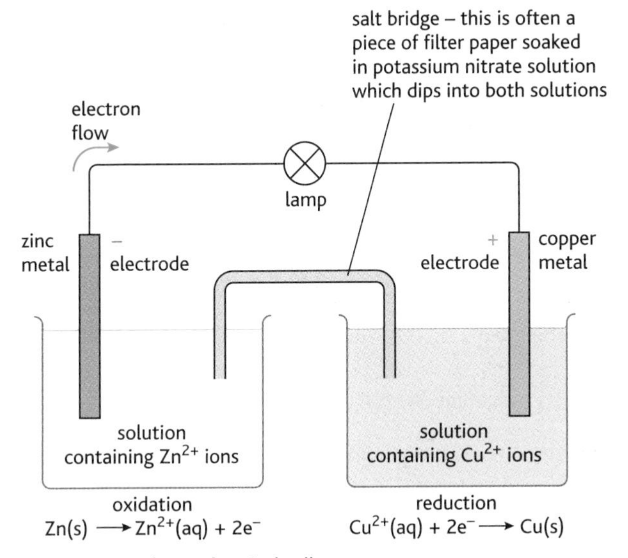

fig. 2.1.10 An electrochemical cell.

In **fig. 2.1.10** the copper electrode slowly increases in mass as copper ions leave the solution around it (copper ions are reduced), while the zinc electrode slowly decreases in mass as the zinc forms zinc ions (it is oxidised). As this happens, electrons flow along the wire joining the two electrodes – this flow of current can be detected using a small lamp.

The lamp in **fig. 2.1.10** can be replaced by a high-resistance voltmeter and the maximum potential difference across the cell – its **electromotive force** or e.m.f. – can be measured. The e.m.f. of a cell can be thought of as the 'push' that it is able to provide to a current flowing through it – an e.m.f. is measured in volts. When you look at cells as energy sources later in this chapter, you will see that the e.m.f. of a cell is related to the maximum amount of useful work we can get out of it.

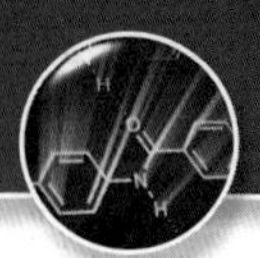

SC If you look at the diagram of a simple electrochemical cell in **fig. 2.1.10**, you will see that a **salt bridge** is present, consisting of filter paper soaked in potassium nitrate dipping into the solutions in both half-cells. What does this salt bridge do? Develop an idea about its role and then do some research to find out if your idea is correct.

Half-cell notation

Chemists use a system of 'shorthand' notation to represent the half-cells that make up an electrochemical cell. The zinc/copper cell would be represented as in **fig. 2.1.11**.

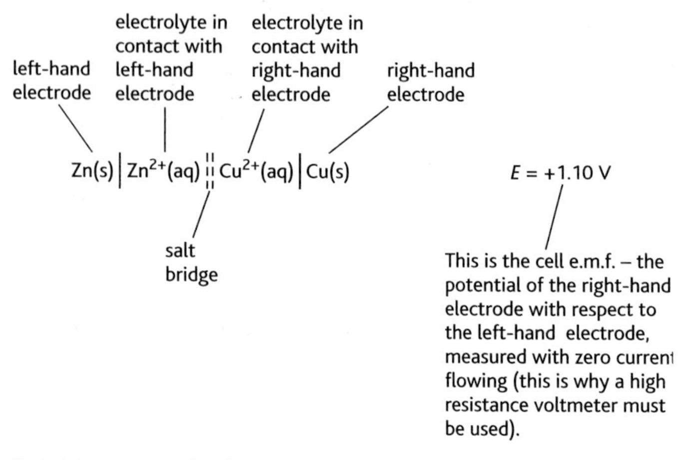

fig. 2.1.11 The notation for representing electrochemical cells.

Notice that when you measure the e.m.f. of this cell, the right-hand electrode is positive with respect to the left-hand electrode, because the electrons flow from the zinc to the copper. By convention, the cell notation always refers to the reaction taking place from left to right, which in this case is:

$Zn(s) \rightarrow Zn^{2+}(aq) + 2e^-$

$Cu^{2+}(aq) + 2e^- \rightarrow Cu(s)$

Overall:

$Zn(s) + Cu^{2+}(aq) \rightarrow Zn^{2+}(aq) + Cu(s)$

$E_{cell} = +1.10V$

If the cell is reversed – so that zinc is now the right-hand electrode and copper is the left-hand electrode – electrons still flow from zinc to copper, so now the cell e.m.f. is *negative*. The cell notation now refers to the process:

$Cu(s) \rightarrow Cu^{2+}(aq) + 2e^-$

$Zn^{2+}(aq) + 2e^- \rightarrow Zn(s)$

Overall:

$Cu(s) + Zn^{2+}(aq) \rightarrow Cu^{2+}(aq) + Zn(s)$

$E_{cell} = -1.10V$

The result of this convention for representing electrochemical cells is that a change which occurs spontaneously has a positive e.m.f., while a change which does not occur spontaneously has a negative e.m.f.

Electrode potentials

When we measure the e.m.f. of a cell, we imagine that it arises from the competition for electrons between the two half-cells. Each half-cell reaction will have its own tendency to attract electrons, a tendency that is measured by the electrode potential of the half-cell. The more positive a half-cell's electrode potential is, the greater its tendency to attract electrons.

This is more clearly seen if we arrange the half-cells in **fig. 2.1.12** in a diagram to show their potentials relative to one another.

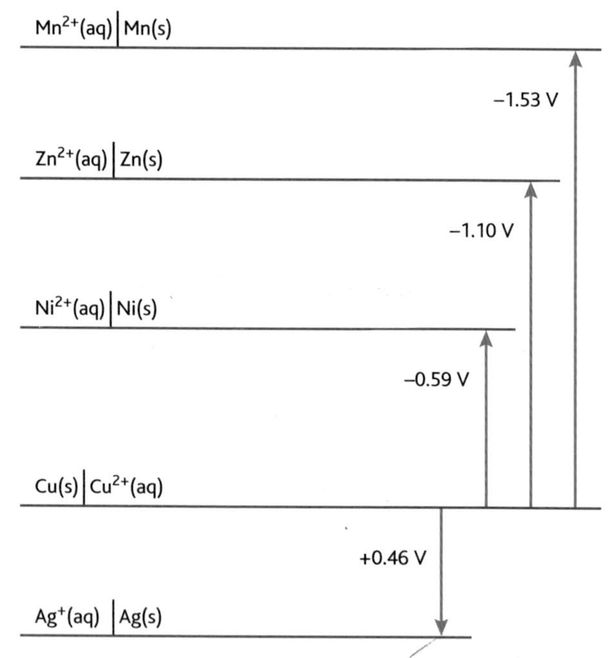

fig. 2.1.12 The relationships between electrode potentials. The relationship of each half-cell is compared to that of the Cu(s)/Cu^{2+}(aq) cell.

In **fig. 2.1.12** the half-cells which tend to attract electrons most strongly are at the bottom of the diagram. So Ag^+(aq) is the strongest oxidising agent out of these substances because it is the strongest competitor for electrons. The manganese half-cell is at the top of the diagram, so Mn(s) is the strongest reducing agent out of these substances because it is the best species at giving away electrons.

When we measure the e.m.f. of a cell, we are comparing the electrode potentials of the two half-cells that make up the electrochemical cell. Clearly, we can never measure the electrode potential of a lone cell – we can only compare it to the electrode potential of another cell. Using this idea and the data in **fig. 2.1.12**, we can calculate the e.m.f. of a cell made up of a $Ni(s) \mid Ni^{2+}(aq)$ half-cell and a $Ag^{+}(aq) \mid Ag(s)$ half-cell:

$$Ni(s) \mid Ni^{2+}(aq) \parallel Ag^{+}(aq) \mid Ag(s)$$

The e.m.f. of this cell is found from:

$$E_{cell} = -(-0.59V) + 0.46V = +1.05V$$

The positive value of the e.m.f. tells us that the reaction:

$$Ni(s) + 2Ag^{+}(aq) \rightleftharpoons Ni^{2+}(aq) + 2Ag(s)$$

proceeds spontaneously in the forward direction. This agrees with our previous deduction that the more reactive metals higher up our table (like Ni) are reducing agents, capable of losing electrons and so being oxidised by ions at the bottom of the table (like Ag^{+}).

Standard electrode potentials

It is impossible to find electrode potentials for individual half-cells. We have to measure the e.m.f. of a cell and attribute part of this to each electrode reaction. To make sure that all of the electrode potentials we use are measured in the same way, we choose to use a reference electrode to compare measurements against – we call this a **reference half-cell**. The reference half-cell is used to measure all other electrode potentials. It is assigned an electrode potential of 0.00V under standard conditions of 1 atm, 298 K (25°C) and solution concentrations of $1.00\,mol\,dm^{-3}$. This measurement of electrode potentials against a standard is similar to the way in which heights above sea level are measured in relation to one agreed point.

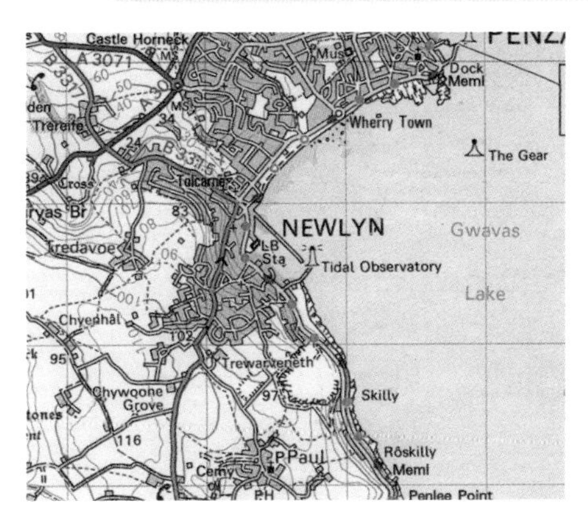

fig. 2.1.13 The reference point against which all other heights are judged is mean sea level at Newlyn, Cornwall, where there is an Ordnance Survey Tidal Laboratory (this is marked on the map). Heights above this level are taken as positive, those below it are negative. A similar principle is used in measuring standard electrode potentials.

The reference half-cell used in practice is the **standard hydrogen electrode**. This consists of hydrogen gas at 1 atm pressure and 298 K bubbling around a platinum electrode in $1.00\,mol\,dm^{-3}$ $H^{+}(aq)$ ions. The symbol used for the standard electrode potential is $E^{\ominus}$. The reaction which occurs in such a half-cell is written:

$$2H^{+}(aq, 1.00\,mol\,dm^{-3}) + 2e^{-} \rightleftharpoons H_2(298\,K, 1\,atm) \qquad E^{\ominus} = 0.00V$$

The double arrows in this equation show that the reaction is reversible – the direction in which the reaction goes will depend on the other half-cell to which the hydrogen half-cell is connected.

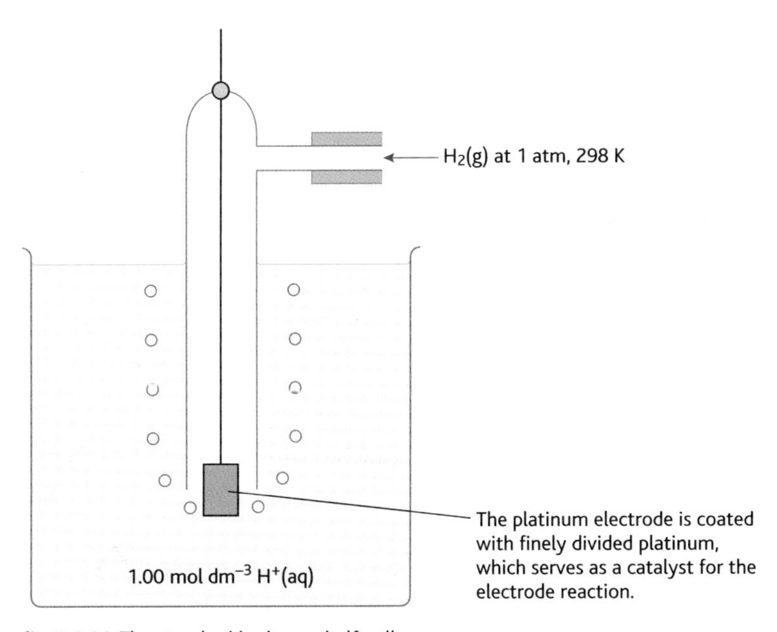

fig. 2.1.14 The standard hydrogen half-cell.

To measure the **standard electrode potential** of another half-cell – say $Cu(s) \mid Cu^{2+}(aq)$ – we need to connect this half-cell to a standard hydrogen half-cell under standard conditions. **Figure 2.1.15** shows this.

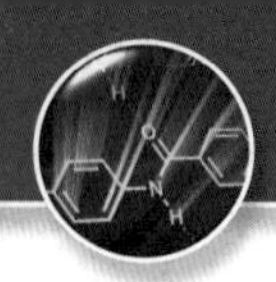

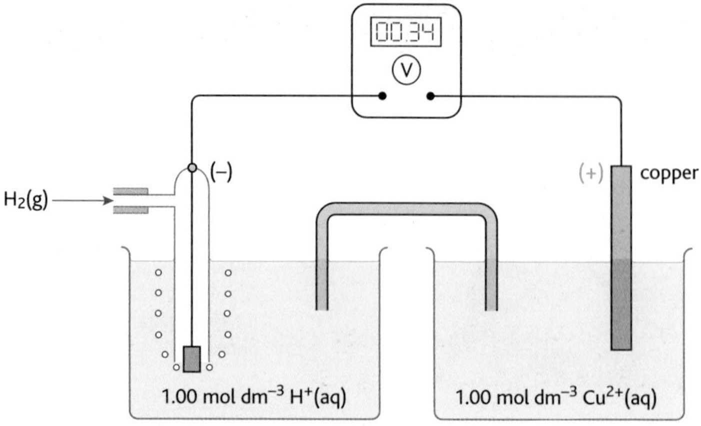

fig. 2.1.15 The electrochemical cell to measure the standard electrode potential of the Cu(s) | Cu^{2+}(aq) half-cell.

Measurement of the cell e.m.f. gives $E^\ominus$ = +0.34V, the copper half-cell being positive with respect to the hydrogen half-cell. Now, the standard electrode potential of the hydrogen half-cell is zero, so the standard electrode potential of the system Cu(s) | Cu^{2+}(aq) must be +0.34V. In other words:

$Cu^{2+}(aq) + 2e^- \rightleftharpoons Cu(s) \quad E^\ominus = +0.34V$

In the same way as for the Cu(s) | Cu^{2+}(aq) system, we can measure the standard electrode potentials of the other systems we considered above:

$Mn^{2+}(aq) + 2e^- \rightleftharpoons Mn(s) \; E^\ominus = -1.18V$

$Zn^{2+}(aq) + 2e^- \rightleftharpoons Zn(s) \quad E^\ominus = -0.76V$

$Ni^{2+}(aq) + 2e^- \rightleftharpoons Ni(s) \quad E^\ominus = -0.25V$

$Ag^+(aq) + e^- \rightleftharpoons Ag(s) \quad E^\ominus = +0.80V$

Notice that, even though we are not measuring the electrode potentials against the copper half-cell, the *differences* between electrode potentials remain the same. So the e.m.f. of the cell:

Ni(s) | Ni^{2+}(aq) ‖ Ag^+(aq) | Ag(s)

is given by:

$E^\ominus = +0.80V - (-0.25V)$

$= +1.05V$

This is exactly the same as the e.m.f. from our previous calculation on this electrochemical cell against the Cu/Cu^{2+} 'standard'.

Conditions and conventions in electrochemistry

The conditions chosen to measure standard electrode potentials are the same as those for other standard measurements (for example $\Delta H^\ominus_{298}$):

- all solutions have unit activity (effectively 1.00 mol dm^{-3} for our purposes)
- all measurements are made at 1 atmosphere pressure
- all measurements are made at 298 K (25°C).

In addition, we must also specify a way of making measurements for half-cells in which there are reactions such as:

$Fe^{3+}(aq) + e^- \rightleftharpoons Fe^{2+}(aq)$

Electrical connection is made through the use of an inert platinum electrode dipping into the solution. The half-cell is then represented as:

Fe^{3+}(aq), Fe^{2+}(aq) | Pt(s)

For any cell containing an inert electrode, the convention is to show the least oxidised (most reduced) species in the half-cell next to the electrode. So the hydrogen half-cell is represented as:

$2H^+$(aq), H_2(g) | Pt(s)

Another convention applies when writing redox reactions for half-cells. These are written in such a way that the species with the higher oxidation number appears on the left-hand side of the half-equation – as in the half-equation involving Fe^{3+} and Fe^{2+} above. The result of this convention is that the standard e.m.f. of any cell can be found from the relationship:

$E^\ominus_{cell} = E^\ominus$(right-hand electrode) – $E^\ominus$(left-hand electrode)

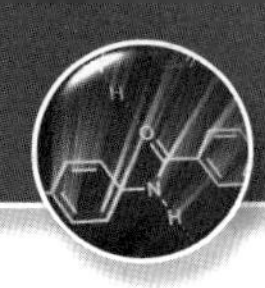

HSW Batteries – easily accessible energy

Alessandro Volta disproved Galvani's ideas about animal electricity by showing that the electricity did not come from the animal tissue, but from the brass and iron coming into contact with each other in the presence of an electrolyte. He invented his famous voltaic piles – stacks of metal discs made of zinc and copper or silver, separated by card, paper or cloth soaked in salty water – which produced electric current. These were the first wet-cell batteries, and the electricity they produced was generated by the potential difference between the different metals in his piles. Ever since the 1780s, when Volta was working, we have developed several different types of battery – they are as useful and popular as ever, but are no longer substantial piles of metal discs.

Batteries store energy in the form of chemicals, the energy being released when a conductor is connected between the terminals of the battery. Batteries are composed of cells, connected together in series to give a larger e.m.f. A car battery consists of six cells, each of which has an e.m.f. of 2 V – so the total e.m.f. of the battery is 6 × 2 V= 12 V.

The lead–acid battery used in a car is an example of a secondary cell. When the chemicals inside one of these batteries have reacted and no more electricity can be drawn from it (it is 'flat'), the battery can be recharged by passing a current through it in the opposite direction. A primary cell cannot be recharged in this way. The half-equations for a lead–acid cell as it discharges are:

Cathode: $PbO_2(s) + 4H^+(aq) + SO_4^{2-}(aq) + 2e^- \rightarrow PbSO_4(s) + 2H_2O(l)$

Anode: $Pb(s) + SO_4^{2-}(aq) \rightarrow PbSO_4(s) + 2e^-$

Overall: $PbO_2(s) + Pb(s) + 4H^+(aq) + 2SO_4^{2-}(aq) \rightarrow 2PbSO_4(s) + 2H_2O(l)$

Notice how this reaction consumes sulfuric acid. When a lead–acid battery discharges, the concentration of sulfuric acid in it decreases. This provides a way of measuring the state of charge of a battery – a hydrometer can be used to measure the density of the electrolyte in it. This gives an indication of the concentration of the acid. The change can be reversed (recharging the battery) by passing an appropriate current through the cell:

$$2PbSO_4(s) + 2H_2O(l) \rightarrow PbO_2(s) + Pb(s) + 4H^+(aq) + 2SO_4^{2-}(aq)$$

Today's car batteries are big and heavy – but they last a long time and can be recharged easily.

There is a wide range of available cells – summarised in **table 2.1.1.** Scientists are working on even better, lighter, longer-lasting cells – you will be looking at these later in the chapter.

Cell	Reactions as cell discharges	Uses
Alkaline dry cell (primary cell – not rechargeable) e.m.f. ≈ 1.54 V	$Zn(s) + 2OH^-(aq) \rightarrow ZnO(s) + H_2O(l) + 2e^-$ (anode) $2MnO_2(s) + H_2O(l) + 2e^- \rightarrow Mn_2O_3(s) + OH^-(aq)$ (cathode)	Radios, cassette players, torches – this type of cell can supply quite large currents for long periods
Nickel–cadmium (secondary cell – rechargeable) e.m.f. ≈ 1.4 V	$Cd(s) + 2OH^-(aq) \rightarrow Cd(OH)_2(s) + 2e^-$ (anode) $NiO_2(s) + 2H_2O(l) + 2e^- \rightarrow Ni(OH)_2(s) + 2OH^-(aq)$ (cathode)	As rechargeable batteries where a lead–acid cell would be too bulky – radios, torches etc.
Mercury (primary cell – not rechargeable) e.m.f. ≈ 1.35 V	$Zn(s) + 2OH^-(aq) \rightarrow ZnO(s) + H_2O(l) + 2e^-$ (anode) $HgO(s) + H_2O(l) + 2e^- \rightarrow Hg(l) + 2OH^-(aq)$ (cathode)	Calculators and cameras – where small size and light weight is important
Silver oxide (primary cell – not rechargeable) e.m.f. ≈ 1.5 V	$Zn(s) + 2OH^-(aq) \rightarrow Zn(OH)_2(s) + 2e^-$ (anode) $Ag_2O(s) + H_2O(l) + 2e^- \rightarrow 2Ag(s) + 2OH^-(aq)$ (cathode)	Watches, miniature cameras and calculators – very small size

table 2.1.1 Some of the many types of cells available.

Competition for electrons

The standard electrode potential of a half-cell is a measure of the oxidising power or reducing power of the species in it – in other words, the ability to compete for electrons. In general the stronger an oxidising agent, the more positive its electrode potential – so ozone gas is a more powerful oxidant than chlorine gas:

$$O_3(g) + 2H^+(aq) + 2e^- \rightleftharpoons O_2(g) + H_2O(l) \quad E^\ominus = +2.08V$$

$$Cl_2(g) + 2e^- \rightleftharpoons 2Cl^-(aq) \quad E^\ominus = +1.36V$$

A strong reducing agent has a large negative electrode potential – so calcium metal is a more powerful reducing agent than lead metal:

$Pb^{2+}(aq) + 2e^- \rightleftharpoons Pb(s) \quad E^\ominus = -0.13V$

$Ca^{2+}(aq) + 2e^- \rightleftharpoons Ca(s) \quad E^\ominus = -2.87V$

From the equations for these reversible reactions it follows that, for example:

$Pb(s) \rightleftharpoons Pb^{2+}(aq) + 2e^- \quad E^\ominus = +0.13V$

$Ca(s) \rightleftharpoons Ca^{2+}(aq) + 2e^- \quad E^\ominus = +2.87V$

since reversing the direction in which we write the reaction (electrons on the right rather than on the left) means that we must reverse the sign of the standard electrode potential too.

Half-reaction	$E^\ominus$/V
$Li^+(aq) + e^- \rightleftharpoons Li(s)$	−3.03
$Na^+(aq) + e^- \rightleftharpoons Na(s)$	−2.71
$Al^{3+}(aq) + 3e^- \rightleftharpoons Al(s)$	−1.66
$Zn^{2+}(aq) + 2e^- \rightleftharpoons Zn(s)$	−0.76
$Fe^{2+}(aq) + 2e^- \rightleftharpoons Fe(s)$	−0.44
$Ni^{2+}(aq) + 2e^- \rightleftharpoons Ni(s)$	−0.25
$Pb^{2+}(aq) + 2e^- \rightleftharpoons Pb(s)$	−0.13
$2H^+(aq) + 2e^- \rightleftharpoons H_2(g)$	0.00
$Cu^{2+}(aq) + e^- \rightleftharpoons Cu^+(aq)$	+0.15
$Cu^{2+}(aq) + 2e^- \rightleftharpoons Cu(s)$	+0.34
$Cu^+(aq) + e^- \rightleftharpoons Cu(s)$	+0.52
$I_2(aq) + 2e^- \rightleftharpoons 2I^-(aq)$	+0.54
$Fe^{3+}(aq) + e^- \rightleftharpoons Fe^{2+}(aq)$	+0.77
$Ag^+(aq) + e^- \rightleftharpoons Ag(s)$	+0.80
$Br_2(aq) + 2e^- \rightleftharpoons 2Br^-(aq)$	+1.09
$Cl_2(aq) + 2e^- \rightleftharpoons 2Cl^-(aq)$	+1.36
$MnO_4^-(aq) + 8H^+(aq) + 5e^- \rightleftharpoons Mn^{2+}(aq) + 4H_2O(l)$	+1.51
$H_2O_2(aq) + 2H^+(aq) + 2e^- \rightleftharpoons 2H_2O(l)$	+1.77
$O_3(g) + 2H^+(aq) + 2e^- \rightleftharpoons O_2(g) + H_2O(l)$	+2.08
$F_2(g) + 2e^- \rightleftharpoons 2F^-(aq)$	+2.87

table 2.1.2 The standard electrode potentials of some half-cells.

Table 2.1.2 shows the standard electrode potentials for a number of half-reactions. Substances to the *right* of the equilibrium sign in the table are reducing agents, becoming oxidised when they react in the direction right to left. The strongest reducing agents are located at the top left of the table – lithium is the strongest in this table. Substances to the *left* of the equilibrium sign are oxidising agents, becoming reduced when they react in the direction left to right. The strongest oxidising agents are located at the bottom right of the table – fluorine is the strongest in this table.

Note that the e.m.f. at an electrode is independent of the number of electrons being transferred – so, for example:

$Zn^{2+}(aq) + 2e^- \rightleftharpoons Zn(s) \quad E^\ominus = -0.76V$

$2Zn^{2+}(aq) + 4e^- \rightleftharpoons 2Zn(s) \quad E^\ominus = -0.76V$

$Cl_2(aq) + 2e^- \rightleftharpoons 2Cl^-(aq) \quad E^\ominus = +1.36V$

$\frac{1}{2}Cl_2(aq) + e^- \rightleftharpoons Cl^-(aq) \quad E^\ominus = +1.36V$

Questions

1 What is meant by these terms:
 a electromotive force
 b standard hydrogen electrode
 c half-cell
 d standard electrode potential?

2 Write half-cell equations for each of the following redox reactions:
 a $2Ag^+(aq) + Cu(s) \rightarrow 2Ag(s) + Cu^{2+}(aq)$
 b $Cl_2(aq) + 2I^-(aq) \rightarrow I_2(aq) + 2Cl^-(aq)$
 c $Zn(s) + Pb^{2+}(aq) \rightarrow Zn^{2+}(aq) + Pb(s)$

3 Explain how the standard electrode potential ($E^\ominus$) of a half-cell relates to the oxidising and reducing power of the species.

4 Calculate the standard electrode potentials for the following cells from the data in **table 2.1.2**. For each cell write down the reaction that occurs when current is drawn from the cell.
 a $Ni(s) \mid Ni^{2+} \parallel Pb^{2+}(aq) \mid Pb(s)$
 b $Pt \mid Fe^{2+}(aq), Fe^{3+}(aq) \parallel Ag^+(aq) \mid Ag(s)$

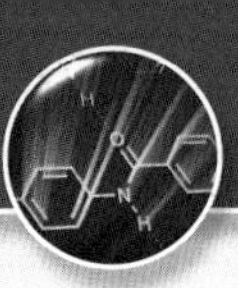

Predicting the thermodynamic feasibility and the extent of reactions [5.3.1d, e, f]

We have just seen how standard electrode potentials can be obtained. These can be used to give guidance about whether particular reactions are likely to happen or not.

Using standard electrode potentials

When comparing two redox systems, a reaction will take place if:

- the stronger reducing agent with the more negative $E^\ominus$ is on the right-hand side of the equation
- the stronger oxidising agent with the more positive $E^\ominus$ is on the left-hand side of the equation.

For a reaction to be feasible, the $E^\ominus_{cell}$ value must be positive, and once the value reaches about +0.6V, the reaction is extremely likely to progress.

$E^\ominus_{cell}$ is directly proportional to the total **entropy** change in a reaction. The more positive the value of $E^\ominus_{cell}$, the more energetically favourable is the reaction.

To make a prediction about a reaction, you need to write down the two half-equations for it. For example, suppose you want to know if aqueous iodine can oxidise bromide ions to aqueous bromine.

Step 1: Write down the equation for the reaction:

$I_2(aq) + 2Br^-(aq) \rightarrow 2I^-(aq) + Br_2(aq)$

Step 2: Split the equation into two half-equations (taking care with the direction of each reaction) and write down the electrode potentials:

$I_2(aq) + 2e^- \rightarrow 2I^-(aq) \quad E^\ominus = +0.54V$
$2Br^-(aq) \rightarrow Br_2(aq) + 2e^- \quad E^\ominus = -1.09V$

(Notice that you reverse the sign of the $Br_2(aq)/Br^-(aq)$ half-cell and write the electrons on the *right* instead of the left.)

Step 3: Add the half-reactions (you should get the equation in step 1) and the electrode potentials together:

$I_2(aq) + 2Br^-(aq) \rightarrow 2I^-(aq) + Br_2(aq)$
$E^\ominus_{cell} = +0.54V + (-1.09V)$
$= -0.55V$

The *negative* value of the cell e.m.f. indicates that the reaction will *not* occur spontaneously – in fact, it will occur in the *reverse* direction and bromine will oxidise iodide ions to iodine:

$Br_2(aq) + 2I^-(aq) \rightarrow 2Br^-(aq) + I_2(aq)$
$E^\ominus_{cell} = +1.09V + (-0.54V)$
$= +0.55V$

fig. 2.1.16 How will these different halogens react in solution? Standard electrode potentials give us one way of predicting the outcome of reactions.

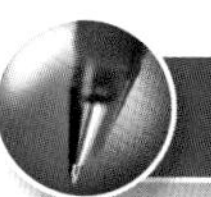

Worked example

Using electrode potentials to predict the likelihood of reactions

Will bromide ions be oxidised by aqueous chlorine?

Step 1: $Cl_2(aq) + 2Br^-(aq) \rightarrow 2Cl^-(aq) + Br_2(aq)$

Step 2: $Cl_2(aq) + 2e^- \rightarrow 2Cl^-(aq) \quad E^\ominus = +1.36V$
$2Br^-(aq) \rightarrow Br_2(aq) + 2e^- \quad E^\ominus = -1.09V$

Step 3: $Cl_2(aq) + 2Br^-(aq) \rightarrow 2Cl^-(aq) + Br_2(aq)$
$E^\ominus_{cell} = +1.36V + (-1.09V)$
$= +0.27V$

The positive cell e.m.f. shows that this reaction is spontaneous in the forward direction – so **chlorine in solution will oxidise bromide ions** and the solution will turn brown.

Is there a rule that helps?

We will use the following half-equations to see if there is a link between the size of $E^\ominus_{cell}$ values and the likelihood of a reaction taking place.

$Zn^{2+}(aq) + 2e^- \rightarrow Zn(s) \quad E^\ominus = -0.76V$
$Fe^{2+}(aq) + 2e^- \rightarrow Fe(s) \quad E^\ominus = -0.44V$
$Cu^{2+}(aq) + 2e^- \rightarrow Cu(s) \quad E^\ominus = +0.34V$
$Fe^{3+}(aq) + e^- \rightarrow Fe^{2+}(aq) \quad E^\ominus = +0.77V$
$Ag^+(aq) + e^- \rightarrow Ag(s) \quad E^\ominus = +0.80V$

For example, for the reaction:

$$Fe(s) + Cu^{2+}(aq) \rightarrow Fe^{2+}(aq) + Cu(s)$$

$$E^{\ominus}_{cell} = (+0.44V) + (+0.34V) = +0.78V$$

This is a fairly large positive value, so the probability of the reaction going ahead is high.

For the reaction:

$$Fe^{2+}(aq) + Ag^{+}(aq) \rightarrow Fe^{3+}(aq) + Ag(s)$$

$$E^{\ominus}_{cell} = (-0.77V) + (+0.80V) = +0.03V$$

This is a much smaller positive value and the reaction is less likely to go ahead compared to the one above.

For the reaction:

$$Fe^{3+}(aq) + Ag(s) \rightarrow Fe^{2+}(aq) + Ag^{+}(aq)$$

$$E^{\ominus}_{cell} = (+0.77V) + (-0.80V) = -0.03V$$

This is a negative value, so the probability of the reaction taking place is relatively low.

When you observe these three reactions, these predictions are found to be true. An $E^{\ominus}_{cell}$ value can tell you more than if the reaction is likely to happen or not – it also gives you some idea of the extent of the reaction. As a general principle:

- if $E^{\ominus}_{cell} > +0.6V$ then the reaction should go to completion
- if $0 < E^{\ominus}_{cell} < +0.6V$ then the products predominate
- if $-0.6V < E^{\ominus}_{cell} < 0$ then the reactants predominate
- if $E^{\ominus}_{cell} < -0.6V$ then there is no reaction.

It is very important to realise that $E^{\ominus}_{cell}$ is related to the *probability* that a reaction will take place. It is not affected by the quantity of materials in the reacting mixture. The reaction will happen – or will not happen – regardless of whether you have a few milligrams of the reactants or several tonnes.

Figure 2.1.17 shows that the rusting of iron is a redox reaction.

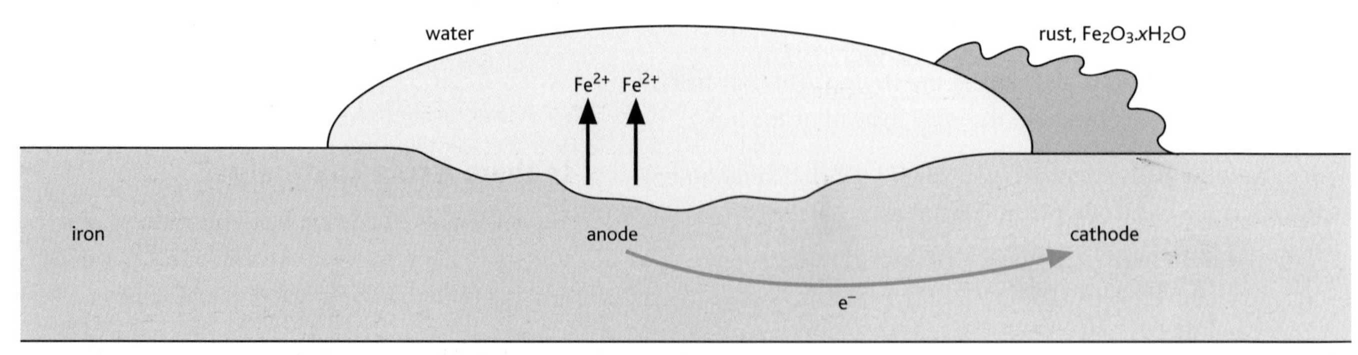

fig. 2.1.17 The rusting of iron in the presence of water is a redox reaction. During rusting Fe^{2+} is oxidised to Fe^{3+}.

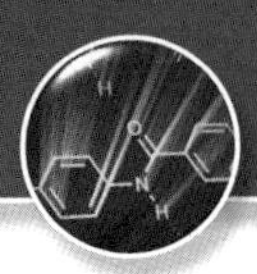

HSW Redox reactions and energy in living cells

As food is broken down in the cells of your body, electrons are removed from glucose and other food molecules. In the mitochondria of the cells these electrons are transported along a series of molecules, which make up the electron transfer chain, in a process called cellular respiration.

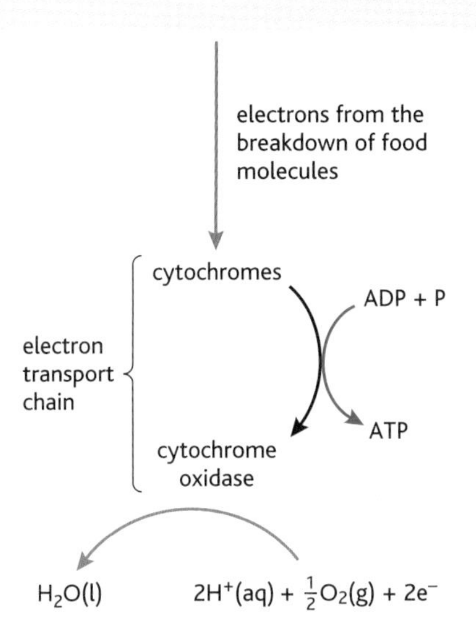

The various components of the chain have different electrode potentials, with the first member having the most negative electrode potential, making it a relatively weak oxidising agent. As the electrons pass along the chain, the electrode potential of the components of the chain increases and their oxidising power increases. At the end of the chain, oxygen oxidises the final member of the chain, forming water in the process. The transport of electrons along the chain releases energy, which is used to drive the synthesis of a molecule called ATP (adenosine triphosphate), the universal 'energy carrier molecule' in living things. This series of redox reactions takes place in almost all living things.

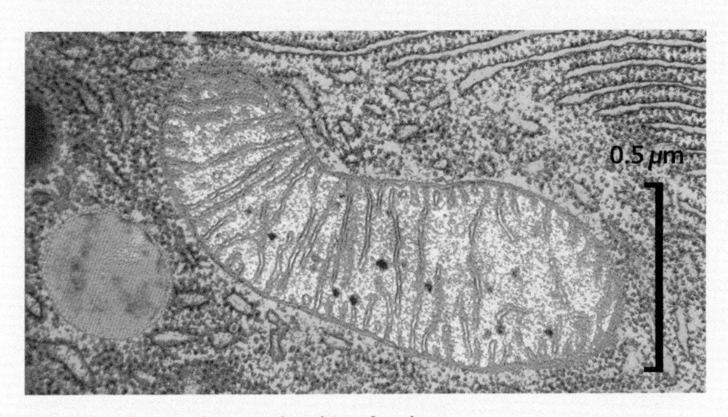

fig. 2.1.18 The electron transfer chain – the redox reactions in this chain take place in the mitochondria, often known as 'the powerhouse of the cell'.

Limitations of standard electrode potentials

Just like enthalpy changes (see chapter 1.2 in *Edexcel AS Chemistry*), standard electrode potentials must be used with care when predicting the likelihood of a reaction. Both ΔH and $E^\ominus$ are concerned with the **energetic stability** of a substance, not its **kinetic stability** – so they can tell you whether a reaction is *possible*, but not whether it happens fast enough to be noticeable! The oxidation of iron(II) salts in acidic solutions is an example of this:

$$Fe^{2+}(aq) \rightleftharpoons Fe^{3+}(aq) + e^- \qquad E^\ominus = -0.77V$$

$$O_2(g) + 4H^+(aq) + 4e^- \rightleftharpoons 2H_2O(l) \quad E^\ominus = +1.23V$$

Overall: $4Fe^{2+}(aq) + O_2(g) + 4H^+(aq) \rightarrow 4Fe^{3+}(aq) + 2H_2O(l)$

$$E^\ominus_{cell} = (-0.77V) + (+1.23V) = +0.46V$$

The positive value suggests that the reaction should proceed spontaneously. However, the oxidation of Fe^{2+} under these conditions is extremely slow because kinetic factors increase the stability of Fe^{2+} in aqueous solutions.

Another important point about standard electrode potentials is that the conditions under which the reaction occurs may be very different from standard conditions (25°C, 1 atm pressure and concentration 1 mol dm^{-3}). Temperature and concentrations are two factors that may be particularly important in influencing a reaction. For example in the equilibrium:

$$Cu^{2+}(aq) + 2e^- \rightleftharpoons Cu(s)$$

increasing the concentration of $Cu^{2+}(aq)$ ions will move the equilibrium to the right (opposing the change), reducing the new number of $Cu^{2+}(aq)$ ions and removing electrons from the system. This is turn makes the electrode potential more positive.

In another example, photographic film uses a light-sensitive emulsion containing silver bromide. The silver in the emulsion comes from silver ingots, which are dissolved in nitric acid:

$$2Ag(s) + 2NO_3^-(aq) + 4H^+(aq) \rightleftharpoons 2Ag^+(aq) + N_2O_4(g) + 2H_2O(l)$$

The half-equations and standard electrode potentials for this reaction are:

$2Ag(s) \rightarrow 2Ag^{+}(aq) + 2e^{-} \quad E^{\ominus} = -0.80V$

$2NO_3^{-}(aq) + 4H^{+}(aq) + 2e^{-} \rightarrow N_2O_4(g) + 2H_2O(l)$
$E^{\ominus} = +0.80V$

So $E^{\ominus}_{cell}$ is 0.0V and this indicates that there should be no reaction between these two substances. However, if the temperature is raised to 60°C, the two electrode potentials become –0.765V for the Ag/Ag^{+} reaction and +0.804V for the $(2NO_3^{-} + 4H^{+})/(N_2O_4 + H_2O)$ reaction, so silver is oxidised under these conditions because it has a more positive electrode potential. Besides raising the temperature, the reaction is also made more favourable by using concentrated acid. This increases the hydrogen ion concentration, pushing the reaction to the right.

So remember that while standard electrode potentials are very useful, they have to be used with care:

- A reaction with a positive $E^{\ominus}_{cell}$ value may not actually take place. Predictions can be made as to the feasibility of these equilibrium reactions but they give no indication of the rate, which may be very slow – because of a high activation energy, for example.
- The conditions used may be different from standard conditions and so the actual electrode potentials will differ from the standard electrode potentials.
- Standard electrode potentials apply to aqueous equilibria. There are many reactions that take place that are not in aqueous solutions.

You saw in **chapter 1.2** that reactions are always possible if ΔS_{total} is positive. It is important to be able to link this to this work on the magnitude and sign of $E^{\ominus}_{cell}$ values.

	Reaction does not take place	Reaction to the left	Equal amounts of reactants and products	Reaction to the right	Reaction to completion
K_c	$< 10^{-10}$	0.01	1	100	$>10^{10}$
$E^{\ominus}$	< –0.6	approx. –0.1	0	approx. +0.1	> +0.6
ΔS_{total}	< –200	approx. –40	0	approx. +40	> +200

table 2.1.3 A summary of the likely results for the reaction of zinc with copper(II) ions. For simplicity units are not shown.

For the reaction of zinc with copper(II) ions, the equation is:

$Zn(s) + Cu^{2+}(aq) \rightleftharpoons Zn^{2+}(aq) + Cu(s)$
$K_c = 1.9 \times 10^{37}$

The equilibrium is well over to the right, so the reaction goes almost to completion.

Using the relationship $\Delta S_{total} = R\ln K_c$ we can work out that $\Delta S_{total} = +707\,J\,mol^{-1}\,K^{-1}$. **Table 2.1.3** summarises the likely results.

Even reactions that have very large K_c values (such as the combustion of sugar) may be slow at room temperature because the activation energy is very high.

Questions

1 a Calculate $E^{\ominus}_{cell}$ for the reaction:
$Cu(s) + 2Ag^{+}(aq) \rightarrow 2Ag(s) + Cu^{2+}(aq)$
b From this decide whether the reaction is likely to be feasible.

2 The equation summarises the reaction of silver(I) ions and iron(II) ions.
$Ag^{+}(aq) + Fe^{2+}(aq) \rightleftharpoons Fe^{3+}(aq) + Ag(s)$
K_c for the reaction is 3.2 at 25°C.
a What does the value of K_c tell you about the position of equilibrium?
b Does it tell you anything about the rate of the reaction?

3 Iron can be protected from rusting (see **fig. 2.1.17**) by putting blocks of zinc on the iron. Explain how this works in terms of electrochemistry.

4 a Use the data in **table 2.1.2** to decide which of these reactions will proceed spontaneously:
A: $Ni^{2+}(aq) + Fe(s) \rightarrow Ni(s) + Fe^{2+}(aq)$
B: $I_2(aq) + 2Fe^{2+}(aq) \rightarrow 2I^{-}(aq) + 2Fe^{3+}(aq)$
C: $Pb^{2+}(aq) + 2Ag(s) \rightarrow Pb(s) + 2Ag^{+}(aq)$
D: $5H_2O_2(aq) + 2Mn^{2+}(aq) \rightarrow 2MnO_4^{-}(aq) + 2H_2O(l) + 6H^{+}(aq)$
b What are the limitations of using standard electrode potentials to predict the feasibility of these reactions?

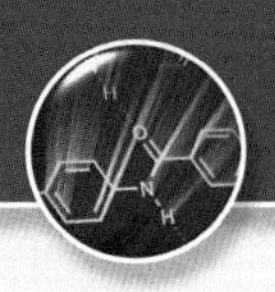

Hydrogen and alcohol fuel cells [5.3.1j]

Tests have shown that although mixtures of hydrogen and air are explosive, the low density of hydrogen means that it is safer than petrol or aviation fuel in an accident. Hydrogen burns in air to form water:

$$2H_2(g) + O_2(g) \rightarrow 2H_2O(l)$$

Small amounts of nitrogen oxides are also formed in a reaction between nitrogen and oxygen in the air at high temperatures.

HSW Is hydrogen the answer to global warming?

Global warming is an increasing problem in the twenty-first century. Scientific opinion suggests, ever more strongly, that the burning of fossil fuels is an important part of the problem. Could burning hydrogen provide a solution?

On 15 April 1988, the first passenger plane flew using a hydrogen-fuelled engine near Moscow. The Tupolev 155 (equivalent to an American Boeing 727) was equipped with two engines – one running on hydrogen, the other on jet fuel. The plane took off and landed on jet fuel, but hydrogen was used during the cruising phase of the flight.

Burning hydrogen as fuel in aircraft and cars is not likely to become the norm. The efficiency of the conversion of energy using a jet engine is only about 20% and although the water in the increased vapour trails produced is clean it too is a greenhouse gas and so adds to global warming!

The future of hydrogen as a fuel in aircraft and cars may well be directed to using hydrogen and oxygen in a **fuel cell**. The reaction is similar to combustion but takes place at a lower temperature, so no nitrogen oxides are formed – and the efficiency is almost 100%. It has been calculated that if 20% of the cars in the USA used hydrogen fuel cells it would cut oil imports by 1.5 million barrels a day.

fig. 2.1.19 There are many advantages to buses powered by hydrogen fuel cells. They are almost silent and produce zero carbon emissions, and are twice as efficient as diesel-powered buses.

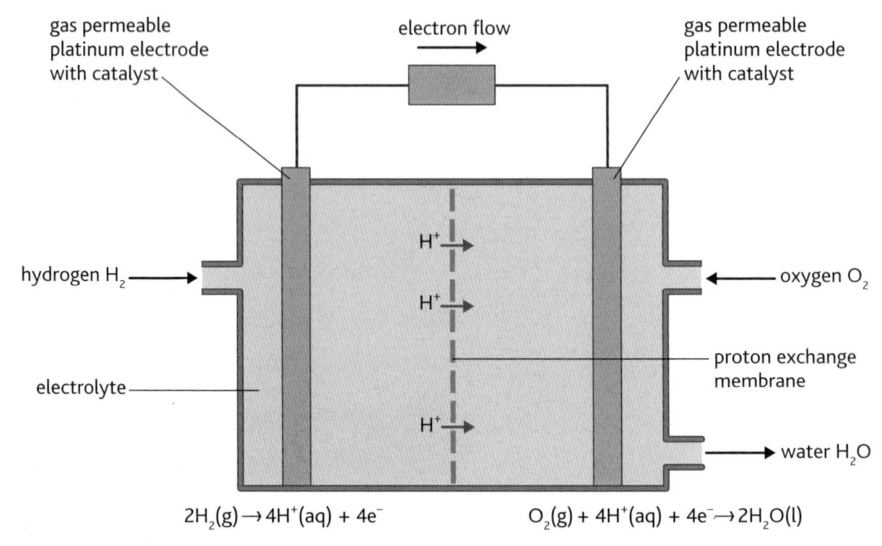

fig. 2.1.20 A fuel cell – reacting hydrogen with oxygen in this controlled way produces a useful source of energy.

How does a hydrogen fuel cell work?

Two electrodes are separated by a membrane, which allows hydrogen ions to pass through but not hydrogen and oxygen molecules. The electrons released by the reaction at the hydrogen electrode are used up at the oxygen electrode. The potential difference set up pushes electrons through the external circuit, where they do work.

The overall reaction for this redox process can be obtained by adding the two half-equations together:

$$2H_2(g) + O_2(g) \rightarrow 2H_2O(l)$$

Hydrogen fuel cells are used in the Space Shuttle, cars, motorbikes and buses and even experimental aeroplanes (see *Edexcel AS Chemistry*, pages 122–3). However, they need a supply of hydrogen. Most of the hydrogen used today is produced either from natural gas or from electrolysis of water. Both methods require energy and produce pollution, adding to global warming.

SC Find out more about the various ways in which hydrogen can be generated and the energy changes involved in the various methods.

Explain how this aspect of hydrogen cells could reduce their usefulness in preventing global warming.

Methanol-based fuel cells

Methanol is a highly flammable, poisonous liquid alcohol. It is made from either non-renewable fossil fuels or agricultural waste such as straw, municipal waste, wood and other **biomass**. It can also be produced by the chemical recycling of carbon dioxide – for example from fossil-fuel burning power plants, or even from the atmosphere itself. Methanol is much easier to store than hydrogen because it doesn't need high pressures or low temperatures – it is a liquid between –97 and +65°C.

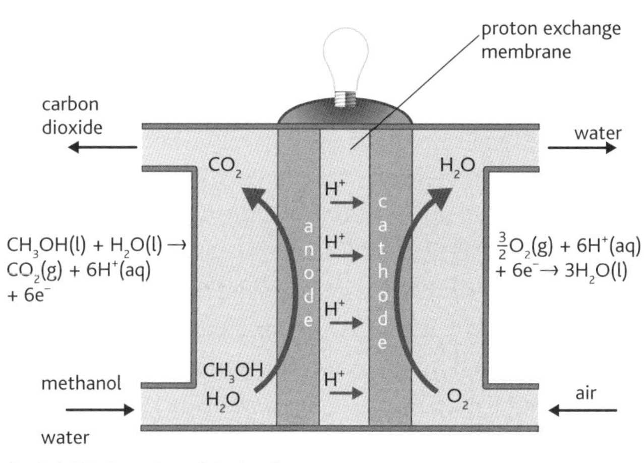

fig. 2.1.21 A methanol fuel cell.

The fuel cell operates at a temperature between 90 and 120°C and uses the oxidation of methanol on a catalyst layer to form carbon dioxide. The proton exchange membrane is often made from Nafion (a polymer with ionic groups attached). Electrons are transported through an external circuit from anode to cathode, providing power to connected devices. The overall equation for the reaction is:

$$CH_3OH(g) + \tfrac{3}{2}O_2(g) \rightarrow CO_2(g) + 2H_2O(l)$$

The amount of energy contained in a given volume of methanol is much greater than that in even highly compressed hydrogen. The waste products from methanol fuel cells are carbon dioxide and water. However, the efficiency of methanol fuel cells is low because the alcohol can pass through the available membrane materials. New polymers being developed should reduce this problem dramatically. Today methanol fuel cells produce only limited power, so they cannot power vehicles, but they are ideal for consumer goods such as mobile phones and laptop computers.

Ethanol-based fuel cell

Ethanol can be used in a fuel cell instead of the more toxic methanol. Ethanol is produced easily by fermentation of sugar cane, wheat, corn or even straw.

In this fuel cell, the oxidation of ethanol involves an expensive, platinum-based catalyst. Alternative catalysts are being developed that use cheaper, more reactive metals – such as iron, cobalt and nickel. A polymer effectively acts as an **electrolyte**. The charge is carried by hydrogen ions. The liquid ethanol is oxidised at the anode in the presence of water, generating carbon dioxide, hydrogen ions and electrons. Hydrogen ions travel through the electrolyte. They react with oxygen from the air and the electrons from the external circuit at the cathode, forming water. Useful power has been obtained from these cells at 25°C.

fig. 2.1.22 The refinery is using the maize to produce ethanol. Growing crops for biofuels takes up some of the carbon dioxide that was discharged into the atmosphere from the fuel used to produce the ethanol and from burning fuels.

On 13 May 2007, a team from the University of Applied Sciences in Offenburg made the first vehicle powered by an ethanol-based fuel cell. However, the development of direct methanol and ethanol fuel cells is generally lagging behind that of hydrogen fuel cells – in part because the hydrogen cell is much closer to providing an everyday solution to the problem. But in the future, alcohol-powered cells may provide the best solution, contributing to a reduction in global warming in both their production and their use.

Questions

1 a Write ionic half-equations for the reactions that take place at the anode and cathode in the ethanol fuel cell.

b Write the overall equation for the reaction.

2 Outline the advantages and disadvantages of hydrogen, methanol and ethanol fuel cells.

How breathalysers work [5.3.1k]

Drinking alcohol and driving a car should not be mixed. More than 20% of the people killed in road traffic accidents are over the drink–driving limit, and 32% have traces of alcohol in their system. The legal blood-alcohol limit for motorists in the UK at the time of writing is 80 mg per 100 ml blood, but there are calls to reduce this limit to match those of many other countries. Several different types of breath analyser (breathalysers) are used by the police to identify people who have been driving under the influence of alcohol.

Dichromate(VI) breath analysers

The first breath analysers consisted of:

- a system to sample the breath of the suspect
- two glass vials containing the chemical reaction mixture
- a system of photocells connected to a meter to measure the colour change associated with the chemical reaction.

The instrument uses the chemical reaction:

$$3C_2H_5OH(g) + 2Cr_2O_7^{2-}(aq) + 16H^+(aq) \rightarrow 3CH_3COOH(aq) + 4Cr^{3+}(aq) + 11H_2O(l)$$

The suspect breathes into the apparatus and the air passes through one tube. Any ethanol in the breath is oxidised to ethanoic acid, reducing the orange potassium dichromate(VI) to green chromium(III). The second tube is used for comparison. The colour change is detected by the meter. An electric current causes the needle in the meter to move from its resting place. The operator then rotates a knob to bring the needle back to the resting place and reads the level of alcohol – the more the operator must turn the knob to return it to rest, the higher the level of alcohol. More modern versions have a digital display.

Fuel-cell breath analysers

Some breath analysers use a fuel cell (**fig. 2.1.24**). As the exhaled air from the driver flows along one side of the fuel cell, a piece of platinum oxidises any alcohol to produce ethanoic acid, hydrogen ions and electrons. An external circuit containing a current meter is attached to both platinum electrodes. The hydrogen ions move through the lower part of the fuel cell and combine with oxygen and the electrons on the other side to form water. The more ethanol that is oxidised, the larger is the electric current. A microprocessor measures the current and calculates the blood alcohol content.

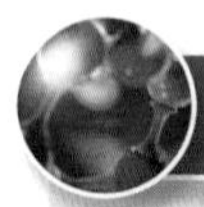

HSW Developing the breathalyser test

fig. 2.1.23 The breathalyser test has helped to revolutionise drink–driving habits and saved many lives.

In 1927 Dr Gorsky, a police surgeon, invited a suspect to blow up a football bladder. He analysed the 2 litres of air collected and found that it contained 1.5 ml of ethanol. As a result, he claimed in court that the man was '50% drunk'!

In 1938 the 'drunkometer' was invented by Professor Harger. He collected breath samples using a fairly large and cumbersome machine and alcohol levels were estimated as a result of colour changes in an acidified potassium manganate(VII) solution.

The first breath analyser which came into regular use was invented in 1954 by Dr Robert Borkenstein, a captain with the Indiana State Police. It used a combination of oxidation and photometry to analyse breath. It was portable and relatively easy to use. Since then further technological refinements have been made to produce very portable and reliable machines.

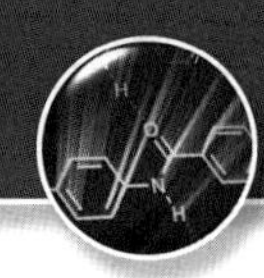

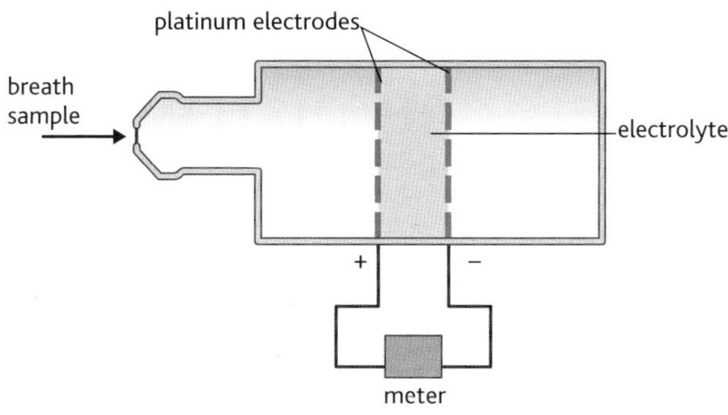

fig. 2.1.24 A fuel-cell breath analyser.

The reactions in the fuel cell are:

$$CH_3CH_2OH(l) + \frac{1}{2}O_2(g) \rightarrow CH_3COOH(l) + 2H^+(aq) + 2e^-$$

$$\frac{1}{2}O_2(g) + 2H^+(aq) + 2e^- \rightarrow H_2O(l)$$

Overall: $CH_3CH_2OH(l) + O_2(g) \rightarrow CH_3COOH(l) + H_2O(l)$

These types of breath analysis machines produce what is known as **preliminary breath testing**. They give police officers an idea of the level of alcohol that a person may have taken in, but they cannot be used in court because there are too many potential sources of error. Further evidence in the form of infrared breath analysis or blood tests will be carried out at the police station to provide results which will be admissible in court.

Infrared breath analysers

Some breath analysers use infrared (IR) spectroscopy (see pages 232–5 in *Edexcel AS Chemistry*). IR radiation is passed through the sample and certain wavelengths associated with the C—O, O—H, C—H and C—C bonds are absorbed. The resulting spectrum allows the concentration of ethanol to be measured accurately by a microprocessor. Other groups, most notably aromatic rings and carboxylic acids, can give similar absorbance readings.

IR spectrometry gives very accurate results but the instrumentation is less portable than for other types of breathalyser. Because the process is both mechanised and accurate, evidence from this type of breathalyser is admissible as evidence in court.

Problems with breath analysers

Breath analysers must be calibrated regularly to ensure that they give accurate readings. Unfortunately, they only give the composition of the air breathed out. It is assumed that this is exhaled from deep within the lungs. However, alcohol may have come from the mouth, throat or stomach. For this reason the operator is trained to leave the suspect for at least 15–20 minutes before taking a further reading.

Find out more about the limitations of breathalysers:

- Why is evidence from many of them not admissible in court?
- How do people try to affect the readings given by breathalyser tests? Explain why most of these methods don't work.
- Why are blood tests so much more reliable than breath tests?

Questions

1 Which type of breath analyser is best for:
   a quick roadside tests
   b a test to produce admissible evidence in courts?
   Explain your answers.

2 Compare the way an ethanol fuel cell breath tester works with the way an IR breath analyser works. Explain the differences in accuracy and validity of results between the two.

2.2 Transition metals and their chemistry

An introduction to transition metals

[5.3.2a, b, c]

What are transition metals?

Many metals you meet on a day-to-day basis are **transition metals**. They are found within the main body of the periodic table (page 270) as members of the d-block of elements. Their character is determined by the numbers of electrons contained in the filling d sub-shells. Some rarer transition metals are shown below the main table itself; these are f-block elements. You will be focusing on the d-block transition metals, particularly those in the first series from titanium to copper.

A simple definition of a transition metal is:

- an element between the s and p blocks of the periodic table.

However, this causes problems when the structure, properties and reactions of the elements are considered, because not all of the metals fit in. To overcome this, a more effective definition is:

- an element which forms one or more stable ions which have incompletely filled d sub-shells.

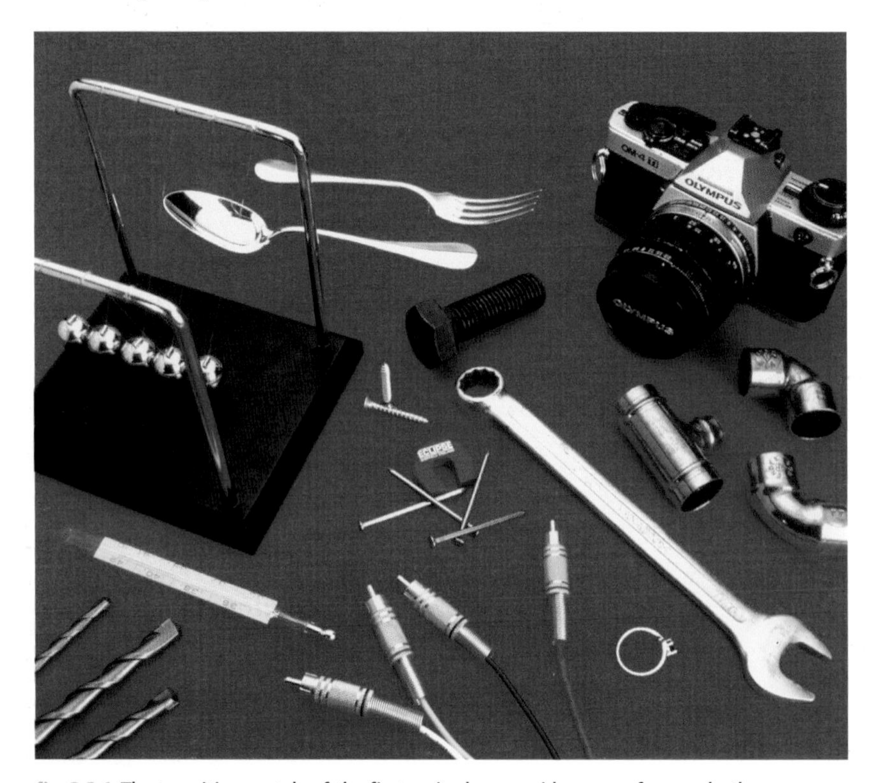

fig. 2.2.1 The transition metals of the first series have a wide range of uses – both as elements and in combination as alloys or compounds.

Based on this definition, the transition metals have a number of properties in common:

- they are all hard metals with high melting and boiling temperatures
- they show more than one **oxidation number** in their compounds
- they tend to form coloured compounds and ions
- many show catalytic activity.

The first, simple definition would include scandium and zinc in the transition metals. Although these two elements are hard metals, they do not show multiple oxidation numbers, they form white rather than coloured compounds and they do not act as catalysts.

Using the second, more effective definition excludes scandium and zinc because:

- a scandium atom loses three electrons, forming a Sc^{3+} ion (with the same electronic configuration, [Ar], as argon)
- a zinc atom loses the outer $4s^2$ electrons, giving a Zn^{2+} ion (with the electronic configuration [Ar] $3d^{10}$).

Neither element forms an ion with an incomplete d sub-shell and so they are not considered to be true transition elements – reflecting the fact that their chemical behaviour differs from that of the other elements in the d block. According to this definition, we shall look at the elements from titanium to copper as the first transition series.

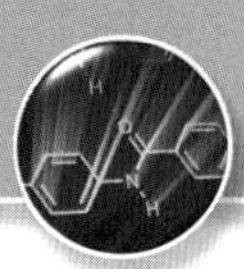

Electron configurations of transition metals

Table 2.2.1 shows atomic number, electronic configurations and the first, second and third ionisation energies of the transition metals together with scandium and zinc.

	Sc	Ti	V	Cr	Mn	Fe	Co	Ni	Cu	Zn
Atomic number	21	22	23	24	25	26	27	28	29	30
Electronic configuration of atom	[Ar] $4s^23d^1$	[Ar] $4s^23d^2$	[Ar] $4s^23d^3$	[Ar] $4s^13d^5$	[Ar] $4s^23d^5$	[Ar] $4s^23d^6$	[Ar] $4s^23d^7$	[Ar] $4s^23d^8$	[Ar] $4s^13d^{10}$	[Ar] $4s^23d^{10}$
First ionisation energy /kJ mol^{-1}	631	658	650	653	717	759	758	737	746	906
Second ionisation energy /kJ mol^{-1}	1235	1310	1414	1592	1509	1561	1646	1753	1958	1733
Third ionisation energy /kJ mol^{-1}	2389	2653	2828	2987	3249	2958	3232	3394	3554	3833

table 2.2.1 The atomic numbers, electronic configurations and ionisation energies of the elements from scandium to zinc.

In a calcium atom (atomic number 20) there are two electrons in the 4s orbital and this is full. Scandium (atomic number 21) has one more electron. However, this does not go into an outer 4p orbital but into a 3d orbital, which is of lower energy than the 4p. The 3d orbitals fill up before the 4p (see pages 64–5 in *Edexcel AS Chemistry*). The full 4s orbital is the outer orbital; the 3d orbitals are inside.

As you move across the transition metals, each successive element has one more electron in its atoms. This extra electron goes into a d orbital. When you reach zinc, all of the five d orbitals are filled with two electrons. So, on moving from zinc to gallium the extra electron goes into the 4p orbital.

When you look at the **ionisation energies** of the transition metals, they help to confirm these electronic configurations. The first electron(s) to be lost are always those in the outer 4s sub-shell, with subsequent electrons lost from the 3d sub-shell. The small fluctuations in the general trend of increasing ionisation energies are explained by the relative stability of full and half-full sub-shells.

So, for example, the second ionisation energy of chromium is higher than that of manganese – although you would expect it to be lower from their positions in the period – because the process:

$$Cr^+(g) \rightarrow Cr^{2+}(g) + e^- \qquad \Delta H = +1592\,kJ\,mol^{-1}$$
$$4s^0 3d^5 \qquad 4s^0 3d^4$$

involves the removal of an electron from the relatively stable half-full 3d sub-shell, whereas:

$$Mn^+(g) \rightarrow Mn^{2+}(g) + e^- \qquad \Delta H = +1509\,kJ\,mol^{-1}$$
$$4s^1 3d^5 \qquad 4s^0 3d^5$$

involves removing a single electron from the 4s sub-shell, exposing the relatively stable half-full 3d sub-shell.

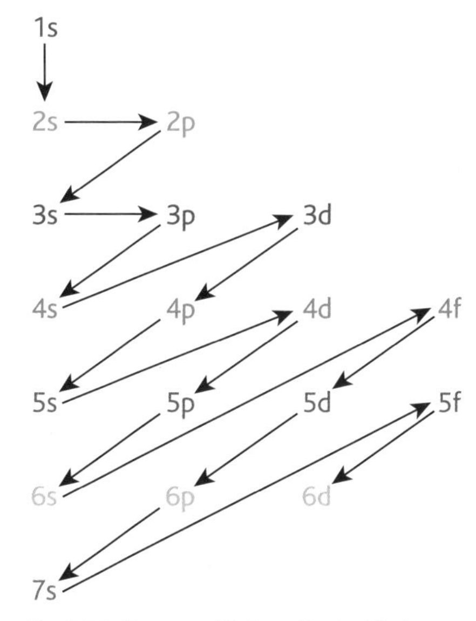

fig. 2.2.2 Electrons fill the orbital with the lowest energy first, which is why the order is not strictly numerical.

SC Plot graphs of the first, second and third ionisation energies for the elements with atomic numbers 20 to 30. Use your graphs to justify the electron configurations of Cr and Cu and their ions, and also to explain why the second ionisation energy of copper is higher than that of zinc.

HSW Transition metals and plants

In the 1980s scientists began to investigate the idea that plants might be used to extract certain metals from the earth. This work has resulted in **phytomining** (plant mining), where transition metals are extracted using plants from areas where they cannot be obtained economically by other methods.

In several states in the USA, *Streptanthus polygaloides* are planted on nickel-rich soils. The plants take up nickel until it makes up as much as 1% of their dry mass. They are harvested and then burned to ash, which is then smelted to produce the metal. The energy produced in burning the plants is used to generate electricity to power the extraction process, with any excess electricity being sold to the local power company. The combination of the money raised from selling the metal extracted and from the electricity generated means that 'metal farmers' can make considerably more money per hectare than the equivalent wheat farmer.

Scientists are also researching the possibility of phytomining metals such as thallium, lead, cobalt and gold. Sometimes chemicals need to be sprayed onto the soil to increase the solubility of these metals to make it easier for the plants to take them up.

There are a number of advantages to phytomining. It can be used to exploit ores and mineralised soils that are uneconomic for conventional mining. **Bio-ores** are almost sulfur free, need less energy for smelting and cause much less acid rain pollution. The metal content of bio-ores is usually much greater than that of a mineral ore. Finally, phytomining is a very 'green' technology when viewed as an alternative to opencast mining of low grade ores.

As with many seemingly perfect schemes, there is a catch. Phytomining is only commercially viable if the price of the metal being extracted remains high. However, the price of a metal at the time of planting can be very different from the price at the time of harvest. Although the dried plants or the ash can be stored and saved until metal prices rise again, the whole process is a gamble – so far, phytomining has not taken off.

Many of the heavy transition metals, such as thallium and lead, are very toxic. If land becomes contaminated by these heavy metals it cannot be used for growing crops or grazing animals. Scientists have identified the use of hyperaccumulator plants to remove these toxic metals from the soil, by **phytoremediation**, making the soil safe to use again. The use of plants to extract metals from the soil will continue for the foreseeable future as a safety measure, if not as a way of making a fortune.

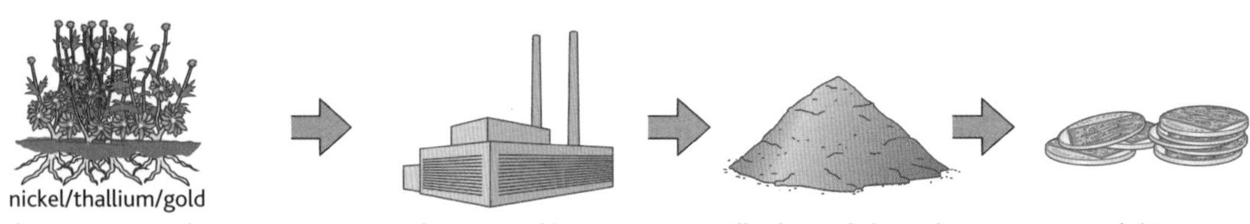

fig. 2.2.3 **A model showing the process of phytomining.**

Questions

1 **a** Write down the electronic configuration of a manganese atom.

b Suggest what change occurs when manganese is oxidised to manganese(II).

2 Technetium, Tc, has atomic number 43 and is below Mn in the periodic table.

a What is the full electronic configuration of Tc?

b Suggest the maximum oxidation number of Tc.

3 'Transition metals are d-block elements'. Do you agree? Give reasons for your answer.

4 What are the main physical characteristics of the transition metals?

5 How do the trends in ionisation energies across the first row of the transition metals support the current model of their electronic configurations?

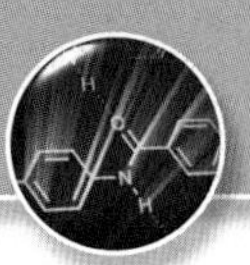

Characteristics of transition metals [5.3.2d (i), (ii), (iii), (iv)]

As you have just seen, one of the defining properties of transition metals is that they have more than one oxidation number in their compounds. The most common oxidation number of most of the elements is +2, due to the loss of the outer two 4s electrons – for example Mn^{2+}, Fe^{2+} and Co^{2+}. However, the underlying d orbitals are very close to the 4s level, and so it is relatively easy to lose electrons from these as well.

This means that several different ions of the same element are possible because the atoms can lose different numbers of electrons – and all these ions are approximately equally stable. The first period of the transition elements always lose their 4s electrons first on forming compounds, and variable numbers of the 3d electrons may be lost as well. The interconversions between one oxidation number and another are an important aspect of transition metal chemistry. This is frequently reflected by colour changes in solutions of the **complex ions**, which help us to see what is happening chemically.

Notice that electrons are never lost to disrupt the inner argon structure. In a number of cases (from Ti to Mn) the highest oxidation number corresponds to all of the electrons beyond the argon core being lost.

Element	Most common oxidation numbers	Electronic configuration of most common ions			Examples of compounds
Ti	+2, +3, +4	[Ar] $3d^1$	[Ar]		Ti_2O_3, $TiCl_4$
V	+1, +2, +3, +4, +5	[Ar] $3d^2$	[Ar]		VCl_3, V_2O_5
Cr	+1, +2, +3, +4, +5, +6	[Ar] $3d^5$	[Ar]		$CrCl_3$, CrO_3
Mn	+1, +2, +3, +4, +5, +6, +7	[Ar] $3d^5$	[Ar] $3d^3$	[Ar]	$MnCl_2$, $KMnO_4$
Fe	+1, +2, +3, +4, +6	[Ar] $3d^6$	[Ar] $3d^5$		FeO, $FeCl_3$
Co	+1, +2, +3, +4, +5	[Ar] $3d^7$	[Ar] $3d^6$		CoO, $CoCl_3$
Ni	+1, +2, +3, +4	[Ar] $3d^8$			$NiCl_2$
Cu	+1, +2	[Ar] $3d^{10}$	[Ar] $3d^9$		Cu_2O, $CuSO_4$

table 2.2.2 The common oxidation numbers for the first row transition elements. The main stable oxidation numbers and their electronic configurations are shown in red. It also shows the electronic configuration of these ions and some examples of compounds with these oxidation numbers.

Variable oxidation numbers

Some general observations on the variable oxidation numbers of the transition metals from titanium to copper are given below:

- +1, +2 and +3 are the most common oxidation numbers for each element
- from titanium to chromium, +3 is the most common
- from manganese onwards, +2 is the most common.

Transition metals usually show their highest oxidation numbers when they are combined with oxygen or fluorine, the most electronegative elements.

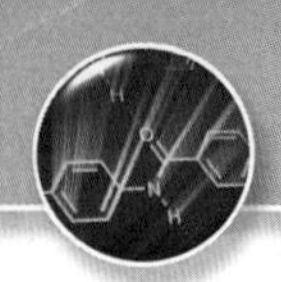

The highest oxidation numbers of all the elements up to and including manganese correspond to the involvement in bonding of all of the electrons outside the argon core – the highest oxidation number of Ti is +4, while that of Mn is +7. Beyond manganese the 3d electrons are held more strongly because of the increasing nuclear charge and so, by and large, the common oxidation numbers involve the 4s sub-shell only.

When the elements exhibit very high oxidation numbers (+4 and above), they do not form simple ions. Either they are involved in covalent bonding (as in Ti_2O_3, CrO_3 and Mn_2O_7) or they form **polyatomic ions** such as CrO_4^{2-} [chromate(VI)] or MnO_4^- [manganate(VII)].

Formation of coloured ions in solution

Most of the transition metals form compounds which dissolve in water to form coloured solutions. Changes in the colour of the solutions can alert us to a change in oxidation number of the ion. Where the ion has a $3d^0$ or $3d^{10}$ arrangement (i.e. the 3d orbitals are empty or completely full) the solution of the ion is colourless.

Table 2.2.3 and fig. 2.2.4 show the colours of different ions and the link with electron configuration.

Hydrated ion	Electron configuration	Colour	Number of unpaired electrons
Ti^{3+}	[Ar] $3d^1$	Violet	1
V^{3+}	[Ar] $3d^2$	Green	2
Cr^{3+}	[Ar] $3d^3$	Green	3
Mn^{3+}	[Ar] $3d^4$	Violet	4
Fe^{3+}	[Ar] $3d^5$	Brown	5
Fe^{2+}	[Ar] $3d^6$	Pale green	4
Co^{2+}	[Ar] $3d^7$	Pink	3
Ni^{2+}	[Ar] $3d^8$	Green	2
Cu^{2+}	[Ar] $3d^9$	Blue	1

table 2.2.3

The colour of the solution is due to electrons absorbing photons of certain frequencies from visible light. The energy of each of these photons matches the energy needed for one electron to jump from a lower to a higher energy orbital. The frequencies of visible light remaining after this absorption give the solution its colour.

fig. 2.2.4 A range of transition metal solutions. From left to right: Ti^{2+}, V^{3+}, VO^{2+}, Cr^{3+}, $Cr_2O_7^-$, Mn^{2+}, MnO_4^-, Fe^{3+}, Co^{2+}, Ni^{2+} and Cu^{2+}.

Forming transition metal complexes

Another characteristic of the transition metals is their tendency to form complex ions (also known as co-ordination compounds). When metals form compounds in the usual way it involves the formation of ionic bonds. Complex formation involves the atoms forming **dative covalent** (also known as co-ordinate) **bonds**. These are bonds in which both bonding electrons come from the same atom.

As a result of their small size, the d-block ions have a strong electric field around them. This field attracts other species that are rich in electrons. Complex ions are formed when the central metal ion is surrounded either by anions or by molecules that act as electron-pair donors. These electron-pair donors are called **ligands**.

A simple example of complex formation involves the common laboratory compound copper(II) sulfate, $CuSO_4$. In aqueous solution, the copper is not present as simple Cu^{2+} ions – the copper ions become bonded to six water molecules to give $Cu(H_2O)_6^{2+}$ which gives the characteristic blue colour (see fig. 2.2.5).

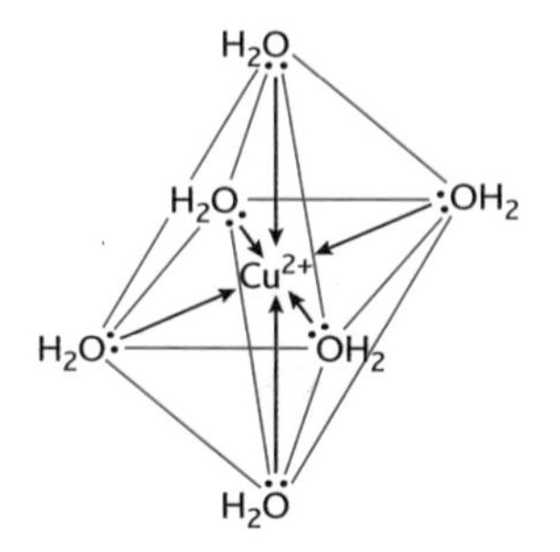

fig. 2.2.5 Cu^{2+} ions in aqueous solution are bonded to six water molecules, forming $[Cu(H_2O)_6]^{2+}$, the hexaaquacopper(II) ion.

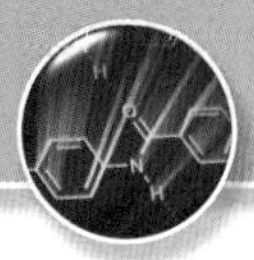

Ligands must be either anions or neutral molecules. Most importantly, they must have a lone pair of electrons available for donation. Common anions that act as ligands include:

- halide ions (F^-, Cl^-, Br^-, I^-)
- the sulfide ion (S^{2-})
- the nitrate(V) ion (NO_3^-)
- the nitrate(III) ion (NO_2^-)
- the cyanide ion (CN^-)
- the hydroxide ion (OH^-)
- the thiocyanate ion (SCN^-)
- the thiosulfate ion ($S_2O_3^{2-}$)
- the ethanoate ion (CH_3COO^-).

Neutral molecules that act as ligands include:

- H_2O
- NH_3
- CO.

The number of water molecules or other ligands involved in a complex ion varies from element to element. For example, nickel(II) ions combine with four cyanide ions to form $[Ni(CN)_4]^{2-}$, whereas cobalt(II) ions combine with six water molecules to produce the $Co(H_2O)_6^{2+}$ ion. Figure 2.2.6 shows the structure of the tetracyanonickel(II) ion and the hexaaquacobalt(II) ion.

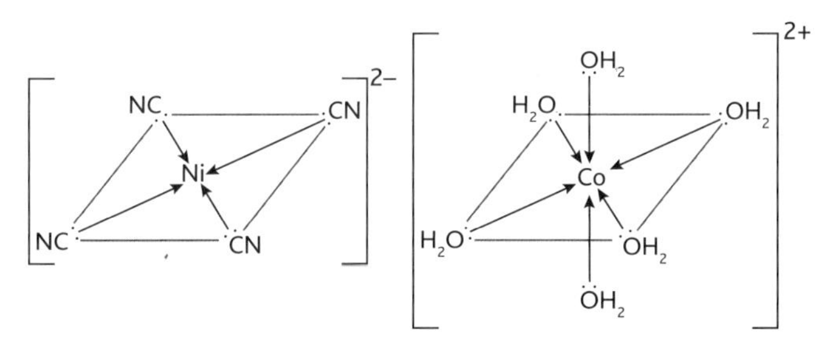

fig. 2.2.6 Two complex ions.

When the formula of a complex ion is given, the metal ion always comes first followed by the ligands. The charge on the complex ion is the sum of the charge on the metal ion and the charge(s) on the ligands. This means that when the ligands are neutral molecules, such as water and ammonia, the charge on the complex ion is the same as the charge on the metal ion, but when the ligands are anions then the charge will be different.

The nickel complex ion in fig. 2.2.6 is an **anionic complex ion** because it would be attracted to an anode during electrolysis. The copper and the cobalt complexes are **cationic complex ions**. The number of lone pairs attached to the metal ion is called the **co-ordination number**. In fig. 2.2.6 nickel has a co-ordination number of 4 and cobalt has a co-ordination number of 6.

Polydentate ligands

All the ligands mentioned so far have been **monodentate** ligands, which means that they join to the metal ion by one atom only. Other ligands are **bidentate** – they join to the metal ion by two atoms; examples are ethanedioate ions and 1,2-diaminoethane.

fig. 2.2.7 Two bidentate ligands – the arrows indicate how the co-ordinate bonds are formed to the metal ion.

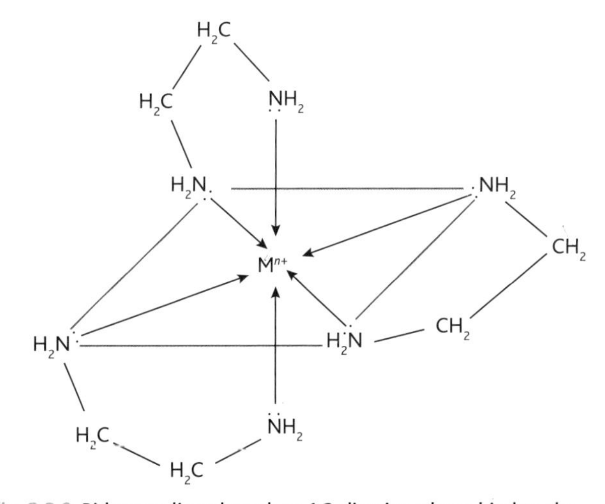

fig. 2.2.8 Bidentate ligands such as 1,2-diaminoethane bind to the metal ion by means of two atoms – here three ligands are shown attached to a metal ion.

Other ligands are **polydentate**, which means that they may attach to the transition metal ion by more than two atoms, leading to the formation of complex ring structures – for example ethylenediaminetetraacetic acid (EDTA)

Chlorophyll, vitamin B_{12} and haemoglobin are all vital biological molecules which contain transition metal ions within complexes.

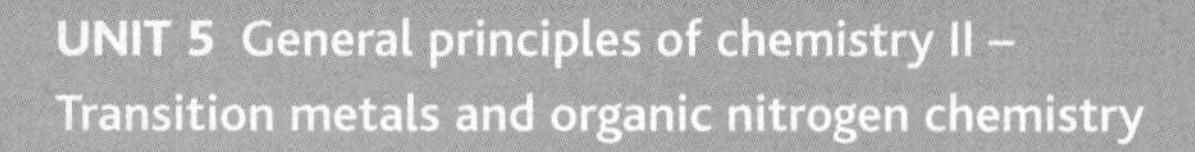

HSW Ethylenediaminetetraacetic acid

One of the most common polydentate ligands is the compound EDTA. The anion $EDTA^{4-}$ has six available donor electron pairs, allowing it to wrap itself around metal ions and form very stable complexes. It is used in trace amounts in foods to prevent spoilage – it forms a complex with any traces of metal ions that might catalyse the reactions of the oils with the oxygen from the air, making the product go rancid. Many shampoos contain this anion to help soften the water, and it may be added in minute amounts to blood for testing to mop up calcium ions and so prevent the blood from clotting. It is also used in sewage plants to help remove unwanted metal ions.

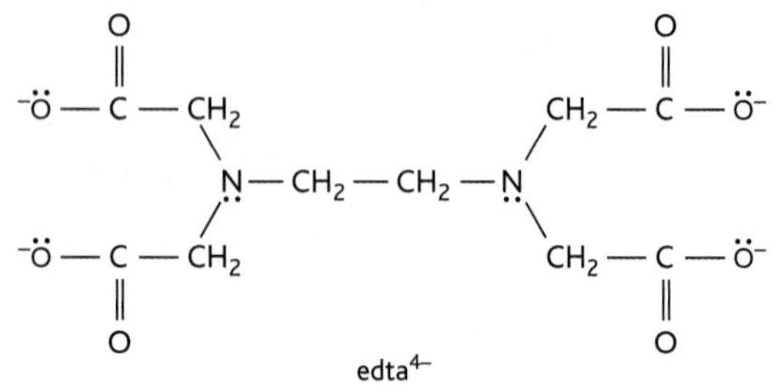

```
      O                                      O
      ‖                                      ‖
⁻:Ö — C — CH₂                        CH₂ — C — Ö:⁻
            \                        /
             N — CH₂ — CH₂ — N
            /                        \
⁻:Ö — C — CH₂                        CH₂ — C — Ö:⁻
      ‖                                      ‖
      O                                      O
                    edta⁴⁻
```

fig. 2.2.9 A polydentate ligand – EDTA.

The geometry of complex ions

The complex ions of the transition metals can form a number of different geometrical shapes including linear, planar, tetrahedral and octahedral. Metal ions with a co-ordination number of 2, such as silver and copper, generally form complexes with a linear structure – for example $[Ag(NH_3)_2]^+$ and $[CuCl_2]^-$:

$$[H_3N \rightarrow Ag \leftarrow NH_3]^+ \qquad [Cl \rightarrow Cu \leftarrow Cl]^-$$

Metal ions with a co-ordination number of 4 produce complex ions with one of two possible geometries – it depends on whether the d orbitals are complete or not. So a transition metal with a co-ordination number of 4 and complete d orbitals shows tetrahedral geometry – cobalt is a good example of this. A co-ordination number of 4 combined with incomplete d orbitals, as in copper, nickel and platinum complexes, sometimes results in square planar geometry (see fig. 2.2.10).

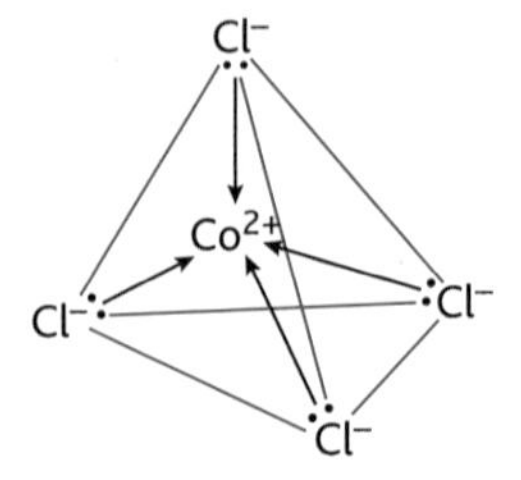

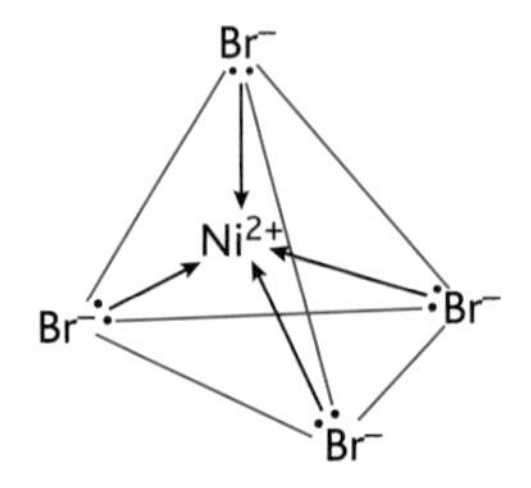

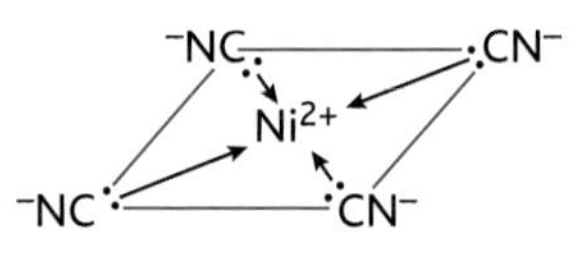

fig. 2.2.10 Examples of complex ions with co-ordination number 4.

The most common co-ordination number for the transition elements is 6, and these complexes are almost all octahedral (fig. 2.2.11).

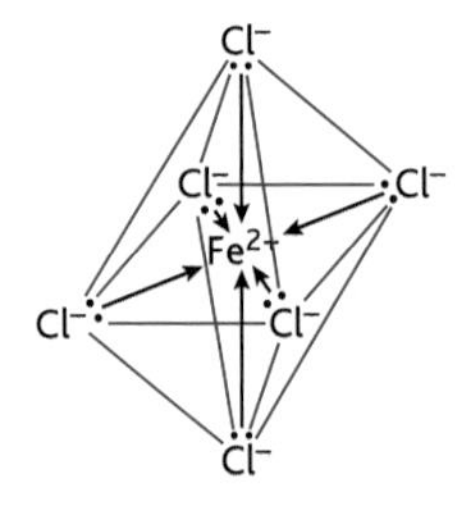

fig. 2.2.11 The octahedral arrangement of ligands in an iron complex – this geometry is common to almost all complexes with a co-ordination number of 6.

Questions

1. Copper(II) compounds are blue but copper(I) compounds are often colourless. Explain this.
2. What is the difference between a covalent bond and a dative covalent bond?
3. a What structural feature must a ligand have?
   b Give one example each of mono-, bi- and polydentate ligands.
   c Explain how each forms a complex ion with a transition metal.
4. Transition metals often form coloured ions in solution. Explain how this can be useful to a chemist.

Using standard electrode potentials to predict the feasibility of forming different oxidation numbers of a transition metal

[5.3.1g, 5.3.2d (i), f (i)]

We found out earlier that reactions are feasible when the $E^\ominus_{cell}$ is positive – and very likely to happen when it is greater than 0.6V. We can use this in studying the reactions of the transition metals – for example, looking at the oxidation numbers of vanadium.

Oxidation numbers of vanadium

As you know, transition metals can exist with different oxidation numbers (fig. 2.2.12). You can demonstrate this very strikingly using vanadium, which shows a sequence of very distinct colour changes in its different oxidation numbers. This is summarised in table 2.2.4.

Zinc acts as a reducing agent in the acidified solution.

You can explain what is happening in these reactions by looking at the standard electrode potentials for the half-equations. For example, in the first stage:

$$Zn^{2+}(aq) + 2e^- \rightleftharpoons Zn(s) \qquad E^\ominus = -0.76V$$

$$VO_2^+(aq) + 2H^+(aq) + e^- \rightleftharpoons VO^{2+}(aq) + H_2O(l) \quad E^\ominus = +1.00V$$

The second of these reactions moves to the right, because it is positive in that direction. The first reaction moves to the left because it will be positive in that direction. So the overall reaction is:

$$2VO_2^+(aq) + 4H^+(aq) + Zn(s) \rightarrow 2VO^{2+}(aq) + 2H_2O(l) + Zn^{2+}(aq)$$

$$E^\ominus_{cell} = (+0.76V) + (+1.00V) = +1.76V$$

Ion	Name of ion	Oxidation number	Colour of solution
V^{2+}	Vanadium(II)	+2	Purple
V^{3+}	Vanadium(III)	+3	Green
VO^{2+}	Oxovanadium(IV)	+4	Blue
VO_2^+	Dioxovanadium(V)	+5	Yellow

table 2.2.4 The oxidation numbers of vanadium.

Dissolving ammonium vanadate(V) (NH_4VO_3) in sodium hydroxide and acidifying the solution with sulfuric acid produces dioxovanadium(V) ions in acid solution, which gives a yellow colour

Shake the solution with granulated or powdered zinc, and you observe a change of colour through green to blue. This is the colour of oxovanadium(IV) ions

The reaction continues to produce a green solution containing vanadium(III) ions

The sequence finishes with the formation of violet vanadium(II) ions

fig. 2.2.12 The colours of vanadium ions in their different oxidation numbers are very distinct and make for some attractive chemistry!

SC Here are the half-equations for the remaining reduction reactions between vanadium ions and zinc – you already have the first two.

$V^{2+}(aq) + 2e^- \rightleftharpoons V(s) \quad E^\ominus = -1.18V$

$V^{3+}(aq) + e^- \rightleftharpoons V^{2+}(aq) \quad E^\ominus = -0.26V$

$VO^{2+}(aq) + 2H^+(aq) + e^- \rightleftharpoons V^{3+}(aq) + H_2O(l)$
$E^\ominus = +0.34V$

Use them to write equations for each stage of the process in the reaction with zinc, giving the standard electrode potential for the cell in each case.

The electrode potentials for the different stages of the reaction can be displayed in graph form, as shown in fig. 2.2.13.

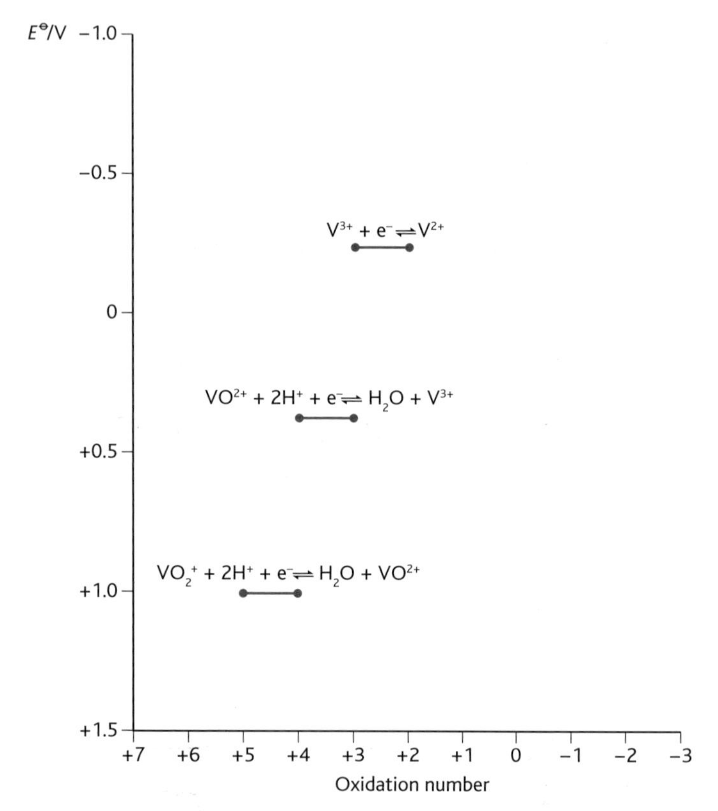

fig. 2.2.13 **Electrode potential chart for vanadium.**

In the sequence of reduction reactions you have looked at above starting with a yellow solution of dioxovanadium(V) ions, you could get vanadium in any of its other oxidation numbers by using suitable reducing agents. For example, suppose you replaced the zinc with tin – what would be the oxidation number of the final product? To answer this, you need to know the standard electrode potential for tin:

$$Sn^{2+}(aq) + 2e^- \rightleftharpoons Sn(s) \quad E^\ominus = -0.14V$$

Now, imagine that information in the electrode potential chart in fig. 2.2.13. For a reduction to happen, the reaction involving vanadium species has to have the more positive $E^\ominus$ value because we want it to go to the right. That means that the tin must have a more negative value.

- In the first reduction from +5 to +4, the tin value is more negative – so that reaction should work and vanadium(IV) be produced.
- In the second reduction from +4 to +3, the tin value is again the more negative, so this reaction should also work.
- In the final reduction from +3 to +2, tin no longer has the more negative $E^\ominus$ value – so it will not reduce vanadium(III) to vanadium(II) under standard conditions.

You have to be careful making these predictions. If dioxovanadium(V) ions are reduced with H_3PO_3:

$$H_3PO_3(aq) + 2H^+(aq) + 2e^- \rightleftharpoons H_3PO_2(aq) + H_2O(l) \quad E^\ominus = -0.50V$$

you might expect reduction right through to vanadium(II) because this electrode potential is more negative than the three values shown in fig. 2.2.13. However, although the reaction is feasible, it stops at oxovanadium(IV) because the final two stages are extremely slow.

This is a reminder that standard electrode potentials are only part of the story. **Activation energy** and kinetic feasibility also affect whether or not a reaction will take place.

Stability of the oxidation numbers

Each of the transition metals may have a number of different oxidation numbers, but these are found in some more frequently than in others. How can you predict or determine the relative stability of the different oxidation numbers? And how can you work out which oxidation number will appear? Manganese is more stable in the +2 number than in the +3 number – but iron is exactly the opposite.

To explain facts like this the answer, again, is to use standard electrode potentials. The $E^\ominus$ values for the reaction $M^{3+}(aq) + e^- \rightleftharpoons M^{2+}(aq)$, where M is a transition metal, are plotted in fig. 2.2.14.

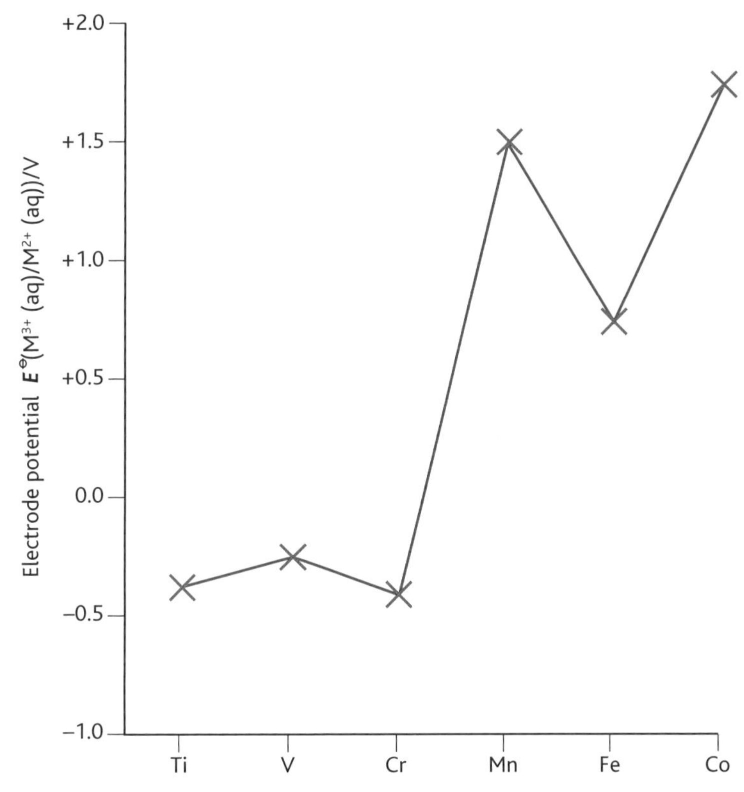

fig. 2.2.14 Graph showing the standard electrode potentials for the M^{3+}/M^{2+} systems for transition metals.

You can see from these data that the +3 oxidation number is more stable than the +2 oxidation number for titanium, vanadium, chromium and iron ($-1 < E^{\ominus}(M^{3+}/M^{2+}) < 1$), while the +2 number is more stable for manganese and cobalt. Why is the +2 number more stable in manganese than the +3 number, while the +3 number is more stable in iron? Think of the arrangement of the electrons in these ions and you will see that Mn^{2+} and Fe^{3+} both have half-full 3d sub-shells – it is this which gives them their relative stability. There are always many different factors to take into account!

Questions

1 During the reduction of dioxovanadium(V) with zinc and dilute hydrochloric acid, the colour change is yellow → green → blue → green → violet. Explain how the first green colour is formed. yellow + blue.

2 Give equations for the reduction reactions between dioxovanadium(V) ions in acid solution and tin, showing the product you would expect at each stage.

3 a Draw the arrangement of electrons in the 3d sub-shells in Mn^{2+}, Mn^{3+}, Fe^{2+} and Fe^{3+}.

b Use these to explain the relative stabilities of the ions.

The chemistry of copper [5.3.2e, f, g(i)]

Copper is a transition metal with electronic configuration [Ar] $3d^{10}4s^1$. It forms compounds with oxidation numbers +1 and +2:

- copper(I) [Ar] $3d^{10}$
- copper(II) [Ar] $3d^9$

Copper is an interesting transition metal. In the +1 oxidation number, where a full set of filled 3d orbitals is present, compounds are often white in colour and do not have the typical properties of transition metal compounds. But in the +2 oxidation number it shows typical transition metal properties. The higher charge on the copper ion gives stronger bonding, compensating for the extra energy required to remove an electron from the 3d orbital in addition to one from the 4s orbital.

The copper story

Copper-containing minerals such as malachite, turquoise and lapis lazuli were known and valued by civilisations as early as 4000 BC. It was discovered that when these minerals are heated in a charcoal fire, they yield the metal copper. The Egyptians exploited mines on the Sinai peninsular and produced thousands of tons of copper around 3200 BC. However, it was the development of bronze, an alloy of 90% copper and 10% tin to give a stronger, harder and more resistant alloy, which had a major effect on civilisations of the time and brought about the dawn of the Bronze Age. The new metal could be forged or cast to produce domestic artefacts, jewellery, tools and weapons, all of which were superior to those that had gone before.

SC Transition elements have played an enormous part in human history, including the Bronze Age and the Iron Age. Produce a summary of human technological history showing how the properties and reactivity of the metals have determined the developments possible. Focus on iron and copper – but look briefly at the 'aluminium age' we live in today.

Copper is an orange-red metal which is relatively unreactive and does not corrode readily, making it ideal for use in copper piping carrying both hot and cold water into homes and offices. It has very high thermal and electrical conductivity and is **ductile**, which is why it is widely used in electrical wiring.

The reactions of copper

Copper has negative electrode potentials for both the Cu/Cu^{2+} and Cu/Cu^+ systems and relatively low reactivity. When exposed to oxygen in the presence of carbon dioxide (the conditions present in the air) copper is oxidised very slowly to form a green film of basic copper(II) carbonate, $CuCO_3{\cdot}Cu(OH)_2$.

Copper reacts with acids only under oxidising conditions – it is not affected by hydrochloric acid. It reacts with moderately dilute and concentrated nitric acid to form copper(II) nitrate, and it is oxidised much more rapidly by concentrated nitric acid:

$$Cu^{2+}(aq) + 2e^- \rightleftharpoons Cu(s) \qquad E^\ominus = +0.34V$$

$$\tfrac{1}{2}O_2(g) + H_2O(l) + 2e^- \rightleftharpoons 2OH^-(aq) \qquad E^\ominus = +0.40V$$

$$NO_3^-(aq) + 2H^+(aq) + e^- \rightleftharpoons NO_2(g) + H_2O(l) \qquad E^\ominus = +0.80V$$

In hot, concentrated sulfuric acid, copper reacts to form copper(II) sulfate, with its familiar bright blue hydrated ions.

$$Cu(s) + 2H_2SO_4(l) \rightarrow CuSO_4(aq) + SO_2(g) + 2H_2O(l)$$

Anhydrous copper(II) sulfate is a white compound – it is only the hydrated form which shows the blue colour.

The most stable oxidation number of copper is +2 and most copper compounds contain the Cu^{2+} ion. Cu^+ also plays a part in the chemistry of copper. We are going to consider the two ions separately.

Copper(I)

Copper(I) is unstable in aqueous solution and exists only:

- at high temperatures
- when insoluble and precipitated
- in complexes.

In aqueous solution, copper(I) compounds **disproportionate** (see page 175 in *Edexcel AS Chemistry*) Disproportionation is the oxidation and reduction of the same element in the same reaction.

For example, $Cu^{+}(aq)$ disproportionates in solution:

$$Cu_2SO_4(aq) \rightarrow Cu(s) + CuSO_4(aq)$$

The electrode potentials explain why:

$$Cu^{2+}(aq) + e^- \rightleftharpoons Cu^{+}(aq) \quad E^{\ominus} = +0.15V$$

$$Cu^{+}(aq) + e^- \rightleftharpoons Cu(s) \quad E^{\ominus} = +0.52V$$

$$2Cu^{+}(aq) \rightarrow Cu(s) + Cu^{2+}(aq) \quad E^{\ominus}_{cell} = +0.67V$$

Copper(I) ions can be stabilised in solution by adding concentrated hydrochloric acid to form a complex:

$$CuCl(s) + Cl^-(aq) \rightarrow [CuCl_2]^-(aq)$$

dichlorocuprate(I)

Copper(I) chloride is prepared by boiling copper(II) chloride with concentrated hydrochloric acid. Initially, the tetrachloro-cuprate(II) complex ion is formed:

$$[Cu(H_2O)_6]^{2+}(aq) + 4Cl^-(aq) \rightarrow [CuCl_4]^{2-}(aq) + 6H_2O(l)$$

This is then reduced to the dichlorocuprate(I) ion by boiling with copper:

$$[CuCl_4]^{2-}(aq) + Cu(s) \rightarrow 2[CuCl_2]^-(aq)$$

When this mixture is poured into cold water a white precipitate of copper(I) chloride is formed:

$$[CuCl_2]^-(aq) \rightarrow CuCl(s) + Cl^-(aq)$$

Copper(I) iodide is precipitated as a white solid (tinged by the iodine also formed) when potassium iodide solution is added to copper(II) sulfate solution:

$$2CuSO_4(aq) + 4KI(aq) \rightarrow 2CuI(s) + 2K_2SO_4(aq) + I_2(aq)$$

or $2Cu^{2+}(aq) + 4I^-(aq) \rightarrow 2CuI(s) + I_2(aq)$

SC Use standard electrode potential data on the reaction of copper(II) ions and iodide ions and predict if the reaction is feasible. Then comment on your findings.

Copper(I) oxide, Cu_2O, is precipitated as a red solid when an alkaline solution of copper(II) sulfate is reduced by glucose (or an aldehyde) – this is **Fehling's test**. The copper(II) ions are complexed with 2,3-dihydroxybutanedioates to prevent precipitation of copper(II) hydroxide:

$$2Cu^{2+}(aq) + 2OH^-(aq) + 2e^- \rightarrow Cu_2O(s) + H_2O(l)$$

Copper(II)

Copper(II) is the more stable state of the element in copper compounds. In an aqueous solution of copper(II) sulfate, copper(II) ions are surrounded by six water molecules to form the octahedral complex hexaaquacopper(II), $[Cu(H_2O)_6]^{2+}(aq)$.

When ammonia solution is added, a precipitate of hydrated copper(II) hydroxide is formed:

$$[Cu(H_2O)_6]^{2+}(aq) + 2NH_3(aq) \rightarrow [Cu(H_2O)_4(OH)_2](s) + 2NH_4^{+}(aq)$$

Adding more ammonia solution causes ligand exchange with four water molecules in the complex being replaced by ammonia molecules. This produces a deep inky-blue complex of tetraaminecopper(II) ions:

$$[Cu(H_2O)_4(OH)_2](s) + 4NH_3(aq) \rightarrow [Cu(NH_3)_4(H_2O)_2]^{2+}(aq) + 2OH^-(aq) + 2H_2O(l)$$

fig. 2.2.15 The difference in the colours of hexaaquacopper(II) ions and tetraaminecopper(II) ions gives a clear indication that ligands have been exchanged.

Questions

1 Give examples of copper complexes that are:
  a linear b tetrahedral c octahedral.

2 a Write an equation for the reaction between copper and concentrated nitric acid.
  b Use the standard electrode potentials in the text to calculate $E^{\ominus}_{cell}$ for the reaction.

3 Explain what is meant by the term 'ligand exchange' when applied to copper complexes.

The chemistry of chromium [5.3.2e, f, g(ii)]

Chromium is a silvery, lustrous transition metal that is hard, rather brittle and very resistant to corrosion. Its atoms have the electron configuration [Ar] $3d^5 4s^1$.

Chromium is very unreactive and it provides an excellent protective covering for other metals. It is also widely used in the production of alloys – stainless steel contains about 18% chromium along with 10% nickel and traces of other elements. Nichrome, as the name suggests, is an alloy of nickel and chromium – this is frequently used to make the wire heating element in a variety of heating devices.

Chromium has various oxidation numbers in its compounds, the most common being +2, +3 and +6:

- chromium(II) [Ar] $3d^4$
- chromium(III) [Ar] $3d^3$
- chromium(VI) [Ar].

One of the most distinctive features of the compounds of chromium is their colour.

fig. 2.2.16 Both emeralds and rubies get their colour from traces of chromium in their structure.

The chromium(II) ion, Cr^{2+}, is pale blue in aqueous solution (see fig. 2.2.17(a)). It is very easily oxidised to the most stable oxidation number of chromium, +3. The Cr^{3+} ion forms a complex ion with water, $[Cr(H_2O)_6]^{3+}$ (hexaaquachromium(III)), and the violet/blue-grey colour of this complex is typical of solutions of many chromium(III) compounds (see fig. 2.2.17(b)). However, the hydrated Cr^{3+} ion often gives a green solution when it is involved in reactions. This is because ligand exchange reactions often take place, with one of the water molecules being replaced by a negative ion from the solution, often a sulfate or chloride ion. It is this complex ion that gives the green colour (see fig. 2.2.17(c)) – when dichromate(VI) ions reduce alcohols, for example. Chromium(III) ions also tend

to exhibit **amphoteric** characteristics, in the same way as aluminium(III) ions. This means that they have both acidic and basic characteristics, and so will react with both acids and bases. For example $[Cr(H_2O)_3(OH)_3]$ reacts with acids to give $[Cr(H_2O)_6]^{3+}(aq)$ and with bases to give $[Cr(OH)_6]^{3-}(aq)$.

The highest oxidation number of chromium is +6, in chromium(VI) oxide, CrO_3, for example. This reacts with water to form a strong acid:

$$CrO_3(s) + H_2O(l) \rightarrow H_2CrO_4(aq)$$

Two of the most common ions in which chromium has oxidation number +6 are the chromate(VI) ion (CrO_4^{2-}), which is bright yellow, and the dichromate(VI) ion ($Cr_2O_7^{2-}$), which is orange (see fig. 2.2.17(d)). These two species exist in an equilibrium that is shifted to the right in acidic solution (with many hydrogen ions) and to the left in alkaline solutions (with fewer hydrogen ions available):

$$CrO_4^{2-}(aq) + 2H^+(aq) \rightleftharpoons Cr_2O_7^{2-}(aq) + H_2O(l)$$

All the chromium(VI) species are strong oxidising agents, particularly in acidic solution.

Note that during the change from chromate(VI) to dichromate(VI) there is no change in oxidation number – it is not a redox reaction.

SC Investigate the standard electrode potentials of chromium ions in their different oxidation states, and use them to show which redox reactions take place and why – and to confirm which of the chromium ions is the most stable.

Chromium compounds also have a variety of practical applications. About 35% of the chromium compounds produced annually are used as green and yellow pigments in the building trade, as protective primers, for producing road markings ('yellow lines') and by artists. A further 25% of chromium compound production, in particular chromium(III) sulfate, is used in tanning animal hides to produce leather for everyday artefacts such as shoes and belts. Most of the remaining chromium compounds are used in various ways to prevent the corrosion of metal surfaces.

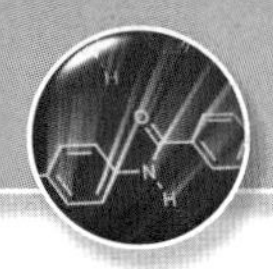

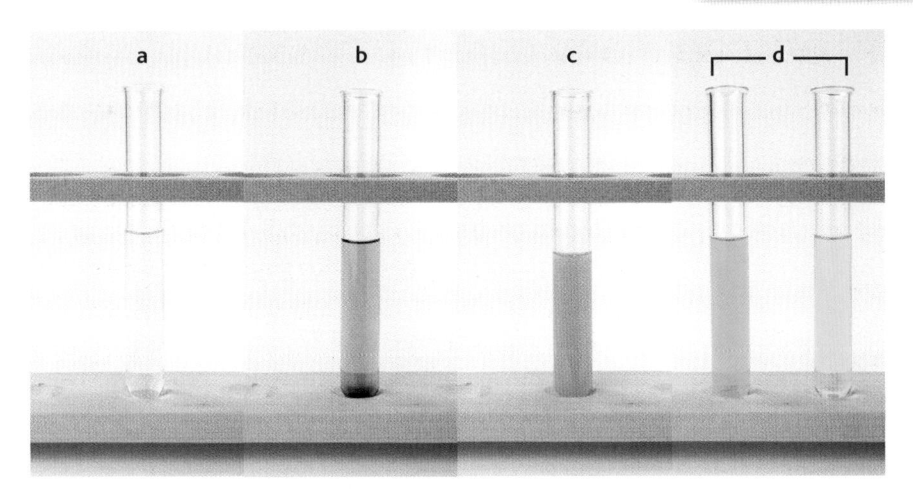

fig. 2.2.17 Chromium compounds produce many brightly coloured solutions.

a When chromium metal is dissolved in a dilute, oxygen-free, non-oxidising acid the pale blue colour of the Cr^{2+} ion is seen.

b The violet colour of this solution of chrome alum $[K_2SO_4 \cdot Cr(SO_4)_3 \cdot 2H_2O]$ is typical of the $Cr(H_2O)_6{}^{3+}$ ion.

c The addition of an alkali to a solution in (b) gives a green solution containing $Cr(H_2O)_2(OH)_4{}^-$ ions.

d The dichromate ion ($Cr_2O_7{}^{2-}$), which predominates in acidic solution, has a reddish-orange colour; the chromate ion ($CrO_4{}^{2-}$) formed in alkaline solutions has a bright yellow colour.

HSW Risk and benefits – chromium and cancer

Most of the commercial uses of chromium involve compounds with the oxidation number +6. Significant amounts of chromium ore contain the Cr^{3+} ion and this is converted into chromium(VI) in a process that involves the ore being crushed and then heated with reactive metals. The process produces much dust, including chromium particles, in the air.

Chromium production soared in the first half of the twentieth century, and by the 1930s it was becoming obvious that workers in the chromium industry were suffering from raised levels of lung cancer compared with workers in similar jobs. The lung tissue of long-term chromium workers was shown to contain up to 10% chromium!

Scientists isolated the form of chromium which caused the problem. Water-soluble Cr(VI) salts, such as sodium and potassium chromate(VI), were the worst offenders. Scientists and technologists worked together to change the way in which the ore was handled, and acceptable exposure levels were set. Once the safe working practices were in place, studies on workers who began their working careers in the chromium industry in the 1960s have shown that the risk of developing lung cancer is not significantly different from that of the general population.

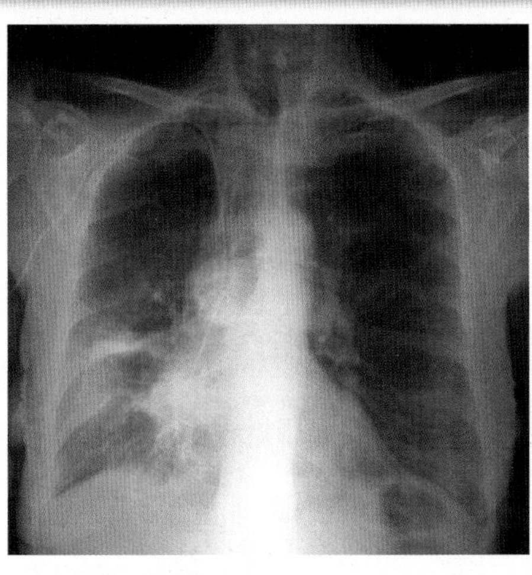

fig. 2.2.18 **Chromium dust can cause a lot of damage inside the body. Many chromium workers have died over the years from lung cancers (like the one which shows up as an orange/yellow area in this coloured X-ray) to provide us with this useful metal.**

Investigate and write a report on the way in which the link between the chromium industry and lung cancer was investigated, and how the industry was made safer.

Questions

1 Chromium(III) is oxidised to chromate(VI) when warmed with hydrogen peroxide in an alkaline solution. Write a symbol equation for this reaction.

2 Research the redox reactions used to produce chromium(VI) from chromium(III) in the processing of chromium ore.

Explaining the chemistry of copper and chromium [5.3.2f (i), (ii), (iii), (iv)]

The colours of complex ions

Solutions of the complex ions of many transition metals are very distinctively coloured – figs 2.2.12 and 2.2.17 show this very clearly. Why is this?

We see the colours of objects around us as a result of the light that falls on them. White light (from the Sun or from an electric lamp) consists of a range of colours, from red through to violet. When light falls on an object, some is reflected, some is absorbed and some – if the object is transparent – may travel through it. Figure 2.2.19 shows how such absorption causes an object to be coloured.

It is transitions between the partly-filled d orbitals in transition metal ions that are the source of their colours. For example, consider an aqueous solution containing Cr^{3+} ions. It is a violet colour because of absorption of light in the middle part of the visible spectrum. The Cr^{3+} ion has the electronic configuration [Ar] $3d^3$. Simple calculations show that movement of an electron between the 3d sub-shell and other sub-shells (such as 4s) cannot account for the absorption of **photons** of visible light – the energy difference between the orbitals is too large. Instead, transitions within the 3d orbitals are responsible for the absorption.

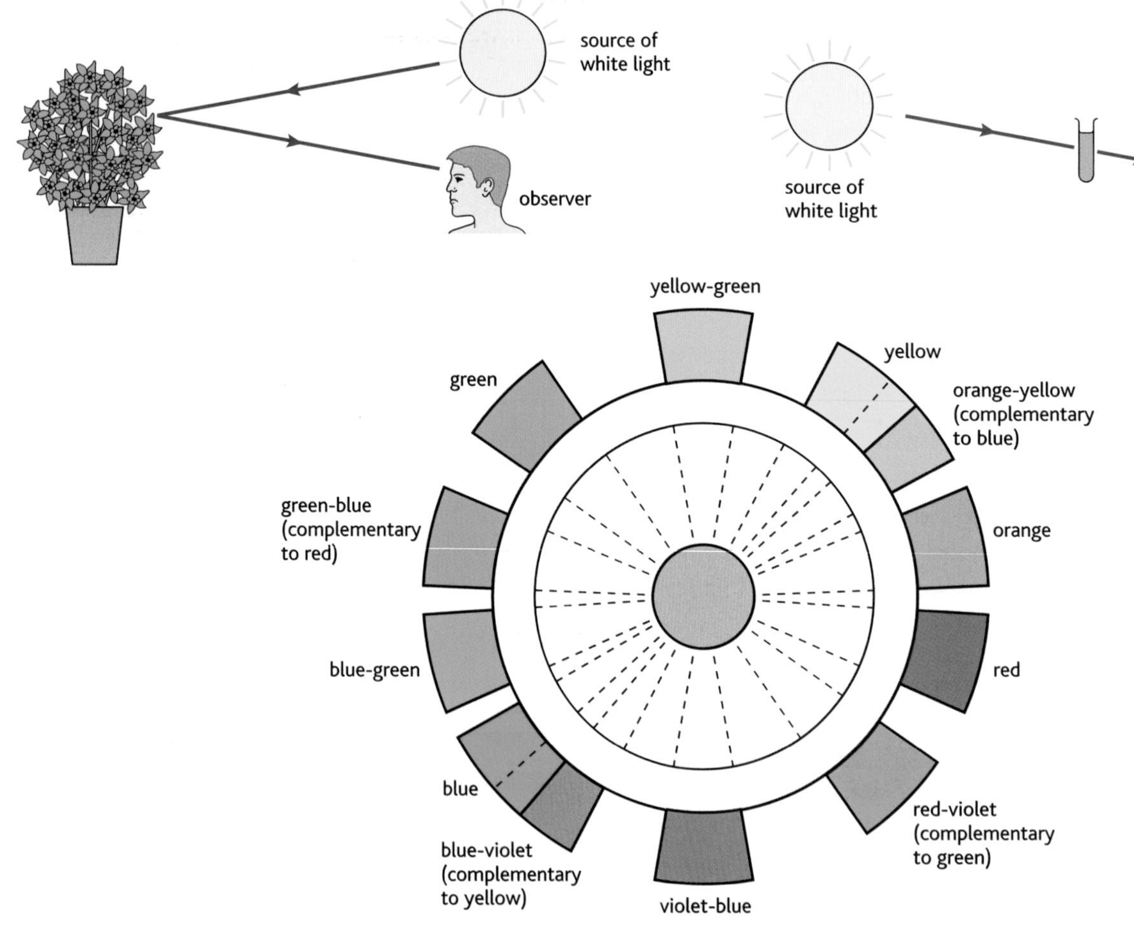

fig. 2.2.19 Opaque objects appear coloured because of the range of wavelengths they reflect; transparent objects appear coloured because of the range of wavelengths they transmit. For example, copper(II) sulfate solution is pale greenish blue because it absorbs light in the red region of the spectrum; these are complementary colours.

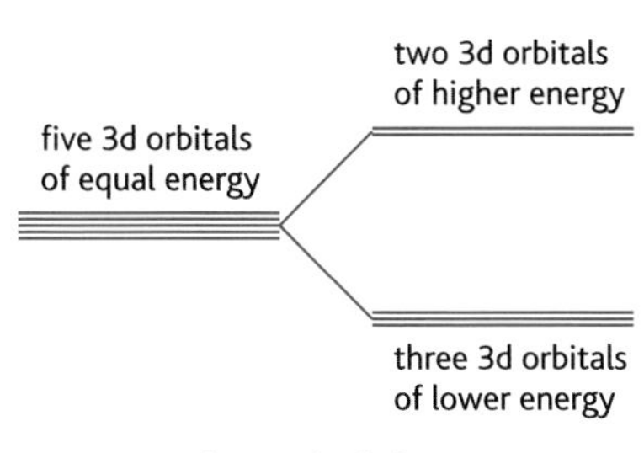

fig. 2.2.20 The octahedral arrangement of water molecules around the Cr^{3+} ion causes the five 3d orbitals to split into two higher energy orbitals and three of lower energy.

An isolated Cr^{3+} ion will have five 3d orbitals, all with the same energy. However, this is no longer true when the ion is surrounded by six water molecules acting as ligands. The orbitals close to the water molecules tend to have a higher energy than those that are further away, with the result that the five 3d orbitals are 'split' into three orbitals of lower energy and two orbitals of higher energy (see fig. 2.2.20). Absorption of visible light can now be explained in terms of a photon causing the single electron to jump from the orbitals with lower energy to those with higher energy.

Although we have looked at a simple example, the theory (**crystal field theory**) can be applied to other ions of the transition elements equally effectively. The presence of greater numbers of electrons gives rise to different combinations of transitions, so that transition metal ions undergo changes of colour when changes of oxidation number occur. As we saw in the case of vanadium, these changes of colour can be quite spectacular.

A closer look at crystal field theory

From the Cr^{3+}example we have seen how an octahedral arrangement of ligands causes a splitting of the 3d orbitals. From the point of view of conservation of energy, it is necessary that this split happens in such a way that the *total* energy of the 3d orbitals remains unchanged. Figure 2.2.21 shows how this happens, as well as the way the orbitals split in the case of tetrahedral complexes.

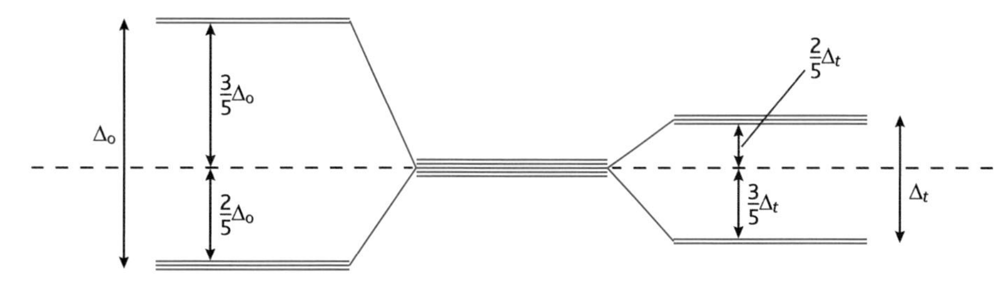

fig. 2.2.21 The splitting of the 3d orbitals in octahedral and tetrahedral complexes.

Questions

1 a Why do transition metals tend to form complex ions?
 b Why do many complex ions produce coloured solutions?

2 Explain why the colour of transition metal ions is different in different complex ions.

Preparing a sample of a complex ion [5.3.2g (iii)]

As you have already seen, chromium ions can exist in several oxidation numbers, some of which can form complexes.

Chromium(III) complexes

In aqueous solution, chromium(III) exists as the hexaaquachromium(III) ion $[Cr(H_2O)_6]^{3+}$.

When concentrated ammonia solution is added, **ligand exchange** takes place and hexaaminechromium(III) $[Cr(NH_3)_6]^{3+}$ is formed:

$$[Cr(H_2O)_6]^{3+}(aq) + 6NH_3(aq) \rightarrow [Cr(NH_3)_6]^{3+}(aq) + 6H_2O(l)$$

When concentrated hydrochloric acid is added to an aqueous solution containing $[Cr(H_2O)_6]^{3+}$, the complex formed is tetrachlorochromate(III), $[CrCl_4]^-$:

$$[Cr(H_2O)_6]^{3+}(aq) + 4Cl^-(aq) \rightarrow [CrCl_4]^-(aq) + 6H_2O(l)$$

These changes are accompanied by the formation of different coloured solutions.

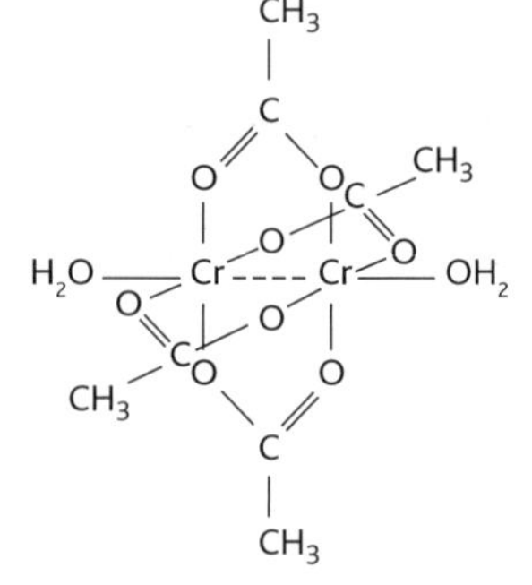

fig. 2.2.22 Structure of chromium(II) ethanoate.

Chromium(II) complexes

Chromium(II) ethanoate, $Cr_2(CH_3CO_2)_4(H_2O)_2$, is unusual. Chromium is usually stable in compounds in which it has oxidation numbers of +3 or +6. But the chromium(II) ion is stable in red-coloured chromium(II) ethanoate. It can be made by reducing a chromium(III) compound using zinc, to form a blue solution. When sodium ethanoate is added to this solution, bright red chromium(II) ethanoate is precipitated. It is a neutral octahedral complex with CH_3COO^- and H_2O as ligands (see fig. 2.2.22).

Preparation of chromium(II) ethanoate

The laboratory preparation of chromium(II) ethanoate requires several hazardous materials including solid sodium dichromate(VI), which is very toxic, an irritant and a carcinogen, concentrated hydrochloric acid, which is corrosive, and zinc powder and hydrogen, which are highly flammable. The procedure is tricky, so check that you are familiar with it before starting.

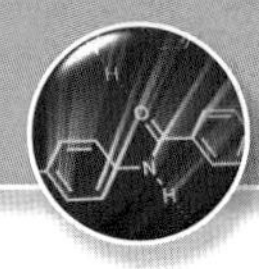

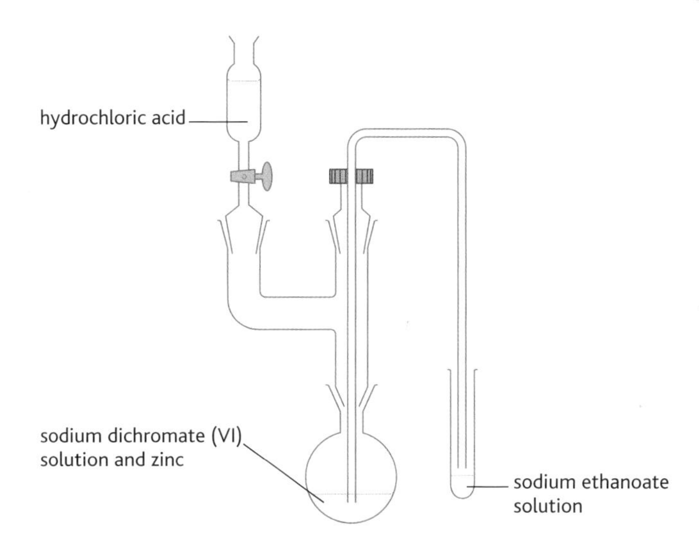

fig. 2.2.23 The apparatus for producing chromium(II) ethanoate.

Preparation of chromium(II) ethanoate

Sodium dichromate(VI) is reduced by hydrogen to produce Cr^{2+}(aq).

When the tap funnel is opened (see fig. 2.2.23), hydrochloric acid runs into the flask where it reacts with the zinc to produce hydrogen:

$$Zn(s) + 2HCl(aq) \rightarrow ZnCl_2(aq) + H_2(g)$$

The hydrogen reduces the orange dichromate(VI) to green chromium(III) and then to blue chromium(II). The redox half-equations are:

$$Cr_2O_7^{2-}(aq) + 14H^+(aq) + 8e^- \rightarrow 2Cr^{2+}(aq) + 7H_2O(l)$$

$$H_2(g) \rightarrow 2H^+(aq) + 2e^-$$

So the overall reaction is:

$$Cr_2O_7^{2-}(aq) + 6H^+(aq) + 4H_2(g) \rightarrow 2Cr^{2+}(aq) + 7H_2O(l)$$

Any surplus hydrogen escapes through the open tap funnel.

When the solution is blue, the tap funnel is closed and the pressure of hydrogen builds up. The contents of the flask are forced up the tube and into the boiling tube containing sodium ethanoate.

A red precipitate of chromium(II) ethanoate is produced. The tube should be removed and sealed immediately, and cooled in running water. If the red precipitate is exposed to air, the Cr(II) ion will be oxidised to Cr(III) and the precipitate will turn grey-green. To separate the brick-red chromium(II) ethanoate you need to filter, wash and dry it in an atmosphere of nitrogen.

$$2Cr^{2+}(aq) + 4CH_3COO^-(aq) + 2H_2O(l) \rightarrow Cr_2(CH_3COO)_4(H_2O)_2(s)$$

Chromium(II) ethanoate is not used widely. It can be used to remove halogen atoms from certain organic compounds and to produce cross-linkages in polymers. It is a good reducing agent because it contains chromium in the 2+ oxidation number, so it is sometimes used in oxygen scrubbers, which remove oxygen from a gas when it is important that none is present in an industrial or chemical process.

SC How is copper(II) ethanoate, which is used as a green pigment, made?

Questions

1 Explain why chromium(II) ethanoate is a neutral complex.

2 Explain the chemistry of the observations made when the red precipitate resulting from the preparation described above is left exposed to air.

Reactions of transition metal ions with aqueous sodium hydroxide and aqueous ammonia [5.3.2j, k]

Sodium hydroxide solution can be used to precipitate metal hydroxides from solution. For example:

$$MgSO_4(aq) + 2NaOH(aq) \rightarrow \underset{\text{white precipitate}}{Mg(OH)_2(s)} + Na_2SO_4(aq)$$

This can be represented by an ionic equation:

$$Mg^{2+}(aq) + 2OH^-(aq) \rightarrow Mg(OH)_2(s)$$

This precipitate does not dissolve in excess sodium hydroxide.

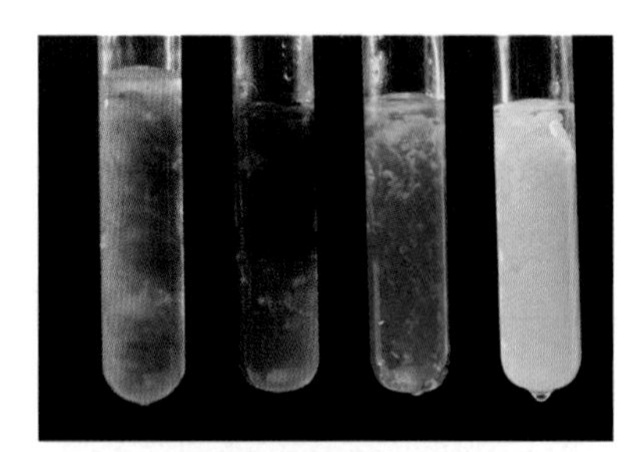

fig. 2.2.24 Precipitates of iron(II), iron(III), copper(II) and nickel(II) hydroxides.

Similarly, when sodium hydroxide solution is added to solutions containing transition metal ions, the metal hydroxides are often precipitated from solution. The colour of a precipitate (fig. 2.2.24) can often be used to help to identify the transition metal ion. In the case of cobalt, the hydroxide precipitate dissolves if excess concentrated sodium hydroxide solution is added.

Different results can be obtained if aqueous ammonia solution is used to precipitate transition metal hydroxides from solution. Ammonia reacts with water in an equilibrium reaction to produce hydroxide ions:

$$NH_3(aq) + H_2O(l) \rightleftharpoons NH_4^+(aq) + OH^-(aq)$$

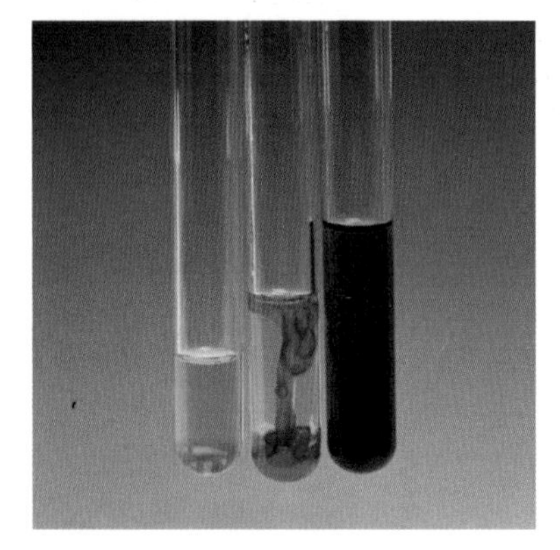

fig. 2.2.25 Copper(II) hydroxide is precipitated when ammonia solution is added to copper(II) sulfate. When excess ammonia solution is added the precipitate dissolves to form a deep blue solution.

Ammonia is a good ligand and can take part in the reaction scheme as well, forming ammine complexes – as shown in fig. 2.2.25.

So, as with sodium hydroxide solution, the transition metal hydroxide precipitate may dissolve if excess ammonia is added – but for a different reason. The results of adding sodium hydroxide solution and aqueous ammonia to solutions of transition metal ions are similar, but not identical – they are summarised in table 2.2.5.

Precipitation of transition metal hydroxides

You may precipitate insoluble metal hydroxides from aqueous solutions of salts containing Cr(III), Mn(II), Fe(II), Fe(III), Ni(II) and Cu(II) ions using both sodium hydroxide and ammonia solution. It is important to add the sodium hydroxide or ammonia solution drop-by-drop to excess so that the true sequence of observations can be made.

As you saw earlier in this chapter, zinc is not classed as a transition metal. Zinc salts form colourless solutions – when sodium hydroxide is added, a white precipitate of zinc hydroxide, $Zn(OH)_2$, forms which dissolves in excess alkali to form a colourless solution containing the $[Zn(OH)_4]^{2-}$ ion. Similarly, with aqueous ammonia the white hydroxide precipitate is formed which dissolves to give a colourless solution containing the $[Zn(NH_3)_6]^{2+}$ ion. These reactions can be useful in identifying unknown salts.

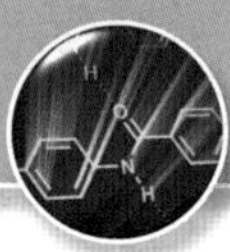

	Cr^{3+}	Mn^{2+}	Fe^{2+}	Fe^{3+}	Ni^{2+}	Cu^{2+}
Colour in aqueous solution	Violet	Very pale pink	Pale blue-green	Yellow	Emerald green	Blue
Reactions with sodium hydroxide solution						
Reaction with small amount of sodium hydroxide solution	Blue-violet precipitate of chromium(III) hydroxide, $Cr(H_2O)_3(OH)_3$	Gelatinous white precipitate of manganese(II) hydroxide, $Mn(OH)_2$, that turns brown on standing in air	Gelatinous pale green precipitate of iron(II) hydroxide, $Fe(OH)_2$, that turns brown on standing in air	Reddish-brown precipitate of iron(III) hydroxide, $Fe_2O_3{\cdot}xH_2O$ also written $Fe(H_2O)_3(OH)_3$	Emerald green precipitate of nickel(II) hydroxide, $Ni(OH)_2$	Gelatinous blue precipitate of copper(II) hydroxide, $Cu(OH)_2$
Reaction with excess sodium hydroxide solution	Precipitate dissolves to form green solution containing $[Cr(H_2O)_2(OH)_4]^-$	Precipitate insoluble in excess	Precipitate insoluble in excess	Precipitate insoluble in excess	Precipitate insoluble in excess	Precipitate insoluble in excess
Reactions with aqueous ammonia solution						
Reaction with small amount of aqueous ammonia	Blue-violet precipitate of chromium(III) hydroxide, $Cr(H_2O)_3(OH)_3$	Gelatinous white precipitate of manganese(II) hydroxide, $Mn(OH)_2$, that turns brown on standing in air	Gelatinous pale green precipitate of iron(II) hydroxide, $Fe(OH)_2$, that turns brown on standing in air	Reddish-brown precipitate of iron(III) hydroxide, $Fe_2O_3{\cdot}xH_2O$ also written $Fe(H_2O)_3(OH)_3$	Emerald green precipitate of nickel(II) hydroxide, $Ni(OH)_2$	Gelatinous blue precipitate of copper(II) hydroxide, $Cu(OH)_2$
Reaction with excess aqueous ammonia	Precipitate dissolves to form yellow solution containing $[Cr(NH_3)_6]^{3+}$	Precipitate insoluble in excess	Precipitate insoluble in excess	Precipitate insoluble in excess	Precipitate dissolves to form lavender blue solution containing $[Ni(NH_3)_6]^{2+}$	Precipitate dissolves to form deep blue solution containing $[Cu(NH_3)_4(H_2O)_2]^{2+}$

table 2.2.5 Summary of the precipitation reactions of transition metal ions with aqueous sodium hydroxide and aqueous ammonia.

Ionic equations for the reactions of transition metal ions with aqueous sodium hydroxide and aqueous ammonia

In this section we are going to explain some of the chemical reactions involved in the precipitation reactions outlined above.

Precipitating metal hydroxides with sodium hydroxide

Iron(II) hydroxide precipitates if sodium hydroxide solution is added to iron(II) sulfate solution:

$$FeSO_4(aq) + 2NaOH(aq) \rightarrow Fe(OH)_2(s) + Na_2SO_4(aq)$$

This can be represented by the ionic equation:

$$Fe^{2+}(aq) + 2OH^-(aq) \rightarrow Fe(OH)_2(s)$$

Similar equations can be written for the precipitation of the other metal hydroxides.

Because transition metal ions exist in aqueous solution as complex ions, the situation may be more complicated than these simple equations suggest. For example, cobalt(II) chloride solution contains hexaaquacobalt(II) ions, $[Co(H_2O)_6]^{2+}$. On adding sodium hydroxide solution, a pink precipitate forms that does not dissolve in excess alkali:

$$[Co(H_2O)_6]^{2+} + 2OH^-(aq) \rightarrow [Co(H_2O)_4(OH)_2](s) + 2H_2O(l)$$

In the case of zinc, addition of excess sodium hydroxide causes the zinc hydroxide precipitate to dissolve and form a zincate ion:

$$Zn(OH)_2(s) + 2OH^-(aq) \rightarrow [Zn(OH)_4]^{2-}(aq)$$

Precipitating metal hydroxides with ammonia solution

Similar ionic equations can be written to represent these reactions. In the case of zinc, the zinc hydroxide precipitate dissolves because it forms a complex ion, with ammonia acting as the ligand:

$$Zn(OH)_2(s) + 4NH_3(aq) \rightarrow \underset{\text{tetraamminezinc(II)}}{[Zn(NH_3)_4]^{2+}(aq)} + 2OH^-(aq)$$

Table 2.2.5 shows that copper(II) compounds behave in a different way with ammonia solution compared to sodium hydroxide solution. In an aqueous solution of copper(II) sulfate, copper(II) ions are bonded to six water molecules to form the octahedral complex hexaaquacopper(II), $[Cu(H_2O)_6]^{2+}(aq)$.

When ammonia solution is added, a precipitate of hydrated copper(II) hydroxide is formed:

$$[Cu(H_2O)_6]^{2+}(aq) + 2NH_3(aq) \rightarrow [Cu(H_2O)_4(OH)_2](s) + 2NH_4^+(aq)$$

Adding more ammonia solution causes ligand exchange, with the four remaining water molecules in the complex being replaced by ammonia molecules:

$$[Cu(H_2O)_4(OH)_2](s) + 4NH_3(aq) \rightarrow [Cu(H_2O)_2(NH_3)_4]^{2+}(aq) + 2H_2O(l) + 2OH^-(aq)$$

Figure 2.2.26 shows the octahedral structure of the copper(II) complex formed. Similar reactions occur with nickel(II) and cobalt(II) compounds.

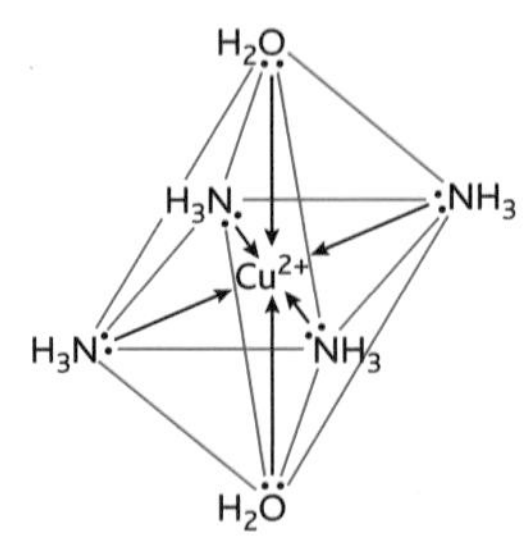

fig. 2.2.26 Ammonia displaces water as a ligand to the Cu^{2+} ion.

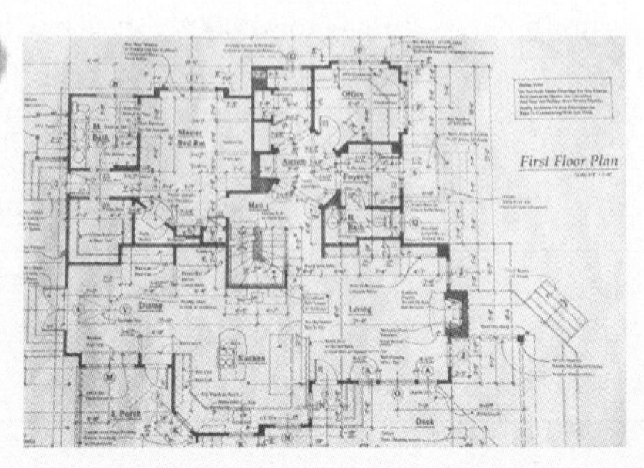

fig. 2.2.27

Investigate the formation of iron complexes with cyanide ions and work out the equations for the reactions. Find out about the compound known as *Prussian blue* and how it was used in the past to produce 'blueprints' like the plan shown above.

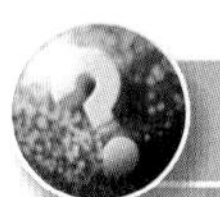

Questions

1 Suggest why the precipitates formed when sodium hydroxide solution is added to solutions containing Mn(II) ions and Fe(II) ions darken on standing in air.

2 You have two unlabelled solutions. One is manganese(II) sulfate and the other is zinc sulfate. Describe how you could find out which is which.

3 Which transition metal ion forms a metal hydroxide that dissolves in excess ammonia solution, but does not dissolve in excess sodium hydroxide solution?

4 Suggest which transition metal ion is present if a white precipitate is formed when ammonia solution is added and the precipitate does not dissolve in excess ammonia.

5 Write a simple ionic equation to represent the reaction between iron(III) ions and hydroxide ions.

6 Write an equation for the reaction that occurs when excess ammonia solution is added to nickel(II) sulfate solution.

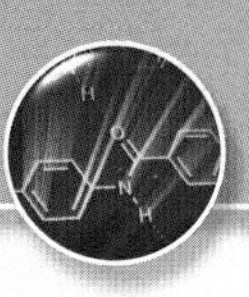

Transition metals as catalysts [5.3.2h, i]

Many chemical reactions occur very slowly or take place only under extreme conditions of temperature, pressure or both. In the research laboratory, these extreme reacting conditions are inconvenient, but if reactions involved in an industrial process do not take place readily then there are many economic and safety implications. For example, a chemical plant that operates at continuous high pressures not only incurs extra capital costs in the engineering of the plant but also presents an increased safety hazard.

The transition metals and their compounds can frequently solve these problems by their ability to act as **catalysts**. There are many examples, including:

- Titanium(III) chloride ($TiCl_3$) in the polymerisation of ethene to poly(ethene) – or polythene:

$$nC_2H_4(g) \rightarrow \left(\begin{array}{cc} H & H \\ | & | \\ -C & -C- \\ | & | \\ H & H \end{array} \right)_n (s)$$

- Iron (Fe or Fe_2O_3) in the Haber process for producing ammonia (see page 206 in *Edexcel AS Chemistry*). Rubidium can be used instead – it is more efficient and enables the reaction to proceed at a lower temperature, but it is more expensive:

$$N_2(g) + 3H_2(g) \rightleftharpoons 2NH_3(g)$$

- Vanadium (V_2O_5 or vanadate, VO_3^-) in the Contact process for producing the sulfur trioxide needed in the manufacture of sulfuric acid:

$$2SO_2(g) + O_2(g) \rightleftharpoons 2SO_3(g)$$

- Platinum gauze in the first stage of the Ostwald process in which ammonia is converted into nitrogen(II) oxide (and ultimately into nitric acid, HNO_3):

$$4NH_3(g) + 5O_2(g) \rightleftharpoons 4NO(g) + 6H_2O(g)$$

Developing new catalysts

Chemists are developing new catalysts to reduce the environmental impact of industrial processes. By using appropriate catalysts, reactions take place at lower temperatures and pressures, using far less electricity (so producing less carbon dioxide).

For example, you studied the Monsanto process for making ethanoic acid from methanol in some detail on page 245 of *Edexcel AS Chemistry*.

$$CH_3OH(g) + CO(g) \rightarrow CH_3COOH(g)$$

The first process developed used cobalt and iodide co-catalyst at 300°C and 700 atmospheres pressure, which needed a big energy input.

Further development led to the use of a rhodium/iodide ion catalyst and then, because rhodium presents some problems, iridium was suggested as an alternative. Iridium is less active than the rhodium complex, but chemists showed that when it combines with ruthenium a very active, specific catalyst is produced. Using iridium reduces costs, releases less carbon dioxide into the atmosphere, reduces the drying columns needed (because less water is used) and decreases the formation of by-products.

HSW Making margarine

Nickel is used as a catalyst in the hydrogenation of vegetable oils to harden them in the process of making vegetable margarine:

$$RCH{=}CHR(l) + H_2(g) \rightarrow RCH_2CH_2R(s)$$

However, the process can lead to the production of many *trans* isomers, known as *trans*-fats, which tend to increase the formation of fatty deposits in human arteries. The *cis* isomers do not have this effect. Scientists are working on a new type of catalyst containing the transition metal palladium as part of a carbon **nanocomposite** material. Its structure means that hydrogenation is very rapid and that the proportion of *trans* isomers in the mixture is reduced.

fig. 2.2.28 The use of transition metal catalysts made the production of margarine from oil possible – and is now improving the product further.

How do transition metals act as catalysts?

Catalysts can be divided into two types:

- **homogeneous catalysts**, which exist in the same phase as the reactants – for example, all dissolved in the same solvent
- **heterogeneous catalysts**, which exist in a different phase from that of the reactants; for example, a solid catalyst in liquid or gaseous reactants – the transition metals frequently act as heterogeneous catalysts.

The reacting mixtures are usually adsorbed (attracted) to the surface of the catalyst. The larger the surface area, the more reactants can be adsorbed. There are a number of models which can help us to understand when catalysis takes place.

Catalysts at work

Using your *Edexcel AS Chemistry* book, remind yourself of how catalysts work. Produce a clear explanation of the difference between heterogeneous and homogeneous catalysts, and investigate the new catalysts which are currently being developed.

Many of the transition metals act as effective catalysts because of their ability to exist in a number of different oxidation states. These variable oxidation states enable them to provide an alternative route for the reaction, thus lowering the activation energy and allowing the reaction to proceed more readily.

Catalysis and the oxidation states of manganese

The decomposition of hydrogen peroxide using manganese(IV) oxide as a heterogeneous catalyst involves the different oxidation states of manganese, and an intermediate compound that cannot be isolated. Investigate this process and write equations for the two stages of the process.

Many of the transition metals act as heterogeneous catalysts in a finely divided state, greatly increasing the surface area for adsorption (see fig. 2.2.29). As the reactant molecules are adsorbed onto the catalytic surface, the bonds between the atoms in them are weakened. This reduces the energy needed to break the reacting molecules apart, lowering the activation energy of the reaction. The adsorbed molecules are held close together on the metal surface, which also increases the likelihood of reaction occurring. Reactants are adsorbed onto the surface of transition metal catalysts at specific points known as active sites. If other molecules bind irreversibly to these sites the catalyst is poisoned and will no longer work.

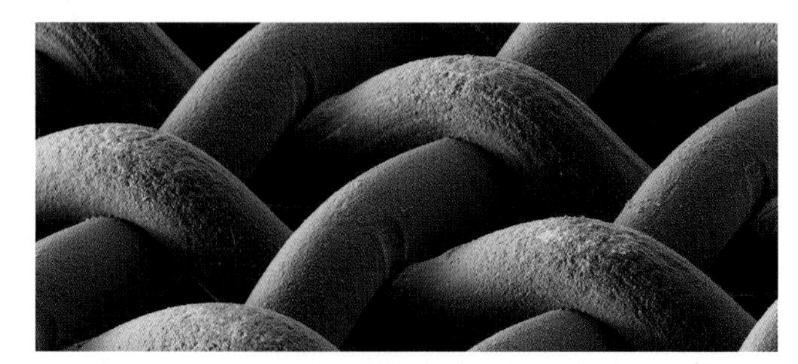

fig. 2.2.29 Transition metals are often pelleted, powdered or made up into incredibly fine wire meshes like the one in this scanning electron micrograph, to provide a large surface area for catalytic reactions. By lowering the activation energy of a variety of reactions – such as the hydrogenation of vegetable oils to make margarine – they enable many everyday substances to be produced at affordable prices.

Catalytic converters

The catalytic converters fitted to car exhausts contain heterogeneous transition metal catalysts. They remove the pollutant gases formed as the fuel is burned. Investigate how catalytic converters work and write a short report on them.

Questions

1. Suggest the advantages of a heterogeneous catalyst in a reaction between two gases.
2. Platinum is used as a catalyst in the oxidation of ammonia in the manufacture of nitric acid. Explain why it is in the form of a very fine mesh.
3. Vanadium(V) oxide acts as a catalyst for the Contact process to make sulfur trioxide. The oxidation numbers of vanadium include +4 and +5. Write equations to show a possible route for this reaction.
4. Suggest why lead additives should never be used in petrol for a car with a catalytic converter.

Modern uses of transition metals [5.3.2l]

HSW Making margarine

In the previous spreads, you have seen examples of how transition metals are used – in many instances as catalysts. In the twenty-first century, although they will remain important as catalysts, transition metals will have many other uses.

Catalysts mimicking living systems

Biological catalysts, or enzymes, are extremely effective and usually work at relatively low temperatures of 30–40°C. Several groups of scientists are developing a new generation of industrial catalysts based on an enzyme model – these are effectively **supramolecules** (very large organic molecules) with associated transition metal ions. A team of scientists in the Netherlands has looked at a number of ways of producing these new-style catalysts. The prospects are bright because these new supramolecular transition metal catalysts make very specific catalysis possible at relatively low temperatures – and they can still be recovered and reused.

Transition metals and cancer treatments

Normal cell division takes place relatively slowly in humans. Cancer occurs when the mechanisms controlling the growth and division of body cells stop functioning, so that cells divide rapidly, forming tumours. One of the main forms of treatment against cancer is chemotherapy, which involves using chemicals to damage or destroy the cancer cells while causing as little harm as possible to healthy tissue.

Because they reproduce so quickly, cancer cells are vulnerable to drugs which interfere with the cell division process. Many of our current chemotherapy drugs damage either the DNA or the proteins involved in cell division. Once this damage has taken place, the cells destroy themselves, removing the cancerous tissue. However, some of your body cells are intended to divide rapidly – for example the immune system, bone marrow and skin and hair cells. These cells are also vulnerable to damage caused by standard chemotherapy, which is why the treatment often makes people feel very ill and can lead to hair loss, weakness, vulnerability to infections, nausea and vomiting.

One commonly used chemotherapy drug is *cis*-platin – a platinum compound that has had a dramatic effect on the treatment of testicular cancer, and subsequently other cancers as well – see fig. 2.2.30.

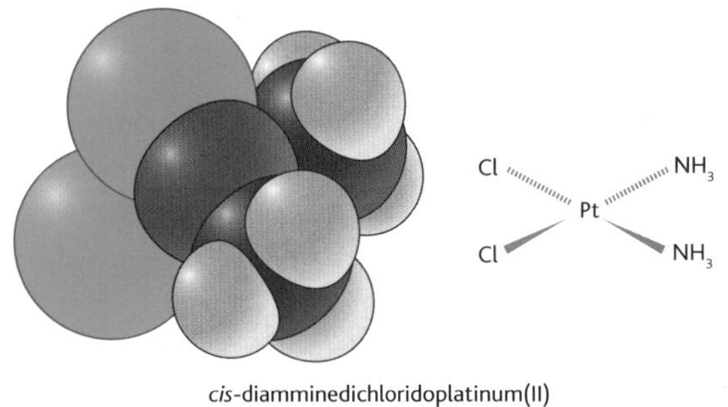

fig. 2.2.30 A *cis*-platin molecule is built around an atom of platinum. The drug slows the growth of cancer cells by binding to the DNA of the nucleus and interfering with the natural repair mechanisms. The red patches on the DNA image show where the *cis*-platin has bound to the DNA.

It can produce some severe side-effects, and also some cancer cells have become resistant to the effects of the drug, which limits its use. It is now mainly used in combination with other drugs – given as a cocktail, they are more effective than any of them alone.

However, a team led by Jan Reedijk in the Netherlands has developed a new version of this drug with far fewer side-effects. It also depends on the transition metal platinum, but it is the *trans* isomer rather than the *cis* isomer (see fig. 2.2.31) that is used.

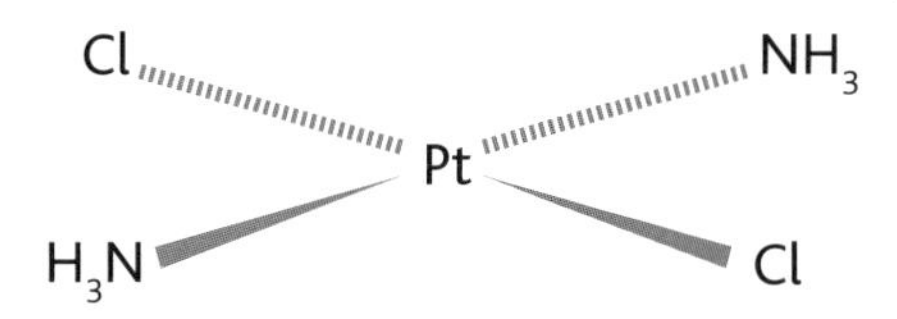

fig. 2.2.31 *Trans*-diamminedichloridoplatinum(II).

On its own, *trans*-platin is largely biologically inactive and does not kill cancer cells. However, combined with other drugs, it seems to enhance their effects without causing serious side-effects, and so this may be another effective use of a platinum-based drug in the future.

Another transition metal increasingly used in chemotherapy is gold. Scientists are looking for ways of getting chemotherapy drugs directly into cancer cells. One potential solution is to package the drugs in gold **nanoparticles** coated in a polymer that is attracted to the cancer cells. The gold particles are small enough to pass through the membranes into the cell, delivering the drug to its target.

Scientists have also shown that heat combined with chemotherapy drugs kills cancer cells more effectively than drugs alone. So trials are being carried out exposing gold nanoparticles in the cancer cells to laser energy, which heats them up. Combined with chemotherapy, the results are promising so far. Heating allows the dosage of chemotherapy drugs to be cut by up to two-thirds while retaining the same efficiency. This makes the treatment cheaper – and reduces the harm done to healthy cells.

All this work is in the early stages, on cells in culture rather than animal models or live patients. Transition metals look as if they will continue to play a major role in our fight against cancer.

Photochromic glasses

Another modern use of transition metals is in the production of **photochromic lenses**. These are spectacle lenses that go darker in sunlight and paler when the light is more diffuse.

fig. 2.2.32 Photochromic glasses can be worn in bright sunlight and in dull light.

Photochromic glass contains silver and copper halides. A redox equilibrium is set up:

$$Cu^+(s) + Ag^+(s) \rightleftharpoons Cu^{2+}(s) + Ag(s)$$

In bright sunlight, the equilibrium moves to the right and clusters of silver atoms are produced – which causes the glass to darken. This is a similar reaction to the one that occurs on photographic film when exposed to light.

If the light level is low, the equilibrium moves to the left, removing the silver atoms and lightening the glass.

SC Investigate and write a report on the use of transition metals in chemotherapy.

Questions

1 What do you think are the potential benefits of developing transition metal catalysts that mimic enzyme systems? Can you suggest any disadvantages?

2 How do you explain the difference in biological activities between *cis*-platin and *trans*-platin?

3 Photochromic lenses can cause problems when there is a sudden change in light levels. Suggest some examples of such a situation.

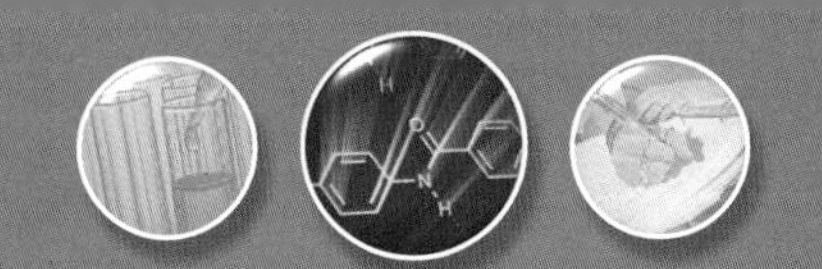

2.3 Organic chemistry: arenes

Evidence for the structure of the benzene ring [5.4.1a]

Benzene is the compound that is the basis of the chemistry of all the **aromatic hydrocarbons** – it was first isolated in 1825. Benzene has a characteristic odour and is immiscible with water. It has the molecular formula C_6H_6 and is a colourless liquid that boils at 80°C and freezes at 6°C.

Many important compounds contain benzene rings with, for example, an alkyl group substituted for one or more of the hydrogen atoms around the carbon ring.

fig. 2.3.1 Many things we use every day, such as dyes, are based on aromatic compounds.

The very high carbon to hydrogen ratio of the C_6H_6 molecule suggests an extremely unsaturated compound. This led chemists to propose a number of **aliphatic hydrocarbon** structures for it – such as those shown in fig. 2.3.2.

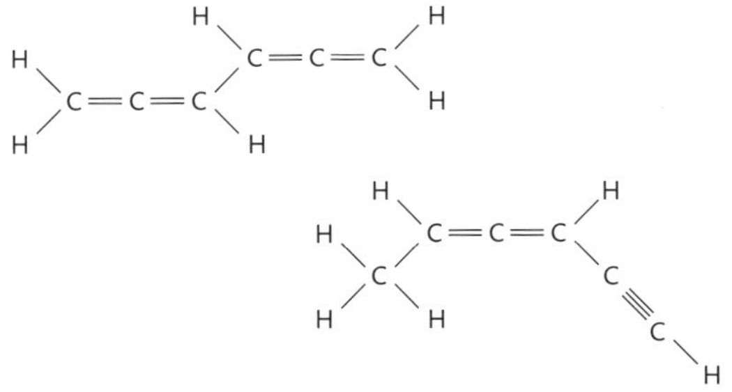

fig. 2.3.2 Displayed formulae for two possible structures for aliphatic benzene.

However, these suggested structures are not very good at explaining the properties of benzene. For example, benzene is very unreactive – failing to react with bromine even when heated. If either of the structures for benzene in fig. 2.3.2 were correct, then the substance would be very reactive because of its double and/or triple bonds. There is a second problem with such suggested structures in that, if they did react with halogens, they would produce many isomers of **monosubstituted compounds** due to the lack of rotation around the multiple bonds, as fig. 2.3.3 shows. The doubts about these aliphatic structures for benzene are further strengthened by the fact that only one form of bromobenzene and chlorobenzene has ever been isolated.

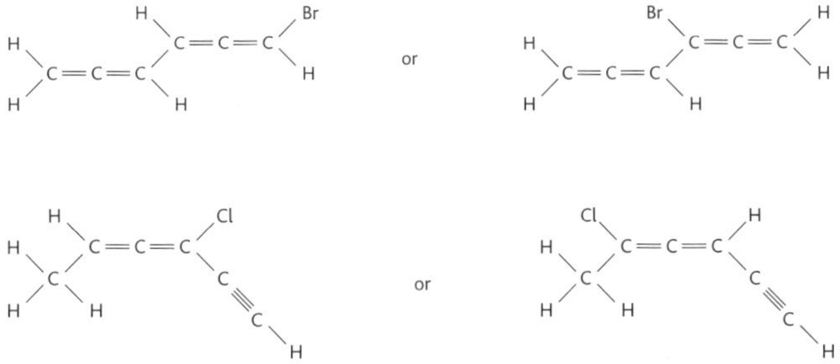

or

or

fig. 2.3.3 The displayed formulae of possible isomers of monosubstituted halogen compounds for the possible benzene structures.

The structure of benzene continued to puzzle chemists for some years, until 1865 when Friedrich August Kekulé first proposed a ring structure for benzene. In this model, benzene consists of a six-membered carbon ring with alternate double and single bonds between the carbon atoms.

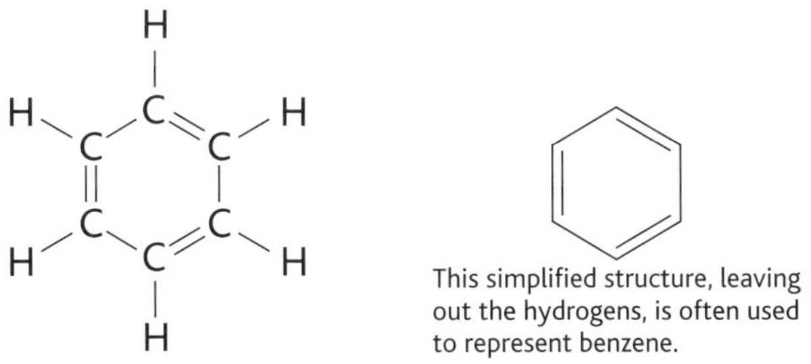

This simplified structure, leaving out the hydrogens, is often used to represent benzene.

fig. 2.3.4 The displayed formula of Kekulé's benzene ring structure.

The **Kekulé model** of benzene was used for many years and was capable of explaining many of the observed properties and reactions of this important compound.

HSW Developing structural models of benzene

- In 1825, Michael Faraday (best known for his work on electricity and magnetism) isolated benzene during the fractional distillation of whale oil.
- In 1834, the molecular formula of benzene was worked out to be C_6H_6.
- Work from 1834 onwards produced much speculation about the structure of the benzene molecule.
- In 1862, a German chemist called Kekulé had what might be called a 'revelation' about the chemistry of the molecule as he dozed by the fire. Gazing into the flickering flames, more than half asleep, he 'saw' snakelike molecules dancing and writhing in the flames until one of them grasped its tail in its mouth, making a ring which whirled round. Kekulé awoke at this point but held the snake dream in his mind. In 1865, he proposed the now famous benzene structure – a six-membered carbon ring with alternating single and double bonds.

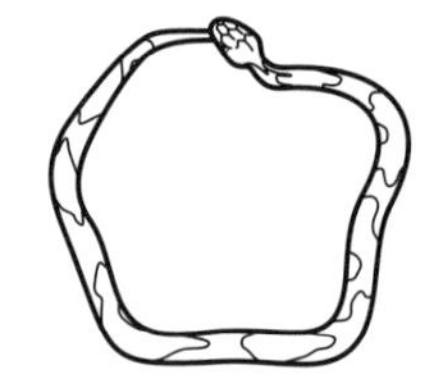
Kekulé and his dream model of benzene

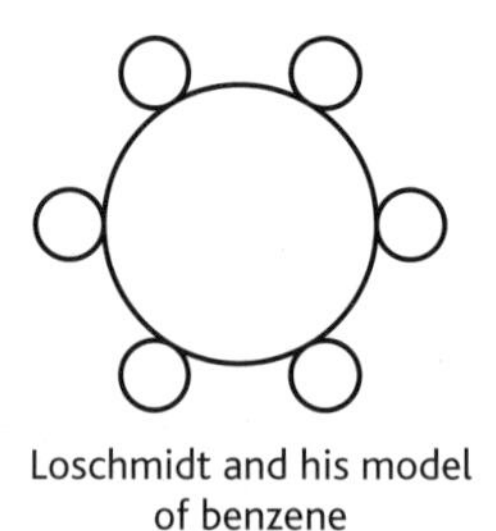
Loschmidt and his model of benzene

fig. 2.3.5 **Kekulé and Loschmidt both contributed to an understanding of the structure of benzene.**

- The first mention of a cyclic structure for benzene was made in a book published in 1861 by Josef Loschmidt, an Austrian school teacher, in which he suggested a six-carbon ring as the most feasible structure for benzene. Kekulé had heard of Loschmidt and his theories, for he refers briefly to them in his paper on the structure of benzene. How much influence Loschmidt's book had on Kekulé's unconscious mind we shall never know.
- In 1883, the first pure sample of benzene was produced by Victor Meyer. He realised that it was pure after it *failed* a test that was commonly used at that time to show the presence of benzene! Careful research showed that all previous samples of benzene had been contaminated with thioprene and that this so-called test for benzene was actually a test for thioprene! Meyer's synthesis of benzene had yielded pure benzene – hence the false test result.

The case against Kekulé's model

The Kekulé model for benzene held good for many years. However, as chemistry became an increasingly precise and quantified science, evidence began to build up that suggested that this structure could not explain all the observed facts. There were two major strands to this evidence.

1 If the structure of benzene included three double bonds, it would be expected that these double bonds would show a similar level of reactivity and a tendency to undergo addition reactions – like the double bonds in the molecules of the alkenes. This is not the case. Most reactions involving benzene are substitution reactions – addition reactions do not occur easily.

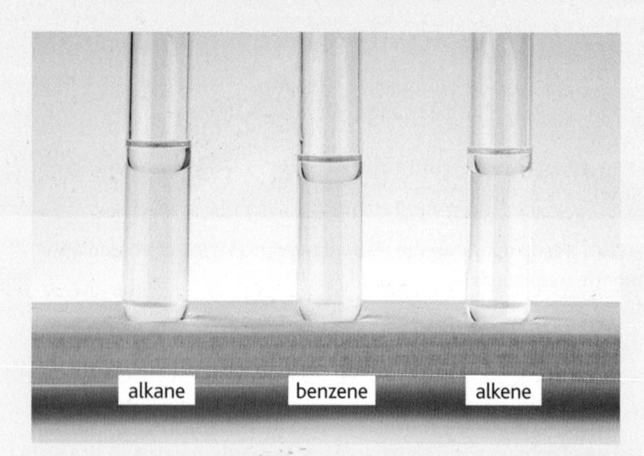

fig. 2.3.6 **After shaking with bromine – only alkenes react readily when shaken with bromine dissolved in tetrachloromethane.**

2 Thermochemical evidence was also against the Kekulé model of benzene. If the enthalpy change for the formation of gaseous benzene from its elements is calculated (using the Kekulé structure) the theoretical value is $+252\,kJ\,mol^{-1}$. In practice it is only $+49\,kJ\,mol^{-1}$. This suggested that the actual structure of benzene is considerably more energetically stable than the Kekulé model would suggest.

Evidence for a new model

In *Edexcel AS Chemistry* we saw how the use of bond energies and the concept of enthalpy changes (both theoretical and real) during the course of reactions can help us to predict the outcome of particular reactions and also to understand the internal arrangement of different molecules (see pages 116–119). It is when considering the structure of benzene that we clearly see how valuable such calculations can be.

For most reactions, an enthalpy change (ΔH) calculated using bond energies turns out to be extremely close to the observed enthalpy change during the reaction. However, this is not the case when we consider the enthalpy change during the hydrogenation of benzene. The theoretical enthalpy change for the hydrogenation of benzene is calculated using the enthalpy change for the hydrogenation of cyclohexene. There is one carbon–carbon double bond in cyclohexene, and when hydrogen is added the enthalpy change is $-120\,\text{kJ mol}^{-1}$.

fig. 2.3.7 **The hydrogenation of cyclohexene.**

The Kekulé model of benzene has three such double bonds. So it seems reasonable to suppose that when these three carbon–carbon double bonds are involved in a reaction with hydrogen the enthalpy change will be three times that of the hydrogenation of cyclohexene. Thus, the theoretical value of ΔH for the hydrogenation of Kekulé benzene would be $3 \times -120\,\text{kJ mol}^{-1}$, or $-360\,\text{kJ mol}^{-1}$. When the hydrogenation is actually carried out, the measured enthalpy change is only $-208\,\text{kJ mol}^{-1}$. This is substantially less in magnitude than the theoretical value and suggests that the addition reaction with hydrogen is not occurring across a normal double bond. This in turn has to lead us to a much more stable structure for the benzene molecule than had been previously proposed.

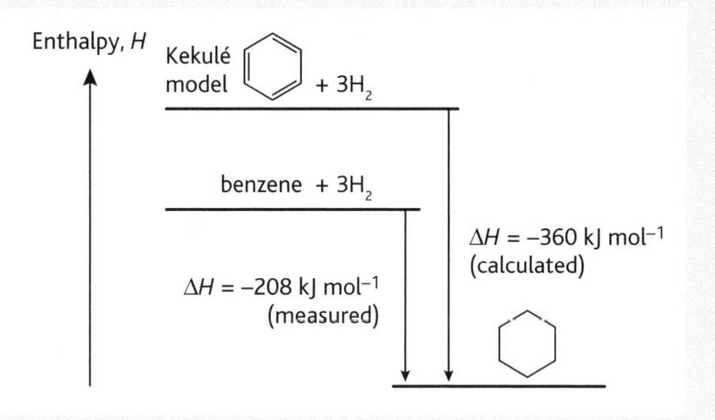

fig. 2.3.8 **The difference between the calculated enthalpy change and the actual enthalpy change for the hydrogenation of benzene can be seen clearly. Data such as this made it imperative that a new model for the benzene model should be found.**

Emerging new techniques also pointed towards the need for a better model. The infrared spectroscopic comparisons between cyclohexene and benzene in **fig. 2.3.9** are revealing. You can see that the spectrum from benzene is simpler. This could be explained if benzene had more symmetry than cyclohexene.

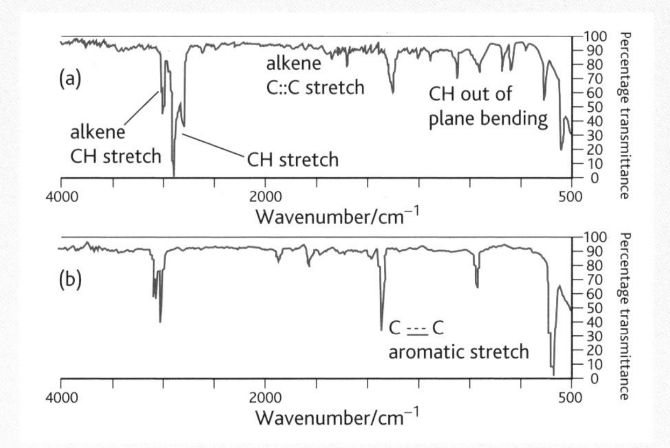

fig. 2.3.9 **IR spectra for (a) cyclohexene and (b) benzene.**

X-ray diffraction studies can be used to measure bond lengths. A carbon–carbon single bond in a cyclohexane ring has a bond length of 0.154 nm. A carbon–carbon double bond in cyclohexene has a bond length of 0.133 nm. If the Kekulé model of benzene structure held true, we would expect there to be two different bond lengths in the molecule – one for the single bonds and one for the double bonds. In fact, X-ray diffraction gives only a single value for the carbon–carbon bonds in the benzene ring. The bond length is 0.139 nm and this suggests that all the bonds are the same in nature, and that the bond length is somewhere between that of a single and that of a double bond.

This is supported by the electron density map of benzene shown in **fig. 2.3.10**.

All the evidence pointed to a symmetrical molecule with six carbon atoms bonded to each other in such a way that all the carbon–carbon bond lengths are identical, with each carbon bonded to a hydrogen atom in such a way that all the bond angles in the molecules are 120°.

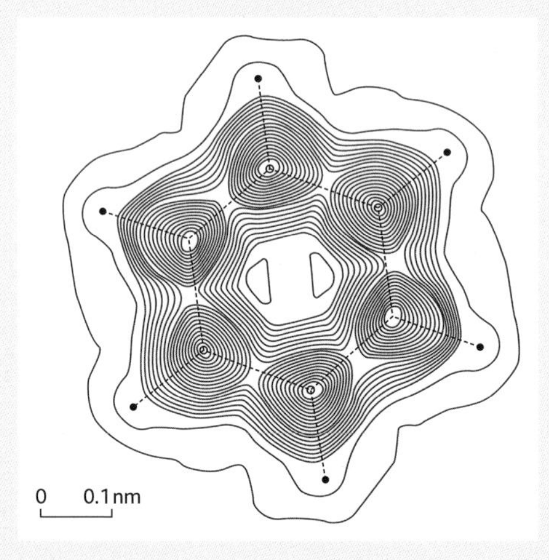

fig. 2.3.10 Electron density map of benzene.

Modern ideas on the structure of benzene

It was the accumulation of such evidence that led Linus Pauling to propose a new way of looking at the structure of the benzene molecule in 1931. Pauling's idea was to treat the molecule as if it was halfway between the two possible Kekulé structures, as shown in **fig. 2.3.11**.

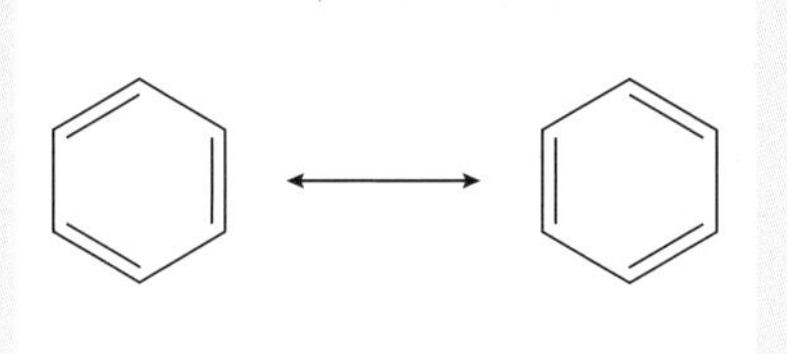

fig. 2.3.11 Pauling proposed representing the structure of 'real' benzene as being a mixture of equal parts of the two possible structures for Kekulé benzene. The double-headed arrow indicates that the real structure is a resonance hybrid of the two structures, not that there is a dynamic equilibrium between the two.

This 'halfway' character, in which the real structure of benzene is called a **resonance hybrid**, explains the observations about the bond lengths in benzene as well as its lack of reactivity, because each bond in the ring is somewhere between a single bond and a double bond. According to Pauling, this idea also explains the extra thermochemical stability of benzene because the resonance hybrid is lower in energy than either of the two Kekulé structures.

This model of the structure of benzene leaves us with a problem when it comes to representing the molecule on paper. The Kekulé model was easy – alternate single and double bonds around the ring. How can we represent bonds which are halfway between single and double bonds all the way round the ring? **Figure 2.3.12** shows the solution.

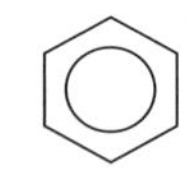

fig. 2.3.12 Symbolic representation of the benzene molecule.

In chemistry texts, you will find benzene represented either by one of the Kekulé structures in **fig. 2.3.11** or by the structure in **fig. 2.3.12.**

This planar structure in which all the bonds between the carbon atoms in the ring are identical means that benzene is a non-polar molecule, and explains its lack of reactivity.

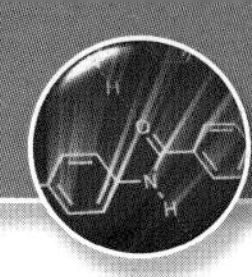

The arrangement of the electrons in benzene

Although **Pauling's model** for benzene provides one way of understanding its properties, a fuller understanding comes from examining the way in which the orbitals belonging to the six carbon atoms in the ring are arranged. Each carbon atom forms a single **sigma (σ) bond** with each of the two carbon atoms joined to it. This forms the skeleton of the molecule, giving a planar hexagonal ring – see **fig. 2.3.13**.

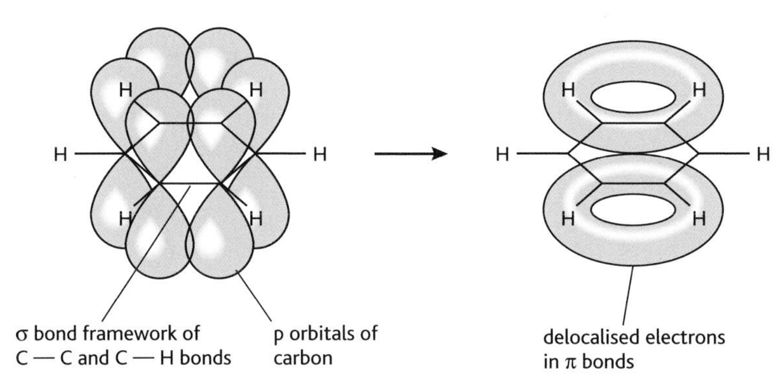

fig. 2.3.13 **The delocalisation of the electrons in the π bonds of the symmetrical six-membered ring structure of benzene explains why the chemical and physical properties of benzene are so different from those of straight-chain compounds containing double bonds.**

Each carbon atom has one p orbital containing a single electron, and it is the lobes of these p orbitals that fuse to form a single ring of charge above and another single ring below the σ-bonded skeleton. Within these extended pi (π) bonds, the electrons are free to move around the entire ring – these electrons are described as being **delocalised**. It is this delocalisation that causes all the carbon–carbon bonds in benzene to be identical, and which also makes the benzene molecule so unexpectedly stable.

We have been using the word 'aromatic' to describe hydrocarbons related to benzene. A much more accurate definition of aromatic hydrocarbons would be *any hydrocarbon system which has a stabilised ring of delocalised π-electrons.*

SC

1 Why were benzene and other similar ring molecules called 'aromatic' hydrocarbons? What are the limitations of this term? Investigate and describe two other molecules with a similar ring structure that are 'aromatic'.

2 It is said that the development of ideas in science depends on both inspiration and perspiration (in other words, hard work!). How does the emergence of our modern model of the structure of benzene confirm this idea?

3 Explain how the development of laboratory techniques and technology has affected the development of our model of the benzene ring.

Questions

1 In practical investigations, benzene attracts and reacts with electrophiles. What structural feature of a benzene molecule attracts electrophiles?

2 a If benzene molecules did have alternate double and single bonds, how would you expect it to react with bromine?
  b Draw the displayed formula of the product.
  c Is the product aromatic? Explain your answer.

Reactions of benzene [5.4.1b (i), (ii), (iii), (iv), (v), 5.4.1d]

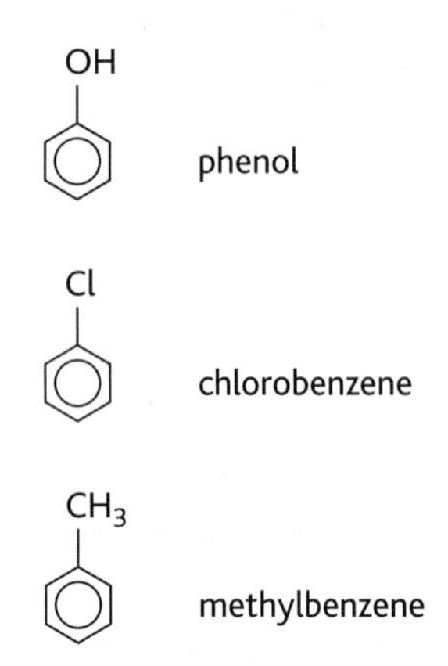

fig. 2.3.14 Naming compounds of benzene: phenol, chlorobenzene and methylbenzene.

Naming aromatic compounds

Most reactions of benzene produce other aromatic compounds. Before looking at the reactions themselves it will be useful to look at the way these other aromatic hydrocarbons are organised and named.

When benzene itself is regarded as a side chain, in the same way as alkyl groups such as the methyl and ethyl groups, it is known as the phenyl group, C_6H_5–. The word 'phenyl' relates back to Michael Faraday's original suggestion of 'phene' as a name for the newly discovered benzene.

The hydrogen atoms in the benzene ring may be substituted by other atoms or groups to give compounds such as phenol, chlorobenzene or methylbenzene (see **fig. 2.3.14**).

Because the benzene molecule is symmetrical, it doesn't really matter where a single substitution product like this is shown. It is simply convention that they are often drawn at the top of the ring. When two or more atoms or groups are substituted into a benzene molecule it becomes more important to indicate where they are positioned. In the examples in **fig. 2.3.15**, there are only two isomers because two of the structures represent the same compound.

The carbon atoms are numbered so that we can give compounds unambiguous names. One of the groups is assumed to be attached to carbon atom 1 in each of the isomers, and the others to different carbon atoms. As you can see from **fig. 2.3.15**, numbering the carbon atoms can be done in either direction because the symmetry of the benzene molecule means that it doesn't matter which way round we count. The important thing is to keep the same group or atom in position 1. However, it is conventional to count in the direction that gives the *lowest* numbers so that, for example, 1,3-dimethylbenzene and 1,5-dimethylbenzene are not mistaken for different isomers – we use the numbering that gives 1,3-dimethylbenzene.

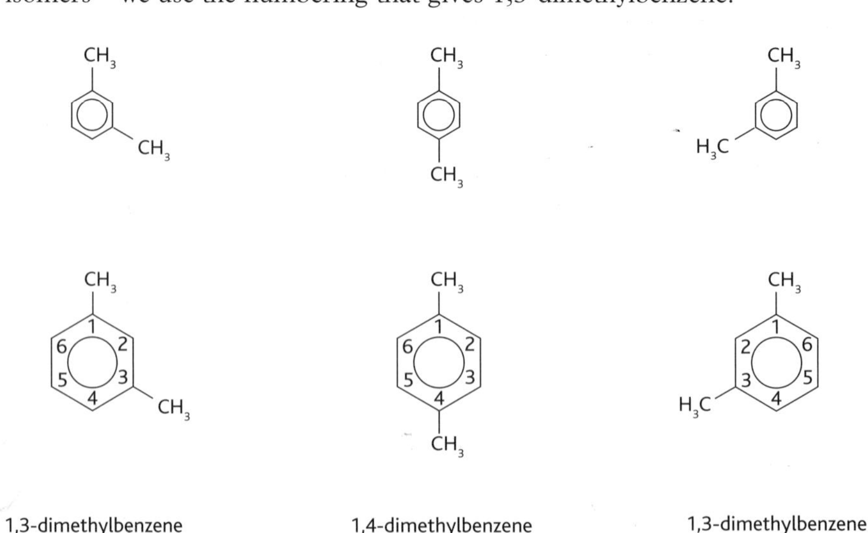

fig. 2.3.15 Naming when there are two substituents attached to a benzene ring.

Combustion of benzene

All hydrocarbons burn in air or oxygen to produce carbon dioxide and water, providing there is sufficient oxygen available. This reaction for benzene is:

$$2C_6H_6(l) + 15O_2(g) \rightarrow 12CO_2(g) + 6H_2O(l)$$

The complete combustion of benzene requires a large volume of oxygen. When benzene burns there may well be insufficient oxygen and some unburned carbon may remain – this will make the flame yellow in colour and smoky.

fig. 2.3.16 Combustion of benzene produces a yellow smoky flame.

Addition reactions of benzene

Reaction of benzene with hydrogen

Alkenes undergo rapid addition reactions with hydrogen in the presence of a nickel catalyst at 200°C (see page 129 in *Edexcel AS Chemistry*). When benzene is mixed with hydrogen in the presence of a **Raney nickel catalyst** (finely divided with a very large surface area and very high catalytic activity) at a temperature of about 150°C it will undergo an addition reaction to form cyclohexane:

$$C_6H_6 + 3H_2 \xrightarrow{\text{Raney Ni, 150 °C}} C_6H_{12}$$

The formation of cyclohexane from benzene is important in the production of nylon.

Reaction of benzene with halogens

We know that alkenes, such as ethene, react readily with bromine in an addition reaction to form a saturated product called 1,2-dibromoethane:

$$H_2C{=}CH_2 + Br_2 \rightarrow H{-}CHBr{-}CHBr{-}H$$

1,2-dibromoethane

This is an electrophilic addition reaction to a double bond (see pages 129–133 in *Edexcel AS Chemistry*). As we saw earlier, benzene does not have double bonds as such and, therefore, we would not expect it to take part in this kind of addition reaction.

On page 118 in *Edexcel AS Chemistry* we saw that alkanes react with halogens, such as chlorine, in substitution reactions. These do not occur as readily as addition reactions with alkenes and usually require additional energy to initiate them, often from ultraviolet light.

It turns out that benzene undergoes addition reactions with bromine rapidly in the presence of ultraviolet light or bright sunlight to form 1,2,3,4,5,6-hexabromocyclohexane:

$$C_6H_6 + 3Br_2 \xrightarrow{\text{light}} C_6H_6Br_6$$

1,2,3,4,5,6-hexabromocyclohexane

The rapidity of the reaction and the requirement for light to overcome the activation energy suggest a **free-radical reaction**. The corresponding chlorine compound 1,2,3,4,5,6-hexachlorocyclohexane is an insecticide called gammexane or lindane.

SC The displayed formula for 1,2,3,4,5,6-hexabromocyclohexane looks simple enough. However, it has eight possible geometric isomers. Research and find out about the reason for these isomers.

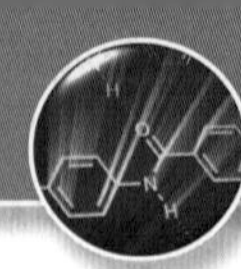

HSW A chemical of the past persists in the future

Lindane (1,2,3,4,5,6-hexachlorocyclohexane) is still used in a few developed countries, but more widely in developing ones. Other organochlorines (including aldrin, dieldrin and DDT) have been banned in many countries because of the link with major health and environmental problems.

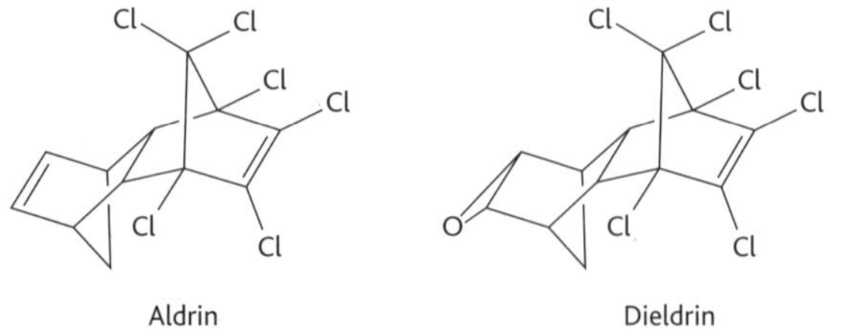

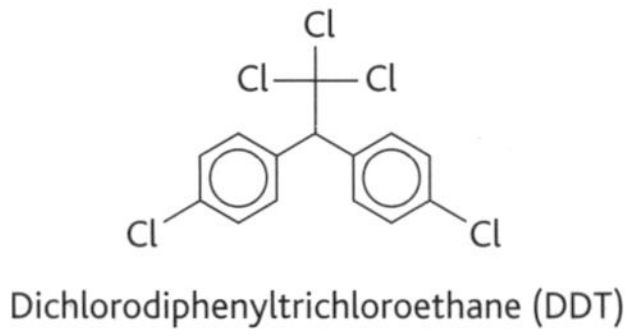

Aldrin

Dieldrin

Dichlorodiphenyltrichloroethane (DDT)

fig. 2.3.17 **Some organochlorine compounds.**

They tend to build up in the fat cells of animals which eat them, getting passed from animal to animal along food chains until they reach toxic levels in larger animals. For example, DDT sprayed on water to kill mosquito larvae and prevent malaria tends to kill large predatory birds, such as herons, at the end of the food chain.

- Lindane was first prepared by Faraday in 1825 but its insecticidal properties were not recognised until the early 1940s, about the same time as DDT was developed.
- Lindane acts on insects largely as a stomach poison, killing those that ingest it. It is not readily broken down and so persists in the environment and organisms, accumulating in animals that eat contaminated prey.
- In developing countries, lindane is used to control a broad spectrum of plant-eating and soil-dwelling pests, public health pests and animal ectoparasites (parasites such as bedbugs, fleas and lice that live outside the body). It is cheap to produce and its persistence in the soil is seen as an advantage.
- Several cases of human poisoning by lindane have been reported. Children are significantly more susceptible than adults to the toxic effects of lindane. In one case, a dose equivalent to 62.5 $mg\,kg^{-1}$ caused death. The highest intake of lindane is likely to occur from consumption of cereals, red meat and tomatoes.

Many of these compounds present us with an ethical dilemma.

fig. 2.3.18 **Using lindane to kill parasites on sheep.**

They bring benefits – for example, more food for people to eat because pests are destroyed, and protection from malaria by killing mosquitoes. On the other hand, they disrupt food chains, poison animals and even people, and can reduce biodiversity. Should they be used or not – and who is to decide?

Substitution reactions of benzene

All the substitution reactions of the benzene ring involve attack on the ring by **electrophiles**.

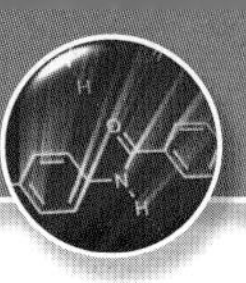

Reaction of benzene with fuming sulfuric acid

Fuming sulfuric acid reacts with benzene to produce benzenesulfonic acid. This reaction is called **sulfonation** – the reaction (shown in **fig. 2.3.19**) takes place at room temperature. It involves hazardous chemicals and so is rarely carried out in school laboratories.

fig. 2.3.19 A summary of sulfonation of benzene. A π complex is formed, and then a σ complex – notice that the reaction is reversible. Sulfonation is used in the manufacture of soapless detergents.

'Fuming' sulfuric acid is concentrated sulfuric acid that contains additional sulfur trioxide, with which benzene reacts more readily because the sulfur trioxide is an effective electrophile.

Reaction of benzene with halogens

Benzene reacts very differently with halogens depending on whether the reaction takes place in the light or in the dark.

As you saw earlier, in ultraviolet light benzene undergoes rapid free-radical addition reactions with chlorine and bromine. In the dark, a very different electrophilic substitution reaction takes place.

Benzene does not react with bromine unless there is a catalyst of iron(III) bromide or iron filings present – this catalyst is called a **halogen carrier**. The benzene, bromine and halogen carrier are refluxed together and bromobenzene is formed:

Iron can be used instead of iron(III) bromide because iron reacts with some of the bromine present to form iron(III) bromide:

$$2Fe(s) + 3Br_2(l) \rightarrow 2FeBr_3(s)$$

The iron(III) bromide polarises the bromine molecule, forming a complex:

$$Br{-}Br + FeBr_3 \rightarrow \overset{\delta+}{Br}{-}\overset{\delta-}{Br}.FeBr_3$$

This acts as an electrophile, forming first a weak π complex, and then the more stable σ complex:

A proton is then lost and this reacts to regenerate the iron(III) bromide, forming HBr:

$$H^+ + FeBr_4^- \rightarrow FeBr_3 + HBr$$

Nitration of benzene

Benzene does not react with concentrated nitric acid. However, it does react with a mixture of concentrated nitric and concentrated sulfuric acid – this is called a **nitrating mixture**. The product is nitrobenzene, providing that the temperature is kept below 55°C:

$$C_6H_6 + HNO_3(conc) \xrightarrow[50°C]{conc\ H_2SO_4} C_6H_5NO_2 + H_2O$$

nitrobenzene

fig. 2.3.20 Apparatus for the nitration of benzene.

Figure 2.3.20 shows the apparatus used to make nitrobenzene. Concentrated sulfuric acid is corrosive; concentrated nitric acid is corrosive and a powerful oxidant.

The round-bottomed reaction flask with the nitrating mixture is held in a beaker of cold water by a clamp. The reaction flask has a tap funnel containing benzene. The benzene is slowly added dropwise to the mixture in the flask. The acids react and give out heat, as does the addition of benzene to the nitrating mixture – this is why the cold water and slow addition are needed. If the temperature rises above 55°C, multiple substitution occurs and 1,3-dinitrobenzene is formed, resulting in a mixture of products:

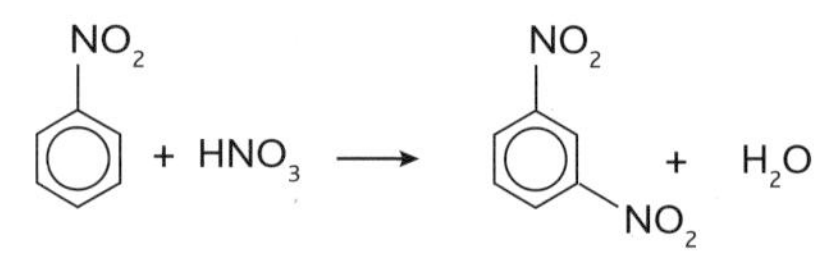

HSW The nitronium ion, NO_2^+

The first step of nitration involves the production of the NO_2^+ (nitronium) cation in the reaction of concentrated nitric acid and concentrated sulfuric acid:

$$HNO_3(l) + 2H_2SO_4(l) \rightleftharpoons NO_2^+(sol) + 2HSO_4^-(sol) + H_3O^+(sol)$$

The nitric acid acts as a base and accepts protons from the sulfuric acid. Evidence for the existence of this cation comes from being able to isolate salts – $NO_2^+ClO_4^-$ for example – and from spectroscopy.

$$C_6H_6 + NO_2^+ \longrightarrow [C_6H_6NO_2]^+ \longrightarrow C_6H_5NO_2 + H^+$$

fig. 2.3.21 The reaction mechanism for the nitration of benzene.

Figure 2.3.21 shows the mechanism of the reaction. There is a ring of negative charge above and below the benzene ring (caused by the delocalised electrons in the extended π bonds) and this allows the nitronium ion to act as an electrophile, forming a weak complex using a pair of electrons from the ring.

In the second stage of the reaction, this intermediate breaks down very rapidly by losing a proton to form the product.

How the use of nitrobenzene has changed

Nitrobenzene is a toxic, flammable, pale yellow, liquid aromatic compound with an odour of bitter almonds. Approximately 95% of nitrobenzene is used in the production of phenylamine ($C_6H_5NH_2$), which is used to make diazonium dyes (page 216).

fig. 2.3.22 Nitrobenzene is used in making rubber, pesticides and pharmaceuticals such as paracetamol. It is used in shoe and floor polishes, leather dressings, paint solvents and other materials to mask unpleasant odours. It is used as a solvent for cellulose and in petroleum refining. It is redistilled to form 'oil of mirbane' which has been used as an inexpensive perfume for soaps.

Although nitrobenzene is not recognised as a carcinogen for humans, there is concern about it forming cancers in mice and rats. A study to evaluate a possible relationship between human cancers and exposure to nitrobenzene was inconclusive. However, prolonged exposure may cause serious damage to the central nervous system, impair vision, and cause liver or kidney damage, anaemia and lung irritation. Inhalation of fumes may induce headache, nausea, fatigue, dizziness, and weakness in the arms and legs, and may be fatal in rare cases. The oil is readily absorbed through the skin and may increase heart rate and cause convulsions and even death. Ingestion may similarly cause headaches, dizziness, nausea, vomiting and gastrointestinal irritation. As a result, it is no longer used in soaps.

Alkylation of benzene

This reaction provides a means of substituting an alkyl group in a benzene ring. It involves refluxing benzene with a halogenoalkane in the presence of a halogen carrier catalyst, such as aluminium chloride:

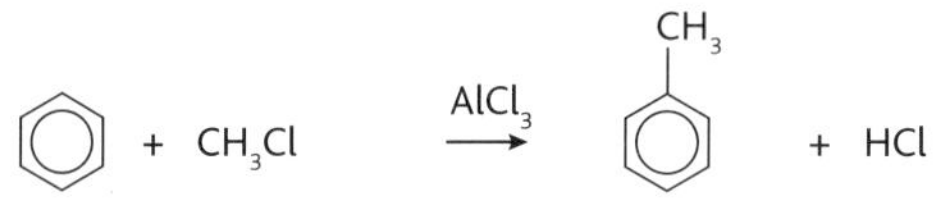

$$C_6H_6 + CH_3Cl \xrightarrow{AlCl_3} C_6H_5CH_3 + HCl$$

benzene + chloromethane ⟶ methylbenzene + hydrogen chloride

The halogen carrier, aluminium chloride, polarises the halogenoalkane molecule, promoting the formation of an electrophilic alkyl cation (**fig. 2.3.23**). This is attracted to the benzene ring, forming first a π complex and then a σ complex. This breaks down to form the alkylbenzene product.

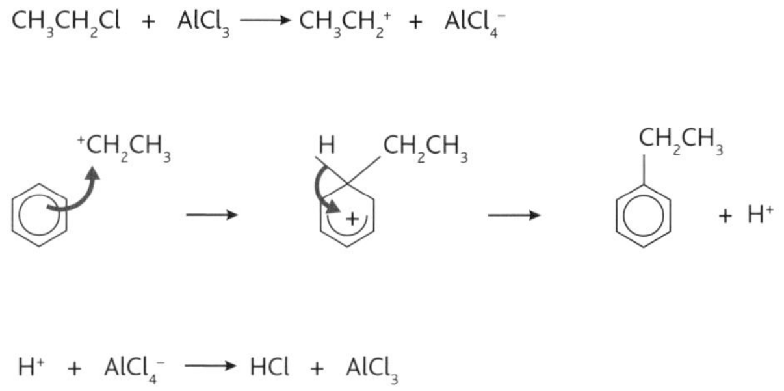

$$CH_3CH_2Cl + AlCl_3 \longrightarrow CH_3CH_2^+ + AlCl_4^-$$

$$C_6H_6 + {}^+CH_2CH_3 \longrightarrow [C_6H_6CH_2CH_3]^+ \longrightarrow C_6H_5CH_2CH_3 + H^+$$

$$H^+ + AlCl_4^- \longrightarrow HCl + AlCl_3$$

fig. 2.3.23 Alkylation of benzene showing the formation and role of the electrophile.

Acylation of benzene

A reaction similar to alkylation takes place when benzene is refluxed with an acyl chloride. The product is a ketone. For example:

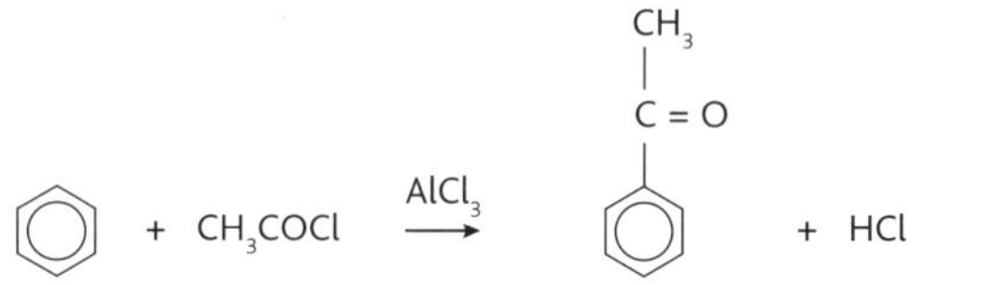

$$C_6H_6 + CH_3COCl \xrightarrow{AlCl_3} C_6H_5COCH_3 + HCl$$

benzene + ethanoyl chloride ⟶ phenylethanone + hydrogen chloride

As **fig. 2.3.24** shows, the mechanism is similar to that for alkylation. The product is more useful this time – it can easily be reduced to form a secondary alcohol, so is often a helpful intermediate in the synthesis of organic chemicals (see **chapter 2.5**).

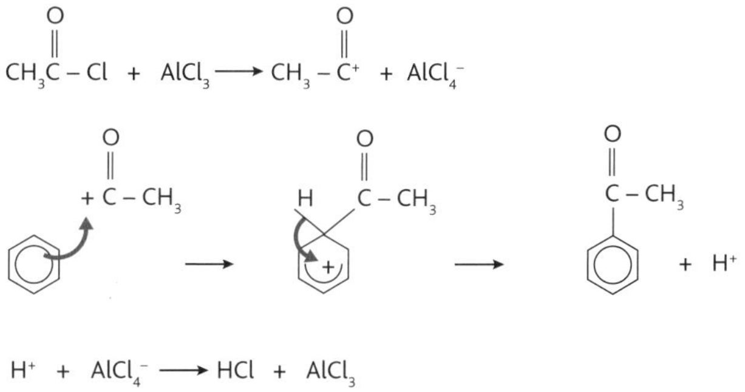

$$CH_3COCl + AlCl_3 \longrightarrow CH_3-C^+{=}O + AlCl_4^-$$

$$C_6H_6 + {}^+COCH_3 \longrightarrow [C_6H_6COCH_3]^+ \longrightarrow C_6H_5COCH_3 + H^+$$

$$H^+ + AlCl_4^- \longrightarrow HCl + AlCl_3$$

fig. 2.3.24 Acylation of benzene showing the formation and role of the electrophile.

Both alkylation and acylation, as described above, are types of **Friedel–Crafts reaction**.

HSW The work of Friedel and Crafts

These reactions of benzene – alkylation and acylation – were discovered by Charles Friedel and James Mason Crafts in 1877. Friedel was a French chemist and mineralogist – he was professor of chemistry at the Sorbonne in Paris. Crafts was an American chemist who came to Europe, first to work with Robert Bunsen in Heidelberg and then with Wurtz in Paris. It was in Paris that Friedel and Crafts first met. Crafts returned to America to work as a professor in two leading universities. In 1874, he returned to Paris to work with Friedel – which eventually led to the development of the Friedel-Crafts reactions.

Development of Friedel–Crafts reactions

Although Friedel–Crafts reactions were first discovered over 130 years ago, it does not mean that chemists are not still making progress in using these reactions in different circumstances. They have found different halogen carriers and different reactants to use them with.

Poly(phenylethene) is an **addition polymer** (see page 134 in *Edexcel AS Chemistry*). It is produced by the addition polymerisation of phenylethene:

poly(phenylethene)

1,4-dimethyl-2,5-dichloromethylbenzene

chain 1

cross-link

chain 2

fig. 2.3.25 **A Friedel–Crafts reaction to produce cross-linked poly(phenylethene).**

An experimental technique has been developed to produce poly(phenylethene) with cross-linking, joining different chains in the structure. Antimony trichloride, $SbCl_3$, is the halogen carrier and the reactants are poly(phenylethene) and 1,4-dimethyl-2,5-dichloromethyl benzene. These are dissolved in a mixture of 1,2-dichloroethane and silicone oil. The reaction is summarised in **fig. 2.3.25.**

The product is colourless and made up of spherical particles of a very high purity and it can be produced in short reaction times. This modification of poly(phenylethene) alters the properties of the polymer, increasing its range of uses.

For example, **aryl ketones** can be prepared by a Friedel–Crafts reaction using atomised aluminium metal powder in solvent-free conditions under microwave irradiation. This produces extremely good yields – the aluminium metal powder can be recovered and reused several times by simple washing with ethoxyethane. This is, therefore, a new and 'greener' method for synthesis of aryl ketones.

SC Friedel and Crafts tried to use their **acylation reaction** to produce benzaldehyde, but without any success. Research why this was not possible and find out how it is now possible using the **Gattermann–Koch reaction**.

The effect of substituent groups on reactions of the benzene ring

When an atom or group replaces one of the hydrogens in a benzene ring, it may affect the electron density of the ring, so we should not be surprised that substituent groups affect the chemistry of benzene.

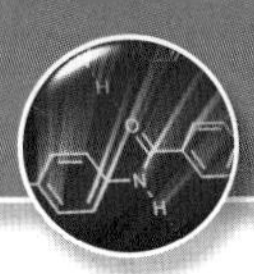

Some atoms/groups allow subsequent substitutions to occur faster than they would compared to benzene itself; other atoms/groups slow down the rate of any further substitution.

The main points about substitutions in benzene derivatives can be summarised as follows.

- Some functional groups donate electron density to the ring, causing electrophilic substitutions to occur faster than they would in benzene itself (see **fig. 2.3.26**).
- Some functional groups withdraw electron density from the ring, causing electrophilic substitutions to occur more slowly than they would in benzene itself.

SC Look at the reactions of alkenes on pages 129–131 of *Edexcel AS Chemistry*. What are the differences and similarities between the reactions of alkenes and those of benzene?

fig. 2.3.26 When a nitro group, NO_2, is substituted into methylbenzene, it tends to be directed to either the 2- or 4- position. The nitration of methylbenzene occurs at room temperature – benzene requires heating at 50°C for about an hour for nitration to occur.

SC The problems caused by DDT and similar chlorinated hydrocarbons are well known.
Suggest why some people advocate the use of these chemicals in parts of Africa.
Suggest some advantages and disadvantages of using such chemicals.

Questions

1 This section has covered the reactions of benzene with hydrogen, bromine and concentrated sulfuric acid.

a Which reactions are addition and which are substitution?

b Draw a table comparing the reactions of the alkanes, alkenes and benzene (arenes) with hydrogen, bromine and concentrated sulfuric acid.

2 Iron filings can be used as a catalyst for the bromination of benzene, although it is not strictly a halogen carrier. Explain why iron filings work.

3 Derivatives of benzene containing more than one substituent are named by numbering the ring with the position of the substituent groups. Hence, A is methylbenzene, B is 1,3-dimethylbenzene and C is 1-butyl-3,4-dimethylbenzene.

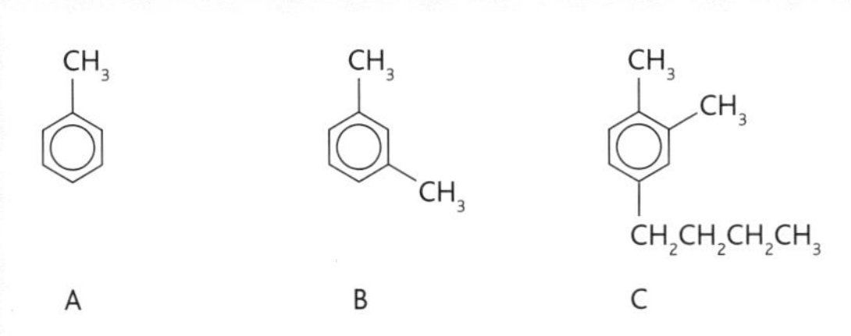

Name the four compounds below.

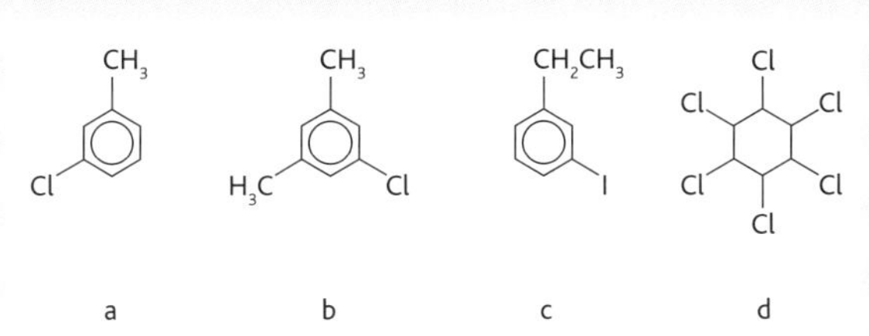

4 Draw the displayed formula of
a butylbenzene **b** 1,2-dimethyl-4-iodobenzene.

5 Benzene reacts, separately, with bromine, fuming sulfuric acid and a nitrating mixture. Using C_6H_6 to represent benzene and C_6H_5 to represent the phenyl group, write equations for these reactions. Give the essential conditions for the reactions.

6 **a** Suggest how cross-linking poly(phenylethene) chains might alter the properties of the polymer.

b Suggest two reasons why cross-linking of poly(phenylethene) chains is likely to be an economic process.

Reactions of phenol [5.4.1e]

Like an alcohol, a phenol contains the hydroxyl, –OH, group.

phenol ethanol

fig. 2.3.27 In an alcohol, the hydroxyl group is attached to a carbon atom in a carbon chain; in a phenol, it is attached to a carbon atom in a benzene ring.

Substitution reactions in phenol

Phenol undergoes similar electrophilic substitution reactions to benzene, where one or more hydrogen atoms may be replaced by, for example –Br, $-NO_2$, $-SO_3H$, R– or RCO–. These reactions occur much more readily in phenol than in benzene. This is because of the –OH group on the benzene ring, the presence of which also directs the incoming groups of the substitution reaction to particular positions around the ring.

Reactions of a phenol

You may carry out reactions with phenol and bromine water and with dilute nitric acid. Make sure your skin does not come into contact with phenol.

Reaction of phenol with bromine water

When bromine water is added to a solution of phenol in water, multi-electrophilic substitution takes place immediately, without heating or introducing a halogen carrier. The bromine water is decolorised. The product is a white precipitate of 2,4,6-tribromophenol, which smells of antiseptic:

$C_6H_5OH(aq) + 3Br_2(aq) \longrightarrow C_6H_2Br_3OH(s) + 3HBr(aq)$

phenol bromine solution 2,4,6-tribromophenol hydrogen bromide

The corresponding chlorine compound produced by reacting chlorine water and phenol is trichlorophenol (TCP), which is widely used as an antiseptic in solution. Phenol itself was originally used as the first antiseptic by Joseph Lister in 1857, but it can damage and burn the skin. Not only is TCP much more effective than phenol at killing bacteria, but it can be used without the unpleasant side-effects.

fig. 2.3.28 Trichlorophenol, produced by the multi-electrophilic substitution of phenol, kills bacteria and can be used safely on the skin.

Reaction of phenol with dilute nitric acid

When dilute nitric acid is added to a solution of phenol, a white precipitate of 2,4,6-trinitrophenol is formed. Again this is a multi-electrophilic substitution reaction. Compared to benzene, phenol does not need sulfuric acid to assist nitration and the nitric acid is dilute. 2,4,6-trinitrophenol is sometimes called **picric acid**. Again, the effect of the –OH group in activating the benzene ring can be seen – and the substitutions happen in the 2, 4 and 6 positions of the ring:

$C_6H_5OH(aq) + 3HNO_3(aq) \longrightarrow C_6H_2(NO_2)_3OH(s) + 3H_2O(l)$

phenol nitric acid 2,4,6-trinitrophenol water

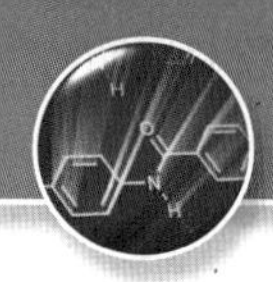

Phenol is slightly acidic and ethanol is neutral. Do some research and find out the reason for this difference.

Why do electrophilic substitution reactions occur easily with phenol?

The hydroxyl group of a phenol has an oxygen atom with two pairs of non-bonding electrons. One pair is drawn into the delocalised π electron system. The benzene ring is now better at attracting electrophiles, so reactions occur more easily. The 2, 4 and 6 positions are particularly susceptible to attack.

HSW Explosives

fig. 2.3.29 Explosives are used for peaceful purposes – in limestone quarries, for example – but all too often they are used for war or similar purposes.

Picric acid was first made in 1742 by nitrating animal horn or natural resin. Its first synthesis, by nitrating phenol, was done in 1841. It was used in World War I as an explosive, but was largely replaced by TNT (trinitrotoluene) and cordite in the twentieth century because it was too sensitive and exploded very easily. Dry picric acid is relatively sensitive to shock and friction, so laboratories that do use it store it in bottles under a layer of water, rendering it safe.

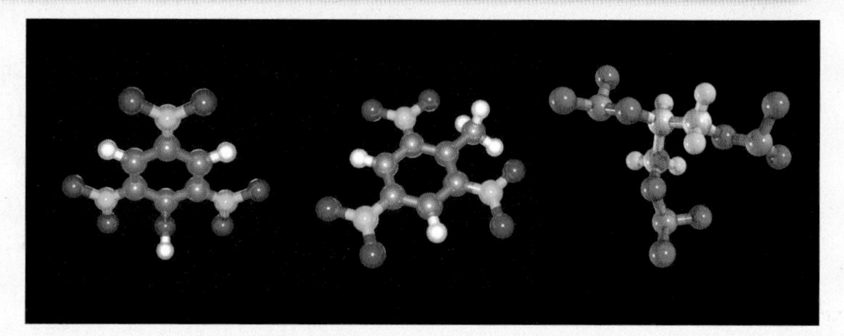

fig. 2.3.30 Computer models of picric acid, trinitrotoluene and trinitroglycerine.

TNT is now one of the most commonly used explosives for military and industrial applications. It is valued because it is insensitive to shock and friction, which reduces the risk of accidental detonation. When TNT explodes the reaction is highly exothermic and there is a large increase in entropy because most of the products are gases. It does not explode easily because the reaction has a high activation energy. It took a long time to realise its power as an explosive just because it is so difficult to detonate.

Nitroglycerine is an explosive liquid that is extremely sensitive to shock. In the early days, when impure nitroglycerine was used, it was very difficult to predict the conditions that would cause it to explode. Alfred Nobel was the first to produce nitroglycerine on an industrial scale when he was working to improve it as an explosive for the mining industry. He made one of his most important discoveries when he found that a mixture of nitroglycerine, an oily fluid and *kieselguhr* could be turned into a paste that could be kneaded and shaped into rods suitable for insertion into drilling holes. He called his paste **dynamite**.

Questions

1 Methylbenzene (toluene), $C_6H_5CH_3$, can be nitrated to form TNT but it takes three stages. Phenol can be nitrated very easily with dilute nitric acid. Explain the difference.

2 Compare the reactions of benzene and phenol with bromine water and nitric acid.

3 Phenylmethanol, $C_6H_5CH_2OH$, has three isomers. Draw their displayed formulae.

2.4 Organic nitrogen compounds: amines, amides, amino acids and proteins

Amines and the formation and reactions of aromatic amines [5.4.2a (i), b (i), (ii), (iii), (iv), c]

There are many important organic nitrogen compounds, including the **amides** and the **amines**. Probably the best known and most widely occurring organic nitrogen compounds are the **amino acids** – the fundamental building blocks of the proteins making up a considerable part of every living organism. We are going to be looking at organic nitrogen compounds in some detail in this chapter.

Ammonia – building block of the organic nitrogen compounds

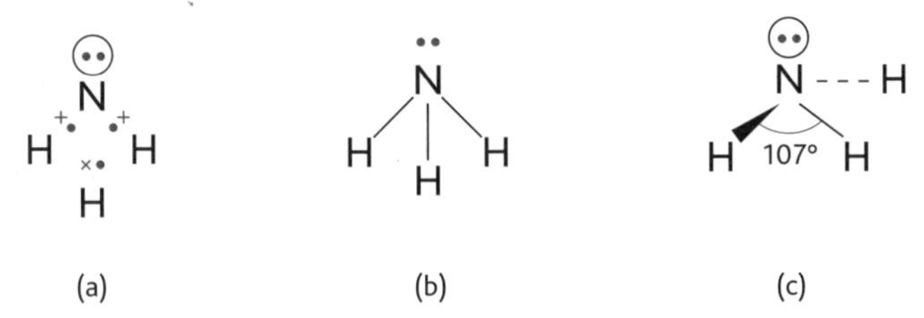

fig. 2.4.1 Different ways of showing the electronic arrangement and shape of the ammonia molecule.

The most important feature of an ammonia molecule is the lone pair of electrons on the nitrogen atom. This accounts for the proton-acceptor property of ammonia that makes it a **base** (see **chapter 1.5**).

The amines

The amines are organic compounds based on ammonia, where one or more of the three hydrogen atoms have been substituted by an **alkyl** or an **aryl** group. They do not occur in the free state widely in the living world, except in putrefying and decaying flesh – they are formed by the action of bacteria on amino acids when proteins break down. Amines also account for some of the body odours detected from humans who do not wash frequently – bacteria on the skin digest amino acids from the sweat of the groin and armpits.

fig. 2.4.2 Di- and trimethylamine occur in rotting fish and account for their very distinctive smell.

Preparing amines

If ammonia is heated under pressure with a halogenoalkane, the ammonia's hydrogen atoms can be replaced by alkyl groups and the resulting products are amines. For example, chloromethane and ammonia react like this:

$$CH_3Cl(g) + NH_3(g) \rightarrow CH_3NH_2(g) + HCl(g)$$
chloromethane ammonia methylamine hydrogen chloride

$$CH_3Cl(g) + CH_3NH_2(g) \rightarrow (CH_3)_2NH(g) + HCl(g)$$
chloromethane methylamine dimethylamine hydrogen chloride

$$CH_3Cl(g) + (CH_3)_2NH(g) \rightarrow (CH_3)_3N(g) + HCl(g)$$
chloromethane dimethylamine trimethylamine hydrogen chloride

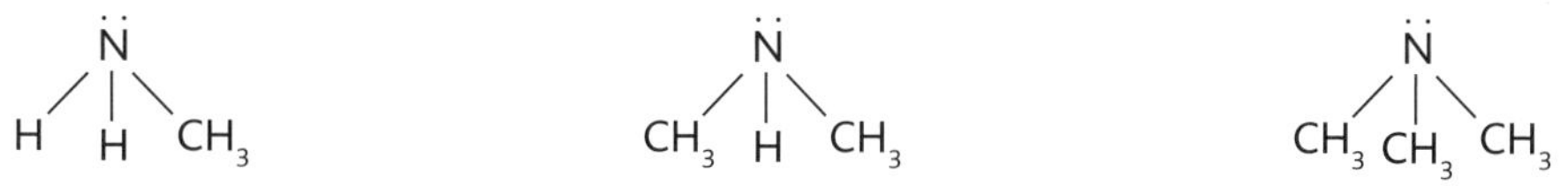

fig. 2.4.3 The structures of three amines – methylamine, dimethylamine and trimethylamine. Notice that they all have a nitrogen atom with a lone pair of electrons.

In *Edexcel AS Chemistry* on page 209 you found out that alcohols can be classified as **primary**, **secondary** or **tertiary alcohols** according to the position of the –OH group in the molecule.

Amines can also be classified using these same terms, but according to the number of hydrogen atoms that have been replaced by alkyl groups:

- if one hydrogen atom has been replaced, e.g. methylamine, then it is a primary amine
- if two hydrogen atoms have been replaced, e.g. dimethylamine, then it is a secondary amine
- if three hydrogen atoms have been replaced, e.g. trimethylamine, then it is a tertiary amine.

Just as with alcohols, whether an amine is primary, secondary or tertiary will affect the properties of the amine.

It is also possible to replace hydrogen atoms in a molecule of ammonia with aryl groups. For example, phenylamine (aniline) is an example of an aryl amine:

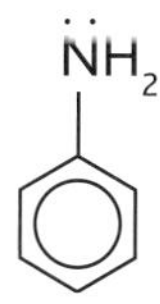

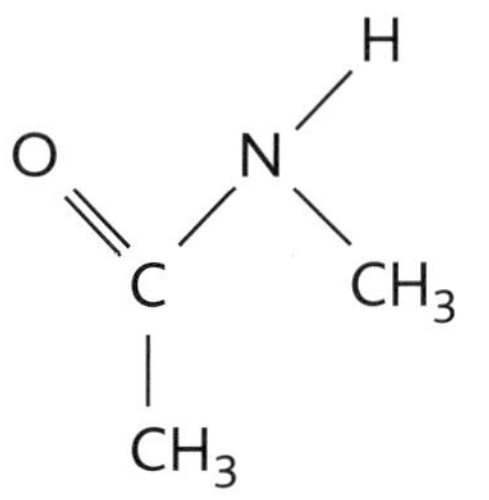

In chapter 1.6, we saw that a hydrogen atom on an **electronegative atom** such as oxygen or nitrogen could be replaced with a CH_3CO– group using ethanoyl chloride. A hydrogen atom on a primary or secondary amine can be replaced in a similar way using ethanoyl chloride or ethanoic anhydride. The product is an amide. For example:

$$CH_3NH_2 + CH_3COCl \rightarrow CH_3CONHCH_3 + HCl$$
methylamine ethanoyl chloride N-methylethanamide

fig. 2.4.4 Structure of N-methylethanamide. 'N-methyl' indicates that the methyl group is attached to the nitrogen atom.

Properties of amines

You may carry out experiments with butylamine and phenylamine to show the common properties of amines. Experiments with butylamine should be carried out in a fume cupboard. Phenylamine is toxic and it may be suggested that you use ethyl 4-aminobenzoate instead.

Reduction of aromatic nitro-compounds to form aryl amines

You will know from chapter 2.3 that benzene and other **aromatic compounds** can be nitrated to form nitro-compounds using a mixture of concentrated nitric acid and concentrated sulfuric acid. These nitro-compounds can produce aromatic amines, which are very useful for making other compounds.

For example, phenylamine can be produced in the laboratory by the reduction of nitrobenzene using a mixture of tin and concentrated hydrochloric acid:

$$C_6H_5NO_2 + 3H_2 \rightarrow C_6H_5NH_2 + 2H_2O$$

The reaction can be carried out in the reflux apparatus shown in fig. 2.4.5.

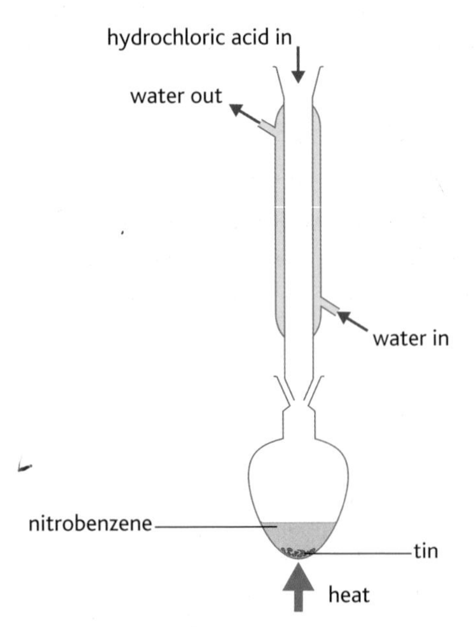

fig. 2.4.5 Reflux apparatus for the conversion of nitrobenzene to phenylamine.

Tin reacts with concentrated hydrochloric acid to produce tin(II) ions and hydrogen:

$$Sn(s) + 2HCl(aq) \rightarrow Sn^{2+}(aq) + 2Cl^-(aq) + H_2(g)$$

The tin(II) ions are oxidised to tin(IV) ions and these reduce the $-NO_2$ group to the $-NH_2$ group:

$$Sn^{2+}(aq) \rightarrow Sn^{4+}(aq) + 2e^-$$

$$C_6H_5NO_2(l) + 6H^+(aq) + 6e^- \rightarrow C_6H_5NH_2(l) + 2H_2O(l)$$

The flask is then cooled and sodium hydroxide solution is added to redissolve the initial precipitate of tin(IV) hydroxide, $Sn(OH)_4$ – tin(IV) hydroxide is amphoteric and with excess alkali produces the soluble $Sn(OH)_6^{2-}$ ion.

Water is then added and **steam distillation** is used to separate the mixture of phenylamine and water. The distillate initially collected is cloudy because it is an emulsion of phenylamine and water. When the distillate is clear, only water is distilling over. Powdered sodium chloride is added to the distillate and the mixture is transferred to a separating funnel. This is shaken well – phenylamine is significantly soluble in water, but very much less so in saturated sodium chloride solution. This process is called **salting out**.

The mixture is then transferred to a separating funnel and ethoxyethane is added and shaken, relieving the pressure in the separating funnel occasionally by opening the tap. The layers are allowed to separate, and then the lower aqueous layer is run off into a small beaker.

The ethoxyethane layer is transferred to a small conical flask, and further extraction of the aqueous layer with further portions of ethoxyethane is carried out. The ethoxyethane extracts are combined – this is an example of **solvent extraction**.

Pellets of potassium hydroxide are added to dry the ethoxyethane extract. Potassium hydroxide is better than calcium chloride or anhydrous sodium sulfate because, being alkaline, it will remove any traces of hydrochloric acid.

The ethoxyethane is distilled off and finally phenylamine is distilled off, collecting the fraction between 180 and 185°C.

In industry a similar method is used to reduce nitrobenzene to phenylamine, though iron is used rather than tin because it is cheaper.

Properties of amines

The lower members of the **homologous series** of alkylamines are gases or volatile liquids. In primary and secondary amines, where there are hydrogen atoms attached to the nitrogen atom, there is the possibility of hydrogen bonding similar to the hydrogen bonding in ammonia (*Edexcel AS Chemistry* page 163). In tertiary amines hydrogen bonding is not possible and the interactions between molecules are only van der Waals forces – therefore, trimethylamine boils at a lower temperature than methylamine or dimethylamine.

Aryl amines, such as phenylamine, are much less volatile and are usually liquids at room temperature. They are also only slightly soluble in water because the aromatic nature of the benzene ring outweighs the tendency of the $–NH_2$ group to form hydrogen bonds. Aryl amines, however, do dissolve in organic solvents.

Like ammonia, lower members of the amines are very miscible with water because hydrogen bonding can occur between water and amine molecules. A solution of an amine in water is alkaline – compare with a solution of ammonia in water. Amines are quite strong bases – the nitrogen atom acts as a proton acceptor:

$$CH_3CH_2NH_2(aq) + H_2O(l) \rightleftharpoons CH_3CH_2NH_3^+(aq) + OH^-(aq)$$

Ammonia, being basic, can form salts with acids, for example:

$$NH_3(aq) + HCl(aq) \rightarrow NH_4^+Cl^-(aq)$$

Amines can form salts with acids in a similar way:

$$\underset{\text{ethylamine}}{CH_3CH_2NH_2(aq)} + HCl(aq) \rightarrow \underset{\text{ethylammonium chloride}}{CH_3CH_2NH_3^+Cl^-(aq)}$$

If dilute hydrochloric acid is added to a solution of ethylamine, you will notice a rise in temperature (**exothermic reaction**) and the loss of the fishy amine smell.

Primary alkyl amines, such as methylamine, are stronger bases than ammonia. This is explained by a shift of electrons from the alkyl group increasing the electron density on the nitrogen atom. This means that the nitrogen atom can hold a proton more strongly.

Phenylamine, an aryl amine, is a weaker base than ammonia. The electrons on the nitrogen atom interact with the delocalised π electrons in the ring and, as a result, the lone pair on the nitrogen atom is less available. This reduces the ease of acceptance of protons by the $–NH_2$ group. Phenylamine will form salts with strong acids, where excess H^+ ions move the equilibrium to the right, producing the salt:

$$C_6H_5NH_2(aq) + H^+(aq) \rightleftharpoons C_6H_5NH_3^+(aq)$$

phenylamine — phenylammonium ion

SC Provide as much evidence as you can to show that amines are stronger bases than ammonia, and that phenylamine is a weaker base than ammonia. You may, for example, choose to investigate the K_a values for the conjugate acid of butylamine, $C_4H_9NH_3^+$, and the conjugate acid of phenylamine, $C_6H_5NH_3^+$.

Ammonia forms complex ions with transition metal cations, for example:

$$4NH_3(aq) + [Cu(H_2O)_6]^{2+}(aq) \rightleftharpoons [Cu(NH_3)_4(H_2O)_2]^{2+}(aq) + 4H_2O(l)$$

Amines form complex ions in a similar way:

$$2\,C_6H_5NH_2(aq) + [Cu(H_2O)_6]^{2+}(aq) \rightleftharpoons Cu(C_6H_5NH_2)_2(H_2O)_4(aq) + 2H_2O(l)$$

HSW Phenylamine and the fight against bacteria

The lone pair of electrons on the nitrogen atom of phenylamine has a tendency to become partly delocalised around the benzene ring. This gives a high electron density around the ring, which in turn means that phenylamine is easily oxidised and readily undergoes **electrophilic substitution** of the benzene ring. What has this to do with the fight against disease-causing bacteria?

4-aminobenzenesulfonamide is a derivative of phenylamine that can be synthesised as a result of the ease of this substitution. This compound and other similar ones derived from it form a class of drugs known as the **sulfonamides**. These were some of the earliest drugs to be really effective in destroying bacteria. They have played a major part in reducing the numbers of deaths from pneumonia, from tuberculosis and from infection after childbirth. They are still in use today.

fig. 2.4.6 4-aminobenzenesulfonamide has given rise to a whole family of antibacterial drugs.

HSW Neurotransmitters

Ecstasy, the illegal drug, is an amine. Its full name is 3,4-methylenedioxymethamphetamine (or MDMA). It produces feelings of energy, euphoria, empathy and love. This is because it causes the release of serotonin and dopamine in the brain. These **neurotransmitters** are involved in the regulation of mood – high levels are associated with positive feelings. The exact mechanism by which the drug causes the release of the neurotransmitters remains unclear.

The drug also interferes with the normal temperature regulation mechanisms of the body. This can lead to dangerous, and even fatal, overheating. Some people drink excessive amounts of water to try to prevent overheating. This can also prove fatal because the water can cause swelling of the brain, which can in turn lead to coma and death.

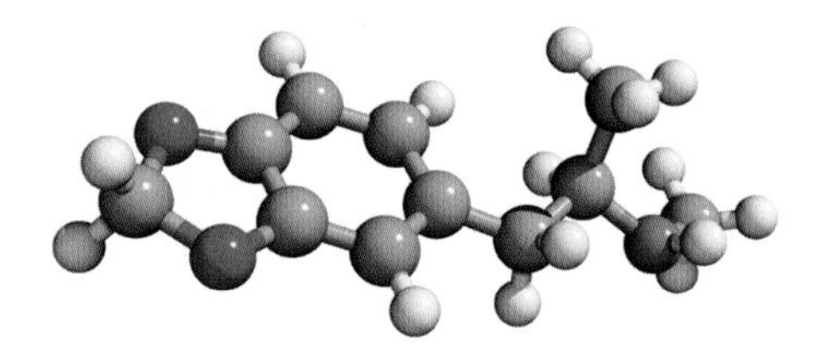

fig. 2.4.7 In this model, showing the structural formula of the drug ecstasy, the atoms are colour coded – carbon (grey), nitrogen (blue), oxygen (red) and hydrogen (white).

Questions

1 Ammonia reacts with chloromethane to form, in theory, methylamine.

- a Why is it difficult to produce a pure sample of methylamine in this reaction?
- b What could you do to improve the yield of methylamine?

2 a What is the molecular formula of the compound shown here?

- b Is it a primary, secondary or tertiary amine? Explain your answer.

3 Explain why the production of phenylamine, $C_6H_5NH_2$, from nitrobenzene, $C_6H_5NO_2$, is considered to be a redox reaction.

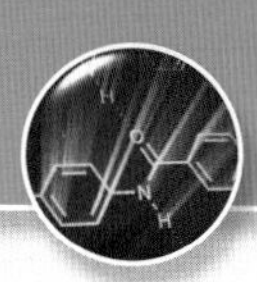

Making paracetamol [5.4.2b (v)]

Earlier, in chapter 1.6, you saw examples of **ethanoylation** – the replacement of a hydrogen atom attached to a nitrogen or oxygen atom by CH_3CO– using ethanoyl chloride or ethanoic anhydride. In this section we are going to consider this reaction further, in particular with respect to manufacturing paracetamol.

Reaction of an amine with ethanoyl chloride

If butylamine is mixed with ethanoyl chloride, smoky white fumes of hydrogen chloride are seen to indicate that a reaction is taking place (fig. 2.4.8). Much of the hydrogen chloride that is made reacts with any remaining base in the reaction mixture – unreacted butylamine, and even the product of the reaction, N-butyl ethanamide.

$$CH_3COCl + H_2N{-}CH_2CH_2CH_2CH_3 \longrightarrow CH_3CO{-}NH{-}CH_2CH_2CH_2CH_3 + HCl$$

fig. 2.4.8 Equation for the reaction of butylamine and ethanoyl chloride.

The amine, with a lone pair of electrons on the nitrogen atom, acts as a **nucleophile** and attacks the positive centre in the ethanoyl chloride molecule (fig. 2.4.9). This positive centre is caused by withdrawal of electrons in the C=O bond because of the greater electronegativity of the oxygen atom – this is further enhanced by the electronegative chlorine atom. The nitrogen atom attacks the carbon atom and this is accompanied by the loss of a chlorine and a proton.

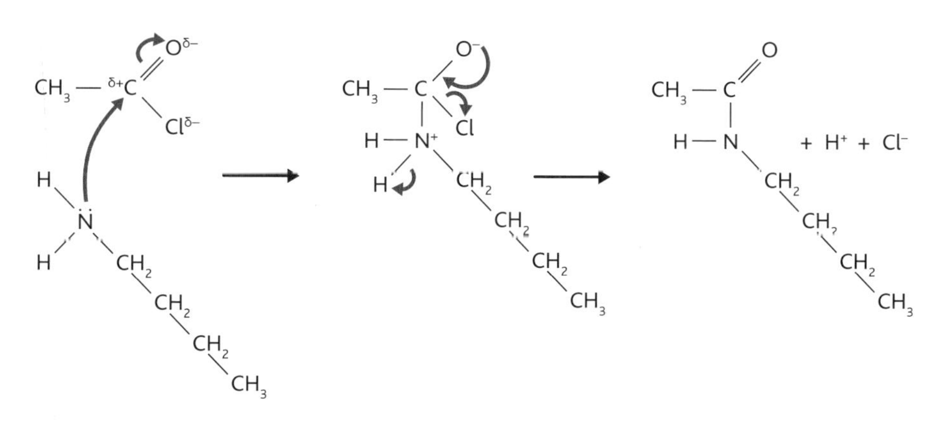

fig. 2.4.9 Ethanoylation of butylamine.

Paracetamol

Paracetamol is a widely used painkiller (analgesic) that is manufactured in huge quantities around the world. It was first synthesised by Harmon Northrop Morse in 1878 using the reduction of 4-nitrophenol with tin in glacial ethanoic acid. However, paracetamol was not used as a painkiller for another 15 years.

In 1948 Julius Axelrod showed that acetanilide, a substance previously studied as a painkiller, breaks down into paracetamol – this also acts as an analgesic but is better tolerated by the body.

However, a relatively small overdose of paracetamol can cause fatal damage to the liver, kidneys and brain – it is effective and safe, but must be used with caution.

Manufacture of paracetamol

Figure 2.4.10 shows the structure of the analgesic paracetamol.

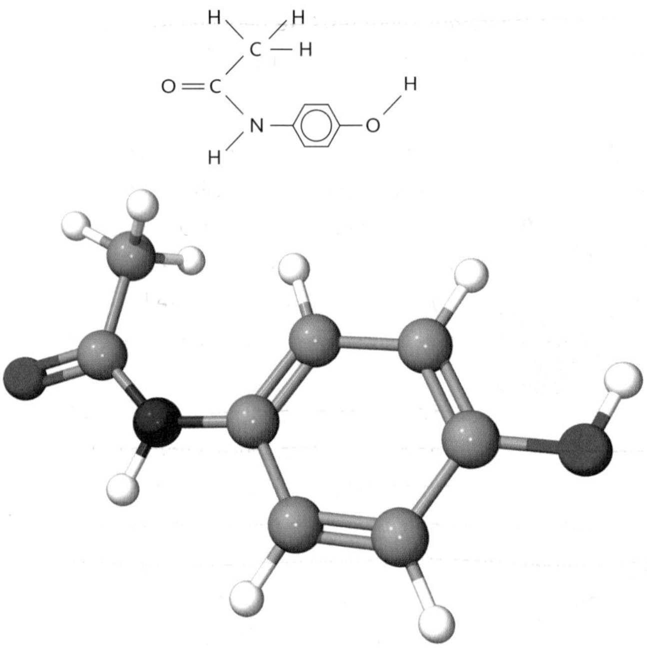

fig. 2.4.10 The displayed formula and structure of a paracetamol molecule.

Paracetamol can be made in a stepwise procedure using phenol as the starting material.

Step 1: Phenol is nitrated using sulfuric acid and sodium nitrate. Unlike the nitration of benzene, it is not necessary to use a mixture of concentrated nitric and sulfuric acids because the ring is activated by the –OH group (**chapter 2.3**). A mixture of two isomers is produced – 2-nitrophenol and 4-nitrophenol:

2-nitrophenol 4-nitrophenol

Step 2: The two isomers are separated by fractional distillation. 4-nitrophenol has a higher boiling temperature than 2-nitrophenol because more effective hydrogen bonding occurs between molecules in 4-nitrophenol.

Step 3: The 4-nitrophenol is reduced to 4-aminophenol using a reducing agent such as sodium tetrahydridoborate(III) in an alkaline medium:

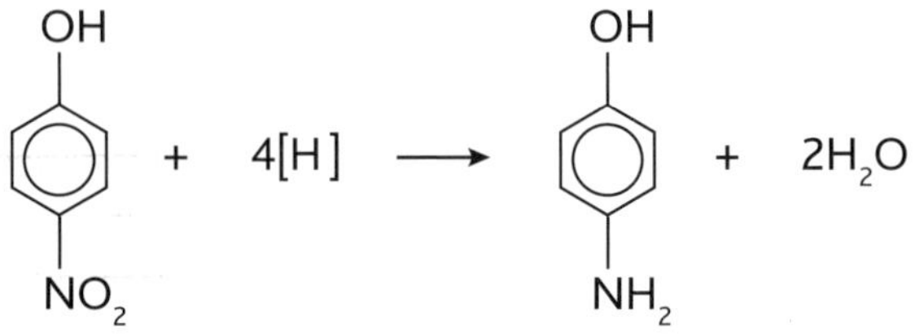

$$HOC_6H_4NO_2 + 4[H] \longrightarrow HOC_6H_4NH_2 + 2H_2O$$

Step 4: The 4-aminophenol is reacted with ethanoic anhydride to give paracetamol:

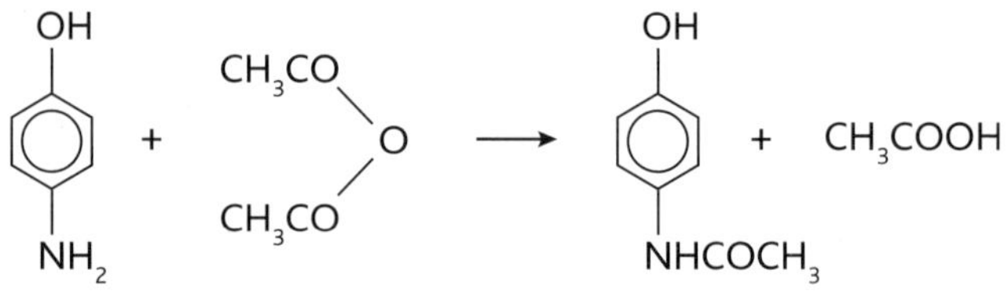

$$HOC_6H_4NH_2 + (CH_3CO)_2O \longrightarrow HOC_6H_4NHCOCH_3 + CH_3COOH$$

The CO.O.CO arrangement should look like this:

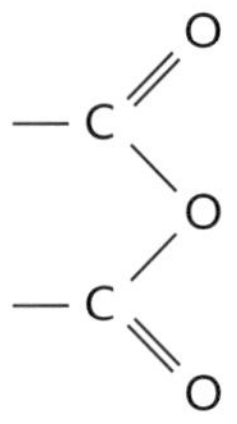

Many processes used for the synthesis of drugs involve molecules with **chiral** centres. This leads to the need to consider the stereochemistry of the products at each stage. There is no chiral centre in paracetamol, or any intermediates, which makes this manufacturing process much simpler.

Why do we need to separate the isomers during the manufacture of paracetamol – and what would be the two products if we did not? Would this matter to its use as a drug?

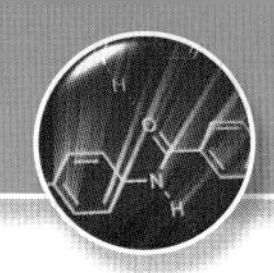

HSW Developing new drugs

As you saw in *Edexcel AS Chemistry*, page 53, every new medicine which comes onto the market is the result of around 10 years of research and development and costs of several hundred millions of pounds. A new drug must be safe and effective, easily taken into and removed from the body, stable so that it can be stored, and capable of being manufactured on a very large scale in a very pure form – not to mention relatively cheap to produce!

Scientists use computer modelling to help them to identify possible new drugs, but they also look for new medicines in the natural world. Any potential new medicine goes through a rigorous process of testing to investigate its effectiveness against disease and its safety. It is tested on cell cultures, tissue cultures and whole organs in the laboratory. These tests are designed to see if the chemical performs as the chemists expect it to. Many chemicals fail at this stage, but those that succeed move on to testing in animals.

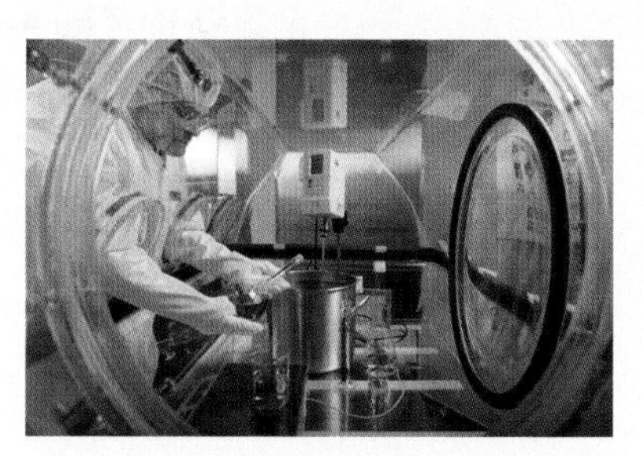

fig. 2.4.11 **It takes years of hard work to develop a new drug which is safe and effective against a disease.**

Mammals are used to see how the chemicals perform in a whole organism. Wherever possible, animal testing is replaced with tissue culture and computer modelling but these have limitations. There is a legal requirement for two different species to be used for testing a drug – a rodent (rats or mice) and a non-rodent. Once a drug passes the animal testing stage successfully, it moves on to human trials. The drug is given first to healthy volunteers, and then to very sick patients. If it gets through these stages successfully then it is tested on a larger group of patients, often as a double-blind trial where neither doctor nor subject knows if they are being given the new drug or a harmless placebo. Even if the drug passes all these tests and reaches the market, it is continually monitored in case any harmful side-effects emerge when it is used by thousands of people.

There are ethical and moral issues with testing human medicines on animals. Investigate this and write two articles – one supporting the use of animal testing in the development of new drugs and the other opposing it.

Questions

1 Using the reaction mechanism given for the reaction of a simple primary amine with ethanoyl chloride, suggest a mechanism for the last stage (step 4) in the production of paracetamol given in this section.

2 In the past, drugs were made simply from natural resources after observing their effects. Why is there such a rigorous testing procedure now in the twenty-first century?

Making an azo dye [5.4.2d]

Phenylamine is an aromatic amine. Aromatic amines take part in many reactions to produce a range of very useful aromatic chemicals. This wide range is possible because phenylamine forms an intermediate diazonium ion.

Reaction of phenylamine with nitrous acid

Nitrous acid is a very unstable compound which exists only in aqueous solution and decomposes at room temperature. This means that any reactions involving it have to be carried out under very carefully controlled conditions – the nitrous acid must be prepared as needed (in situ) by mixing ice-cold solutions of sodium nitrite and dilute hydrochloric acid:

$$NaNO_2(aq) + HCl(aq) \rightarrow NaCl(aq) + HNO_2(aq)$$

The reaction vessel must still be kept in an ice–water mixture because the reactions must be carried out at 5°C or lower.

The reaction of nitrous acid with a primary amine produces an alcohol:

$$\underset{\text{ethylamine}}{CH_3CH_2NH_2(aq)} + HNO_2(aq) \longrightarrow \underset{\text{ethanol}}{CH_3CH_2OH(aq)} + N_2(g) + H_2O(l)$$

A similar reaction takes place with an aromatic amine, such as phenylamine, and nitrous acid at temperatures above 5°C to produce a phenol:

$$\underset{\text{phenylamine}}{C_6H_5NH_2(aq)} + HNO_2(aq) \longrightarrow \underset{\text{phenol}}{C_6H_5OH(aq)} + N_2(g) + H_2O(l)$$

However, if the temperature is below 5°C, an aromatic amine will form a diazonium compound with nitrous acid:

$$\underset{\text{phenylamine}}{C_6H_5NH_2(l)} + HNO_2(aq) + HCl(aq) \longrightarrow \underset{\text{benzenediazonium chloride}}{C_6H_5\overset{+}{N}{\equiv}NCl^-(aq)} + 2H_2O(l)$$

If aliphatic amines are used, diazonium salts are not produced because they are so unstable – they decompose almost immediately they are formed, even at very low temperature. Aromatic diazonium salts, such as benzenediazonium chloride, are more stable, at least up to temperatures of 10°C or so. This additional stability results from the delocalised electrons of the benzene ring (**fig. 2.4.12**).

The high reactivity of diazonium salts can be put to use in a wide variety of situations in organic synthesis procedures. The use of diazonium ions is based on two aspects of their chemistry – attack of the diazonium benzene

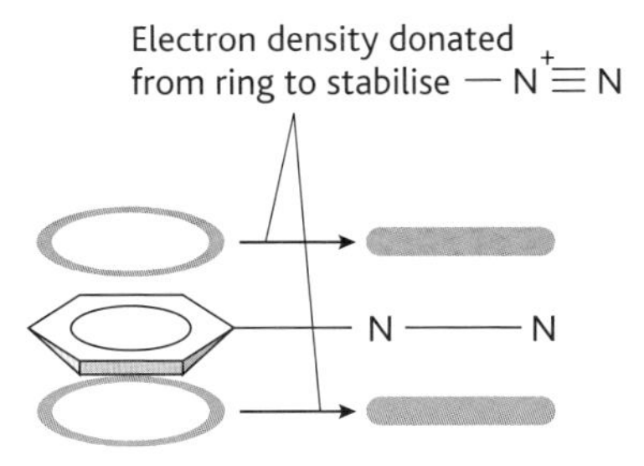

fig. 2.4.12 Benzenediazonium chloride is a very useful compound despite the fact that it decomposes above 10°C and is explosive in the solid form!

ring by nucleophiles, and the ability of diazonium salts to couple with other molecules. When nucleophiles react with the benzene ring, nitrogen gas is evolved – this does not happen when diazonium ions couple with other molecules.

Diazonium compound reactions with nucleophiles

The diazonium group opens the benzene ring to attack by nucleophiles, such as amines. Nitrogen gas is evolved in these reactions. Normally benzene reacts with electrophiles, so a reaction which enables nucleophilic species, such as halide ions or amines, to attack the benzene ring is extremely useful. Some of the more common syntheses in which benzenediazonium chloride is used are summarised in fig. 2.4.13.

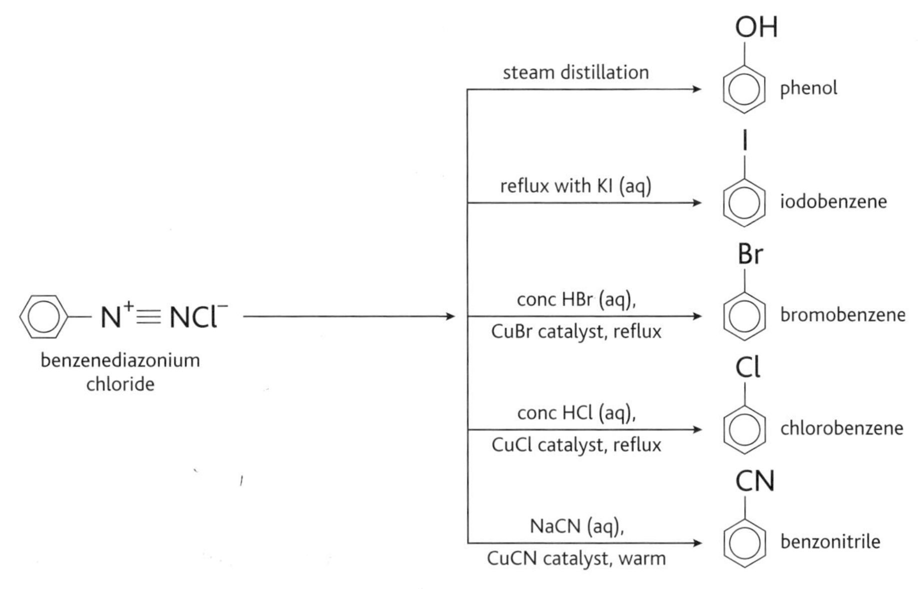

fig. 2.4.13 The reactions of benzenediazonium chloride with a variety of nucleophilic reagents.

Diazonium compound reactions as electrophiles

The benzenediazonium ion carries a positive charge and so is a strong electrophile. It reacts readily in cold alkaline solution both with aromatic amines and with phenols – they attack giving brightly coloured diazo-compounds. These reactions, joining two aromatic rings together, are known as **coupling** reactions. Remember, in coupling reactions no nitrogen gas is produced.

$$C_6H_5{-}N^+{\equiv}NCl^- + C_6H_5{-}OH \longrightarrow C_6H_5{-}N{=}N{-}C_6H_4{-}OH + HCl$$

benzenediazonium chloride + phenol → (4-hydroxyphenyl)azobenzene

fig. 2.4.14 Reaction between benzenediazonium chloride and phenol to give (4-hydroxyphenyl)azobenzene.

The diazo-compounds produced are complex molecules with a minimum of two aromatic rings joined by an N≡N coupling. Unlike the diazonium compounds from which they are made, diazo-compounds are extremely stable and unreactive.

'Direct red 39', bluish red

'Direct blue 2'

fig. 2.4.15 Skeins of wool dyed in a variety of azo-dyes, and formulae of some azo-dyes with their names.

The bright colours are the result of the extensive delocalised electron systems that spread across the entire molecule through the N≡N coupling. The main commercial use of these compounds is as colourfast dyes (**fig. 2.4.15**).

HSW Dyes and dyeing

Throughout history, people have used dyes to colour clothes, decorate their faces and so on. Mummies have been found in Egypt dressed in clothes dyed over 4000 years ago and it is known that the Phoenician, Greeks and Romans also used various natural dyes. Although these natural dyes provided a way of colouring, they had disadvantages – they washed out easily, for example. A **mordant** can be used to fasten the dye molecules onto cloth, which means that the dye washes out less easily. But this doesn't always work, particularly if natural fibres such as wool or cotton are being dyed. Natural dyes are often pale colours and they may also fade in sunlight.

In 1856 William Perkin revolutionised dyeing with the discovery of azo-dyes. The first, called Perkin's Mauve, was discovered by accident. Natural dyes have now been almost entirely superseded by synthetic products, except for a few specialised uses such as Logwood, which is the only natural dye still in large-scale use for dyeing silk and wool, and also secondary cellulose acetate and nylon.

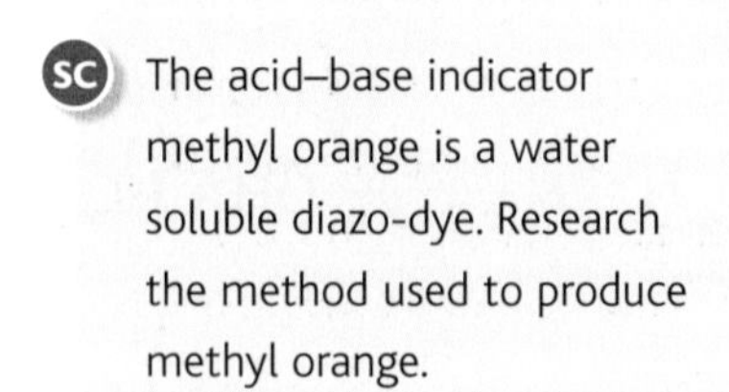

SC The acid–base indicator methyl orange is a water soluble diazo-dye. Research the method used to produce methyl orange.

Questions

1 Draw the structures of the amine and the phenolic compound used to make this azo-dye:

2 The indicator methyl orange is an azo-dye produced from these two compounds:

a Draw the structure of methyl orange.

b Describe the conditions needed to carry out this reaction.

New polymers [5.4.2e, f (i), (ii), g, h]

You will know from earlier work in this chapter that acyl chlorides, such as ethanoyl chloride, react with ammonia to produce an amide. It is also possible to join amide molecules together to form **condensation polymers** called **polyamides**.

HSW New polymers

In chapter 1.6 you found out that it was possible to form a polymer called a polyester in a series of condensation reactions. One such polyester is sold under the trade name Terylene. British chemists John Rex Whinfield and James Tennant Dickson created this first polyester fibre in 1941. Their work was inspired by earlier work by Carothers (see below) who had previously invented nylon – the first synthetic polyamide. Polyamides are produced by a series of condensation reactions involving molecules with two $-NH_2$ groups and two $-COCl$ groups or two $-COOH$ groups.

fig. 2.4.16 How a strand of nylon can be drawn from the interface of the two immiscible solutions.

Benzene derivatives and the production of nylon

In 1934, Wallace Hume Carothers, working for the massive chemical company DuPont, invented nylon. He patented it the following year, but as a result of a series of tragedies and misunderstandings in both his professional and personal life he committed suicide in 1937, the year before his revolutionary new fibre went on the market.

fig. 2.4.17 Wallace Hume Carothers.

Nylon is a totally synthetic fibre, a condensation polymer like many natural fibres but made up of two monomers that are both synthesised from benzene derivatives, and thus originally from coal or oil. Nylon has a multitude of uses in modern daily life. Apart from the obvious use in nylon stockings and tights, the polymer is used in combination with many other fibres to add durability. For example, many carpets have a mixture of wool and nylon, the nylon giving them far better resistance to the wear and tear of family or office life. Nylon is also used to make ropes because they don't rot. The nylon bearings in many machines don't wear – the list is almost endless.

The monomer units of one of the most common forms of nylon are 1,6-diaminohexane and hexanedioic acid (fig. 2.4.18). The polymer made is called nylon-6,6 indicating 6 carbon atoms in each of the monomers.

$$n\ \mathrm{HOOC{-}(CH_2)_4{-}COOH} + n\ \mathrm{H_2N(CH_2)_6NH_2} \longrightarrow \left[\mathrm{-CO{-}(CH_2)_4{-}CO{-}NH{-}(CH_2)_6{-}NH-}\right]_n$$

hexanedioic acid — 1, 6-diaminohexane — nylon 6,6

fig. 2.4.18 Nylon is produced as a result of a complex series of reactions (see page 220), resulting in strong nylon fibres which may be put to a wide variety of uses.

The two monomers are obtained from benzene via phenol or cyclohexane as shown in **fig. 2.4.19**.

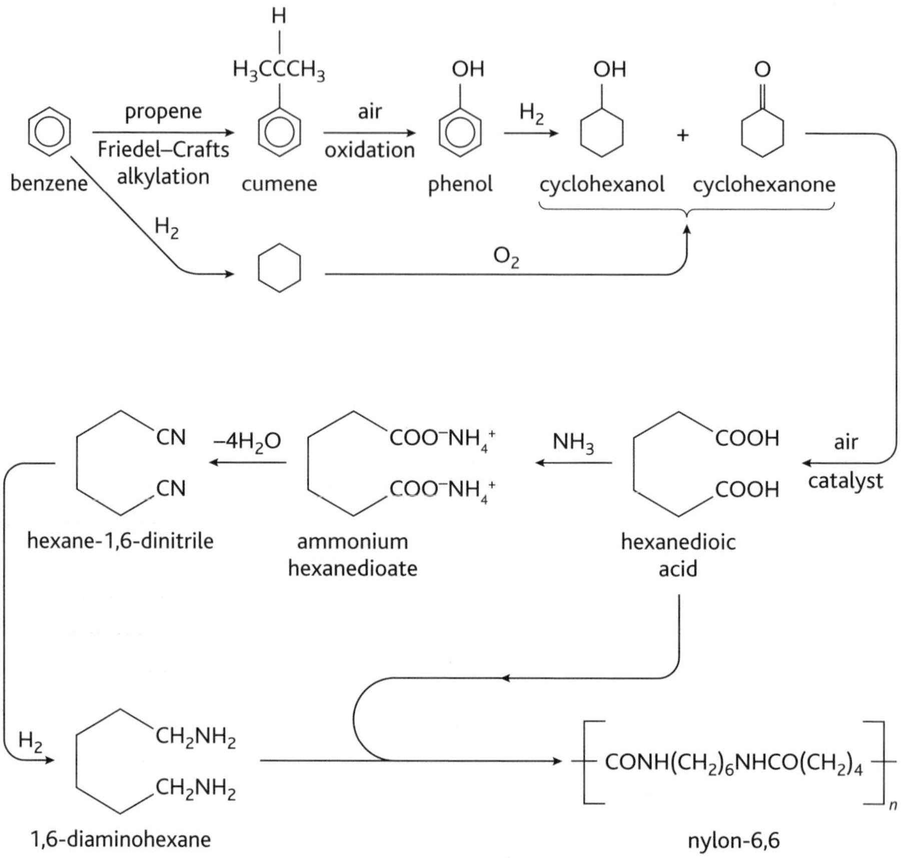

fig. 2.4.19 Diagram to show the stages of nylon-6,6 production from benzene.

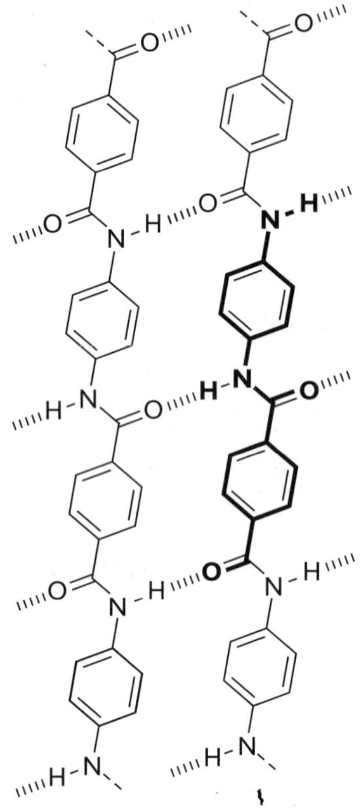

fig. 2.4.20 The structure of Kevlar: it consists of long polymer chains of paraphenylene terephthalamide – sometimes called para-aramid fibres – with strong hydrogen bonds between the chains.

fig. 2.4.21 A police officer with a bulletproof vest made of Kevlar.

Kevlar

Since the development of nylon by the DuPont Company in the 1940s, polyamides have become very important in everyday life. In 1965 Stephanie Kwolek and Roberto Berendt, again working for DuPont, produced a new polyamide called Kevlar® – its structure is shown in **fig. 2.4.20**. This was first used in the 1970s as a replacement for steel in racing car tyres.

This polymer resists high temperatures and has low thermal conductivity. It also has a high tensile strength and is flame-, chemical- and cut-resistant. These properties make it ideal for making many things, for example:

- sealants and adhesives
- transmission belts and hoses for vehicles
- structural parts in aircraft, space vehicles and boats
- data transmission cables
- replacement for asbestos in brake linings
- tyres for aircraft and other vehicles
- bullet- and stab-proof vests
- linings for aircraft fuel tanks
- ropes.

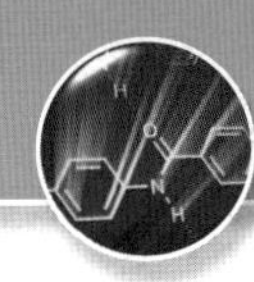

Addition polymerisation

Alkenes can form addition polymers (see page 134 in *Edexcel AS Chemistry*) – for example, ethene makes poly(ethene):

$$3\ \mathrm{H_2C{=}CH_2} \longrightarrow -\mathrm{CH_2{-}CH_2{-}CH_2{-}CH_2{-}CH_2{-}CH_2}-$$

The monomer molecule has a double bond, which is lost when the chain is formed. All the links between alkene molecules are formed in addition reactions, where no other product is formed.

You have already met some important polymers like this, such as poly(propene) and poly(chloroethene) but there are other important addition polymers as well, such as poly(propenamide) and poly(ethenol).

Poly(propenamide)

This is an addition polymer – although the 'amide' in the name might suggest that it is a condensation polymer. It is formed by polymerising 2-propenamide (sometimes called acrylamide):

$$n\ \mathrm{H_2C{=}CH{-}C({=}O)NH_2} \longrightarrow \left[\mathrm{H_2C{-}CH(C({=}O)NH_2)}\right]_n$$

2-propenamide → poly(propenamide)

The polymer chains in poly(propenamide) can be readily cross-linked with other chains. The resulting polymer with cross-chains produces a gel which has a huge capacity to absorb water. It can be used for making soft contact lenses. In the non-cross-linked form it is used as a thickener and a filler in facial surgery.

fig. 2.4.22 Contact lenses are made of poly(propenamide).

SC The monomer 2-propenamide can be made in two ways – either conventionally with a copper salt as a catalyst, or using a biological catalyst. Research these methods and find their advantages and disadvantages.

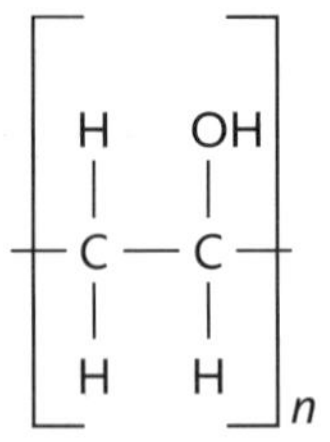

$$\left[\ \begin{matrix} H & OH \\ | & | \\ -C & -\ C- \\ | & | \\ H & H \end{matrix}\ \right]_n$$

fig. 2.4.23 Repeat unit of poly(ethenol).

Poly(ethenol)

Poly(ethenol) is an addition polymer – it is sometimes called poly(vinyl alcohol). It has the repeat unit shown in **fig. 2.4.23**.

It is not manufactured in the usual way by polymerising a monomer, but from another polymer by the process of ester exchange. In the first stage of manufacture, ethenyl ethanoate is polymerised to form poly(ethenyl ethanoate):

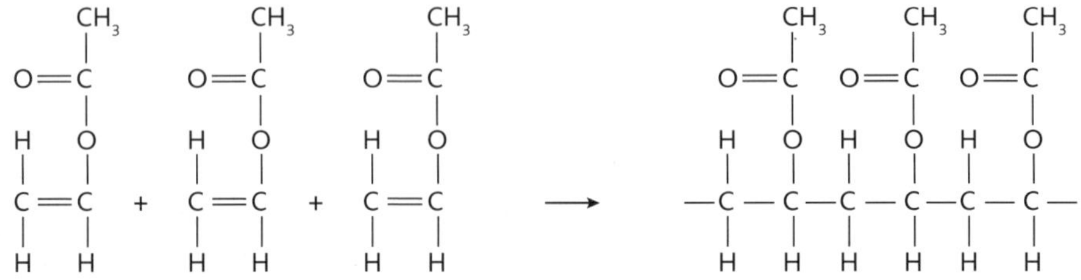

$$3\ CH_2{=}CH{-}O{-}C({=}O){-}CH_3 \longrightarrow {-}CH_2{-}CH(OCOCH_3){-}CH_2{-}CH(OCOCH_3){-}CH_2{-}CH(OCOCH_3){-}$$

ethenyl ethanoate → poly(ethenyl ethanoate)

The new polymer is then reacted with methanol to give poly(ethenol) and methyl ethanoate. The extent of ester exchange can be controlled simply by altering the temperature:

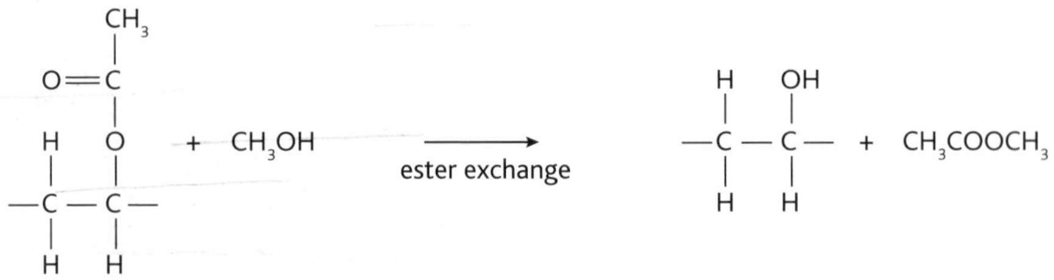

$${-}CH_2{-}CH(OCOCH_3){-} + CH_3OH \xrightarrow{\text{ester exchange}} {-}CH_2{-}CH(OH){-} + CH_3COOCH_3$$

poly(ethenyl ethanoate) → poly(ethenol)

fig. 2.4.24 One of the uses for poly(ethenol) is to make these disposable laundry bags.

The solubility of the final polymer depends on the percentage of ester groups that have been replaced – it ranges from effectively insoluble, through soluble in hot/warm water, to soluble in cold water.

Poly(ethenol) is used to make disposable laundry bags for use in hospitals (**fig. 2.4.24**). Bedding and clothes that might be infected with microorganisms from patients are put into a laundry bag and then the bag can go directly into the washing machine without the laundry workers touching the possibly infected fabrics. During the washing cycle the bag dissolves completely and the clothes are washed clean.

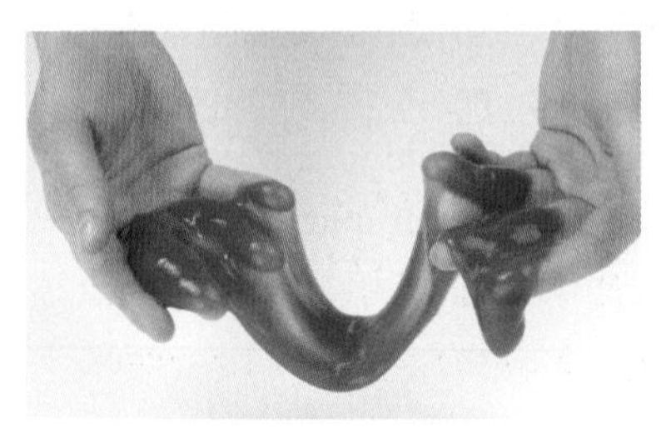

fig. 2.4.25 Using more borax gives a greater number of cross-links, giving the slime a higher viscosity and making it more difficult to pour.

Poly(ethenol) is also used to make the liquid detergent capsules (liquitabs) that contain measured quantities of detergent for use in washing machines. The bags break down and dissolve in water because the chains contain many OH groups. These can form hydrogen bonds with molecules in the water.

Poly(ethenol) can also be used to produce an interesting slimy material (**fig. 2.4.25**). When sodium tetraborate, $Na_2B_4O_7$, is dissolved in water, it hydrolyses to form a boric acid–borate ion solution with a pH of about 9. Borate ions interact with the OH groups in the polymer chains and form cross-links between the poly(ethenol) chains, forming a slimy material – a **viscoelastic gel**.

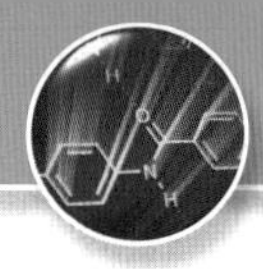

Properties of polyamides

Polymers such as the addition polymers poly(ethene) and poly(propene) are very resistant to a range of chemicals including acids, alkalis and organic solvents.

Figure 2.4.26 shows parts of two polymer chains. The top chain is a natural polymer called gutta-percha and the bottom chain is rubber. Both come from the sap of a tree and the two are isomers. In the gutta-percha (the *trans* isomer), the CH_3 groups are alternately on opposite sides of the molecule; rubber is the *cis* isomer and they are on the same side of the chain. In both cases, the carbon and hydrogen atoms are not easily attacked so the chain is not broken down easily. In general, addition polymers of just carbon and hydrogen atoms are difficult to break down and can cause long-term pollution problems (see pages 138–139 in *Edexcel AS Chemistry*). However, in a polyamide there are links set up between the molecules when they join that are vulnerable to attack under acidic or alkaline conditions.

fig. 2.4.26 **Both the oxygen atom and the nitrogen atom are electronegative and so can interact with adjacent chains and other molecules. These interactions can lead to breakdown of the chain under certain conditions.**

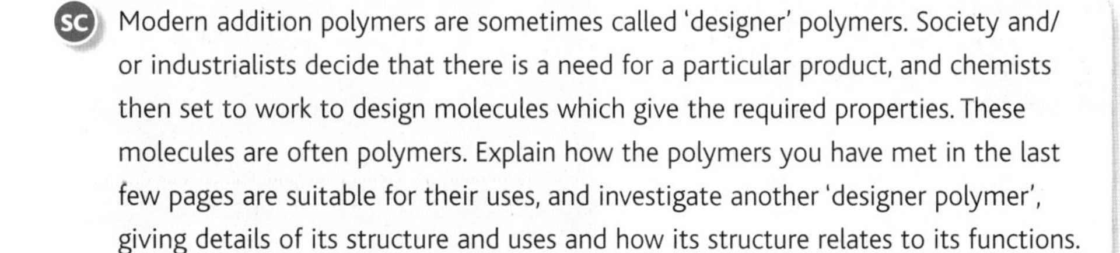

SC Modern addition polymers are sometimes called 'designer' polymers. Society and/or industrialists decide that there is a need for a particular product, and chemists then set to work to design molecules which give the required properties. These molecules are often polymers. Explain how the polymers you have met in the last few pages are suitable for their uses, and investigate another 'designer polymer', giving details of its structure and uses and how its structure relates to its functions.

Questions

1 As suggested by the formulae of the polymers, draw the displayed formulae of the monomers used to produce:
a poly(ethenol) **b** Kevlar.

2 Compare the properties of the synthetic polymers poly(ethenol) and poly(propene).

Amino acids [5.4.2a (ii), i (i)]

We have studied compounds containing the –COOH group (carboxylic acids) and compounds containing the $–NH_2$ group (amines). We are now going to study organic compounds which contain both of these groups attached to the same carbon atom – these compounds are called amino acids.

There are 20 amino acids found in nature. Your body can make 10 of these, but the others have to come from the food you eat. You rely on a regular supply of these amino acids, the building blocks of **proteins**, because your body cannot store amino acids.

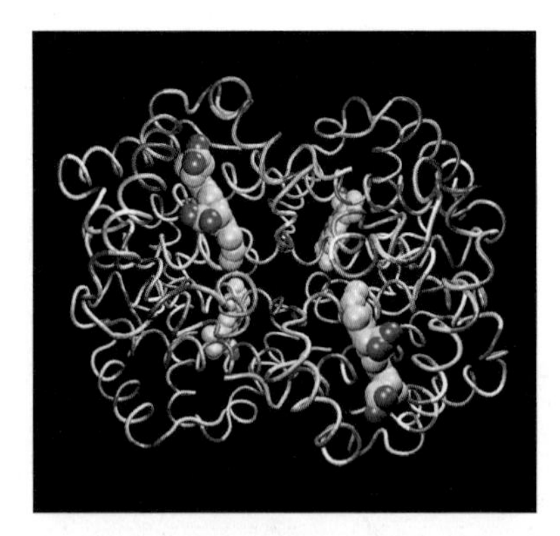

fig. 2.4.27 Hundreds or even thousands of amino acids joined together make up the complex protein structures of the body, such as this haemoglobin molecule.

Proteins make up the **enzymes** which control your body chemistry, as well as much of the cell structure. Tissues such as muscles are made up mainly of amino acids in the form of proteins. Every protein consists of a very long chain of amino acids in a particular sequence, which is determined by your genetic material. The proteins you eat are broken down into amino acids by digestion in your body.

The naturally occurring amino acids can polymerise in almost any order directed by the cell. This means that the potential variety of proteins is enormous. For example, in a protein containing 5000 amino acid units there are more than 10^{6000} possible combinations!

The structure of amino acids

All amino acids have the molecular form shown in fig. 2.4.28. Some specific examples are shown in fig. 2.4.29.

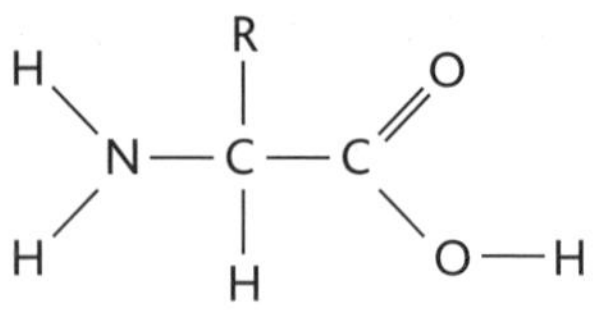

```
H        R      O
 \       |     //
  N  —   C  —  C
 /       |      \
H        H       O—H
```

fig. 2.4.28 The displayed formula of a general amino acid. In the simplest, glycine, R is a single hydrogen atom. In all other amino acids R represents a carbon-based group which may be polar – for example CH_2OH – or non-polar – for example CH_3.

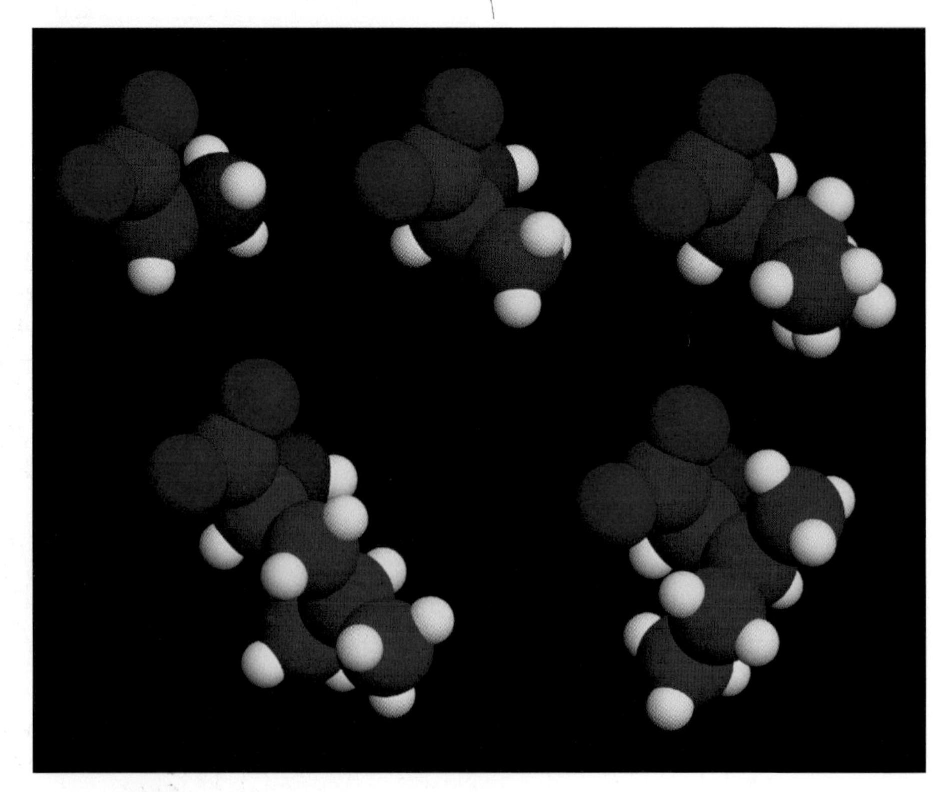

fig. 2.4.29 Models of five amino acids.

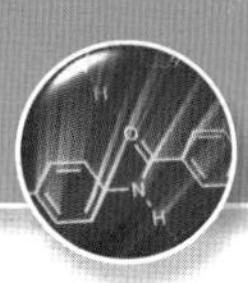

Table 2.4.1 lists some of the common amino acids. They are usually given trivial names, rather than systematic names, because the latter would be long and complex. Each amino acid can also be represented by a three-letter shorthand. This is often the first three letters of the trivial name – for example, 'gly' stands for the simplest amino acid, glycine.

Formula

Formula (backbone)	Side group (attached to the central C)	Amino acid
$H_2NCHCOOH$	H	glycine
$H_2NCHCOOH$	CH_3	alanine
$H_2NCHCOOH$	CH_2–$CH(CH_3)_2$	leucine
$H_2NCHCOOH$	CHC_2H_5–CH_3	isoleucine
$H_2NCHCOOH$	CH_2–C_6H_5 (benzene ring)	phenylalanine
$H_2NCHCOOH$	CH_2OH	serine
$H_2NCHCOOH$	$CHOH$–CH_3	threonine
$H_2NCHCOOH$	CH_2SH	cysteine
$H_2NCHCOOH$	CH_2–C_6H_4–OH (benzene ring)	tyrosine
$H_2NCHCOOH$	$(CH_2)_3$–CH_2NH_2	lysine
$H_2NCHCOOH$	CH_2COOH	aspartic acid
$H_2NCHCOOH$	CH_2–CH_2COOH	glutamic acid

Name	Abbreviation	Side chain	R_f value *
Glycine	gly	Non-polar	0.26
Alanine	ala	Non-polar	0.38
Leucine	leu	Non-polar	0.73
Isoleucine	ile	Non-polar	0.72
Phenylalanine	phe	Non-polar	0.68
Serine	ser	Polar	0.27
Threonine	thr	Polar	0.35
Cysteine	cys	Polar	0.08
Tyrosine	tyr	Polar	0.50
Lysine	lys	Basic	0.40
Aspartic acid	asp	Acidic	0.24
Glutamic acid	glu	Acidic	0.30

* The R_f values given are those determined in a solvent containing butan-1-ol, ethanol and water.

table 2.4.1 **A list of some of the amino acids giving their structure, trivial names and abbreviations.**

HSW When amino acid metabolism goes wrong

In the human body, excess amino acids are either broken down in the liver, or converted from one type to another. Phenylalanine is an essential amino acid, but if levels get too high it can be harmful. Excess phenylalanine is converted into tyrosine by the enzyme phenylalanine hydrolase. However, some babies are born with low levels of this enzyme. Phenylalanine builds up in the blood and eventually this leads to developmental problems – the brain can be irreversibly damaged, leaving the child with severe learning difficulties. This condition is known as PKU (phenylketoneuria).

All babies in the UK are screened for this condition very shortly after birth. If levels of phenylalanine are kept low through a carefully managed diet, the baby will develop normally and grow up to be a healthy adult. Interestingly, as an individual matures the brain loses its sensitivity to phenylalanine and affected adults can eat a relatively normal diet.

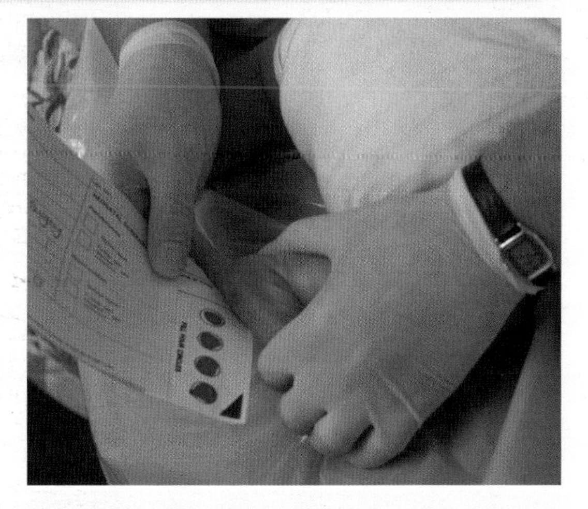

fig. 2.4.30 **A simple blood test for newborn babies enables PKU to be detected and dealt with before it causes any lasting damage.**

The peptide group

When two amino acid molecules join together, they undergo a condensation reaction and lose a molecule of water. The resulting bond is called a **peptide bond** (fig. 2.4.31) and the molecule is called a **dipeptide**.

$$NH_2-CH(R)-C(=O)-OH + H-NH-CH(R')-C(=O)-OH \rightarrow NH_2-CH(R)-C(=O)-NH-CH(R')-C(=O)-OH + H_2O$$

fig. 2.4.31 Forming a dipeptide from two amino acids.

For example, if a glycine molecule and an alanine molecule combine, the result can be gly-ala or ala-gly, depending on which way round they combine. If 50–100 amino acids join together in a similar way then a **polypeptide** is formed.

Acidic and basic properties of amino acids

Amino acids have interesting properties because they contain both an acidic group (–COOH) and a basic group ($-NH_2$). In an alkaline solution, the amino acid molecule loses a proton and forms a negative ion:

$$H_2N-CH(R)-C(=O)-O^-$$

In an acidic solution, the molecule receives a proton and forms a positive ion:

$$H_3N^+-CH(R)-C(=O)-OH$$

In a neutral solution, amino acids exist with both the positive and negative ions within the same molecule. This species is called a **zwitterion**, from the German word for 'hybrid':

$$H^+ + :NH_2-CH(R)-C(=O)-O-H \rightarrow H_3N^+-CH(R)-C(=O)-O^-$$

proton gained   proton lost

Amino acids can exist as zwitterions when in the solid state. This explains their relatively high melting temperature when compared with similar compounds and why they dissolve in water. Amino acids do not dissolve in organic solvents.

Questions

1 a Write equations for the reactions that take place when alanine is added to:
(i) dilute hydrochloric acid
(ii) sodium hydroxide solution.

b Draw the structure of alanine in neutral solution.

2 a Draw the structures of the two dipeptides formed when leucine and serine combine to form a dipeptide.

b Use the three-letter shorthand to represent these dipeptides.

3 a Identify four of the amino acids in fig. 2.4.29 using information in table 2.4.1.

b Write down the molecular formula of the fifth one.

c Which two are isomers?

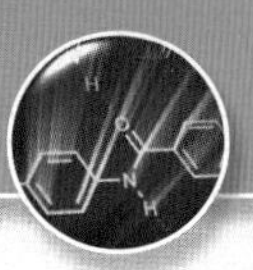

Separation of amino acids [5.4.2i (ii), (iii), (v)]

In the previous section we found out that proteins are made up from chains of amino acids in a particular sequence – but how can we identify the amino acids in a particular protein?

Hydrolysis of proteins

Digestive enzymes in the stomach and small intestine catalyse the **hydrolysis** of proteins into amino acids. The same thing can be done in the laboratory by refluxing a protein with dilute hydrochloric acid – this hydrolysis can take up to 24 hours. The individual amino acids formed can then be identified by paper chromatography.

Separation of amino acids by paper chromatography

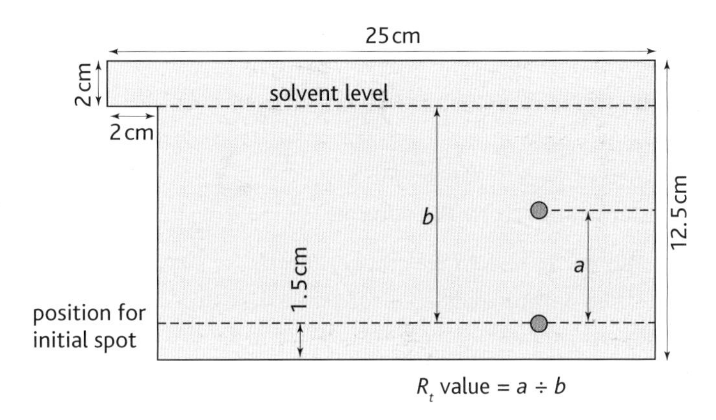

fig. 2.4.32 Apparatus for simplified chromatography.

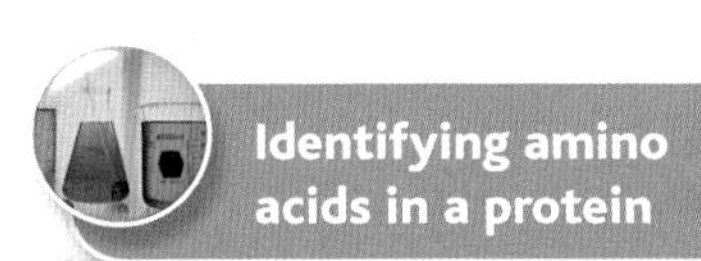

Identifying amino acids in a protein

You may hydrolyse a protein such as egg white or gelatin and attempt to identify the amino acids produced using paper chromatography. The spots are colourless so you may use ninhydrin as a detecting agent. You may also calculate R_f values to help you to identify the amino acids.

The amino acids from the hydrolysis are first neutralised. A spot of the mixture is then put on the bottom of a square piece of chromatography paper (fig. 2.4.32) – it is important that you touch the chromatography paper only along the top edge. Leave the paper to dry.

At this stage you will not be able to see the spots of amino acids because the mixture is colourless. Make the paper into a cylinder and fasten it at the top with a plastic paper clip (fig. 2.4.33). Put some solvent in a large beaker – the solvent should be a mixture of butan-1-ol ($12\,cm^3$), ethanoic acid ($3\,cm^3$) and water ($6\,cm^3$). Stand the cylinder of chromatography paper in the beaker and cover the beaker to prevent the solvent from evaporating.

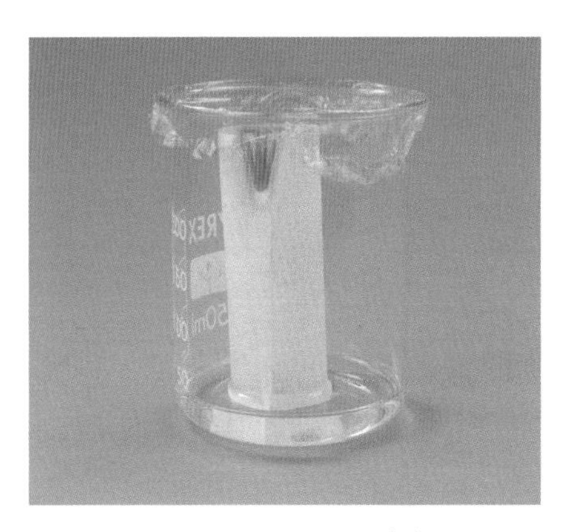

fig. 2.4.33 Separating amino acids by paper chromatography.

After about 20 minutes, remove the paper and mark the position of the solvent front, then dry the paper. At this stage the spots are still invisible. Spray the chromatography paper with ninhydrin in a fume cupboard and then dry it in a hot oven. The amino acids will show up as purple spots. The spots can be preserved by spraying the paper with a mixture of methanol, copper(II) nitrate solution and nitric acid, and exposing it to concentrated ammonia fumes in a fume cupboard. Figure 2.4.34 shows the chromatogram from the products of the hydrolysis of a simple protein.

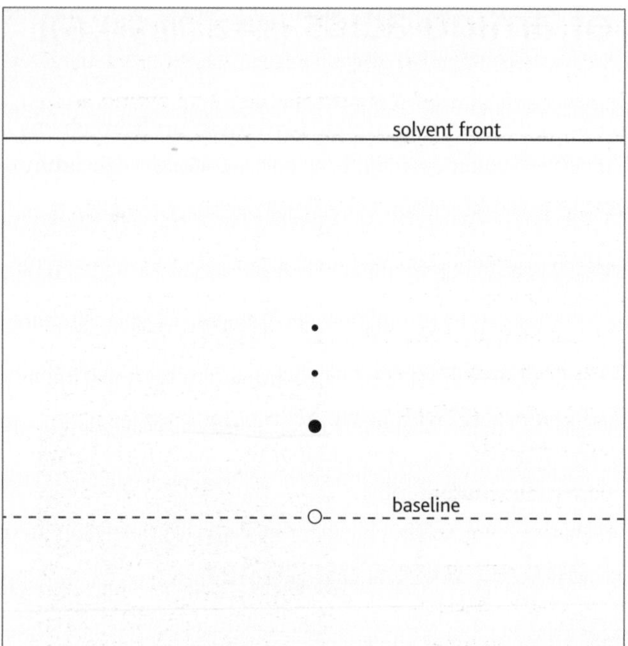

fig. 2.4.34 The three spots on the chromatogram show that there were at least three amino acids in the protein.

You can work out the R_f value of each amino acid using:

$$R_f = \frac{\text{distance moved by the amino acid}}{\text{distance moved by the solvent}}$$

Using the results in fig. 2.4.34 we can get the R_f values of the three amino acids – see question 1 at the end of this section.

Improving the method

If two amino acids have similar R_f values in the butan-1-ol, ethanol and water solvent, it will not be easy to distinguish their spots because they will be very close together. However, it might be possible to repeat the process using a different solvent in which the amino acids have different R_f values, hopefully more widely separated. R_f values are also affected by the type of chromatography paper.

There is also an extension of this technique called two-way paper chromatography. Figure 2.4.35 shows how this might be done.

The mixture of amino acids is applied to baseline (1). The paper is then made into a cylinder and the bottom placed in a solvent of butan-1-ol, ethanol and water in a similar way to that shown in fig. 2.4.34. The positions of the resulting spots – a single spot and a large spot, possibly made from two amino acids very close together – is shown in fig. 2.4.35 (a). The paper is then rotated though 90° and the process is repeated with a different solvent mixture (phenol and water here). This moves at right angles to the original solvent. Figure 2.4.35 (b) shows that the large spot has been separated because two spots are clearly seen. The R_f values are different in the different solvents, making identification of all three amino acids much easier.

(a)

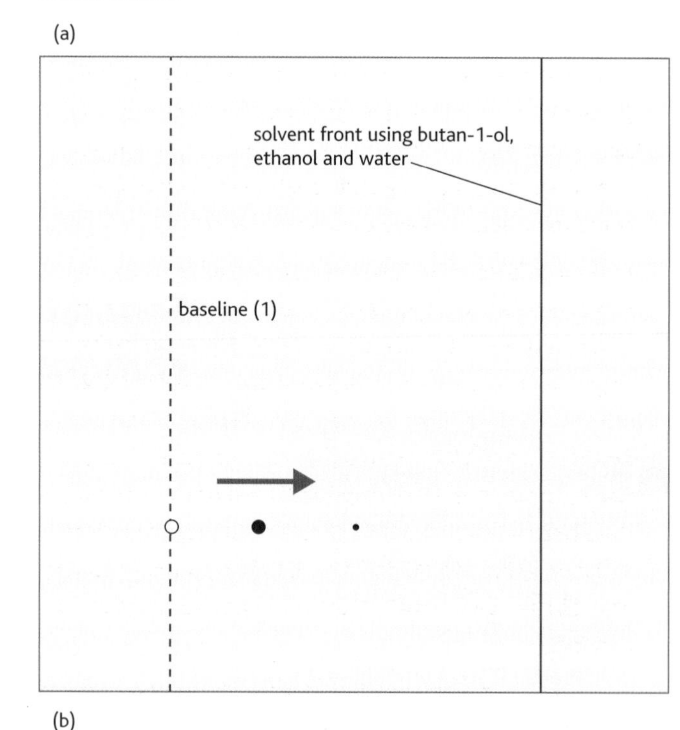

(b)

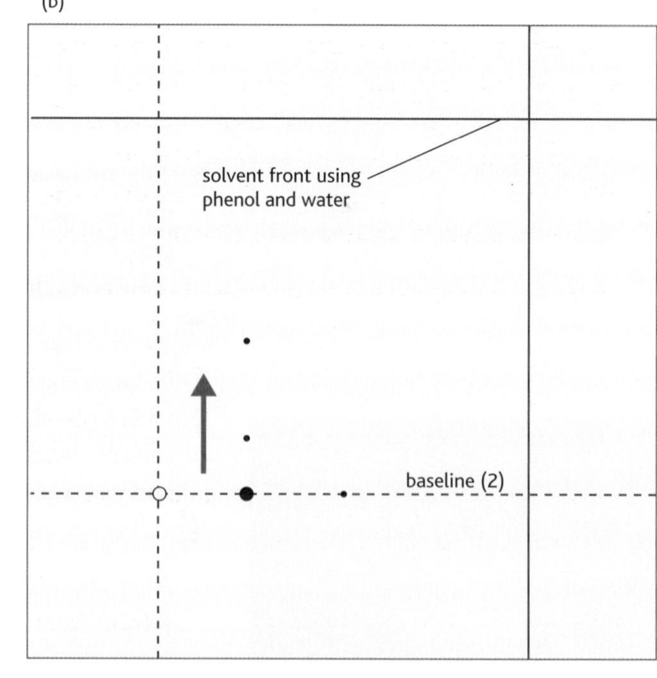

fig. 2.4.35 Two-way paper chromatography: (a) after first separation rotated 90 degrees clockwise; (b) after second separation.

Paper chromatography has its limitations. When more amino acids need to be separated, gel **electrophoresis** is often used. This is a more

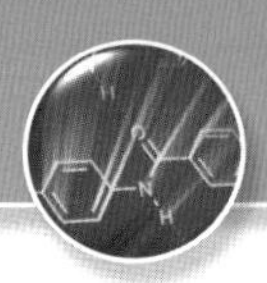

sophisticated form of chromatography that depends on the same principles. It is carried out in a special gel and after the amino acids have been spotted onto the plate an electric current is passed through the gel. The amino acids move through the gel depending on the charges on the functional groups.

HSW Fingerprints

Whenever you touch something with bare skin you leave a fingerprint – a combination of sweat, with its salt and protein components, and grease from the surface of the skin. A fingerprint is unique to an individual, so the police need clear images of fingerprints left at the scene of a crime to be able to make a link with suspects.

Forensic scientists use fingerprints collected from the crime scene as an important tool in identifying suspects. Silver nitrate was the first chemical to be used to make fingerprints on porous substances show up clearly. A 2% solution of silver nitrate in methanol was used. The silver nitrate reacts with the chloride content of the (hidden) fingerprint to form silver chloride. On exposure to light, the silver chloride blackens because silver chloride is light sensitive – the silver(I) ions are oxidised to silver. However, after about a week much of the detail has been lost in this method.

The reaction of amines with ninhydrin to form the coloured reaction product known as Ruhemann's purple was discovered by Siegfried Ruhemann in 1910. However, the use of ninhydrin to identify fingerprints was not developed until 1954. It reacts with the amine part of the amino acids in the protein traces of a fingerprint. Ninhydrin is still the most widely used substance for developing latent fingerprints on paper but its use with some high quality papers, for example bank notes, is limited. However, it has been used to locate fingerprints produced 30 years previously.

fig. 2.4.36 **A fingerprint found using ninhydrin can be compared with the fingerprints of suspects.**

A ninhydrin solution of approximately 0.5% weight per unit volume is required to develop fingerprints on paper. Many different solvents have been used, some both flammable and toxic – they have even included CFCs (*Edexcel AS Chemistry* pages 228–9). Commonly used safer alternative solvents include heptane and ether.

The ninhydrin treatment of fingerprints is the same regardless of the solvent used. The paper is immersed in the solvent containing ninhydrin and developed in the dark for 24 hours. Heating the paper to speed up the development is not recommended.

A permanent record of a fingerprint can be obtained by photographing under white light with a green-yellow filter fitted to the camera.

Optical activity of amino acids

You will know that molecules with a chiral centre have L and D enantiomers (**fig. 2.4.37**) and that these show optical activity (**chapter 1.6**) and rotate the plane of plane-polarised light.

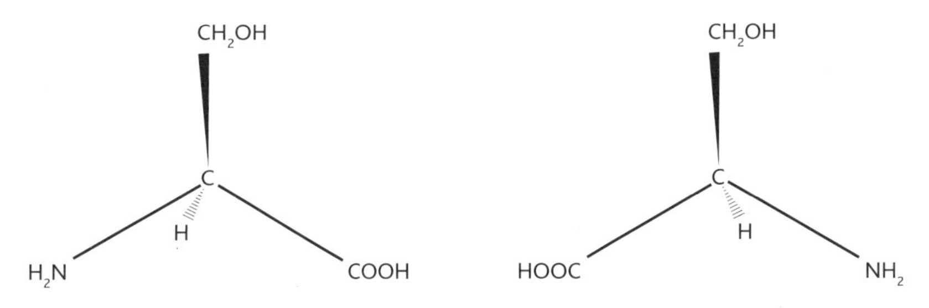

fig. 2.4.37 **Optical isomers of serine.**

Most amino acids form optical isomers – if you refer back to **table 2.4.1** you will see that, with the exception of glycine, they all contain at least one chiral centre. Glycine is the exception because it has two hydrogen atoms attached to the central carbon atom, giving it symmetry, so glycine is not optically active.

All the amino acids found in living organisms are L enantiomers, but the direction in which they rotate plane-polarised light varies. Some are + enantiomers (causing clockwise rotation) and others are – (anticlockwise). It is impossible to predict the directional effect that a particular amino acid will have on plane-polarised light.

Laboratory synthesis of amino acids tends to produce a racemic mixture of L and D enantiomers.

Using a polarimeter

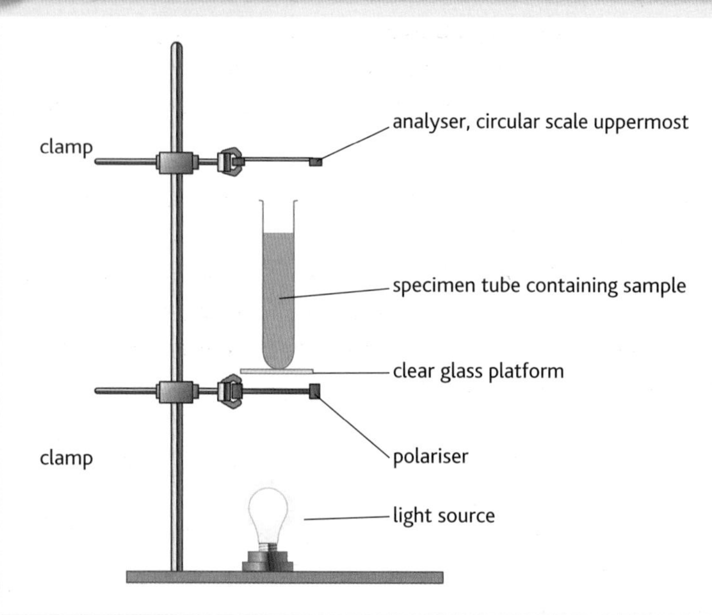

fig. 2.4.38 A simple polarimeter.

Two pieces of Polaroid can be used in a simple polarimeter (**fig. 2.4.38**).

Without the sample in place, rotate the top piece of Polaroid® until the light is extinguished and you cannot see through when viewing from the top. The bottom piece of Polaroid restricts the light passing through to one plane. When the top piece of Polaroid is aligned so that light passes through only when it is at right angles to the previous one, no light will pass through.

When the sample tube is half-filled with the sample, look through the top – you should see light. Rotate the top Polaroid until the light is extinguished, measure the angle of rotation and note whether it is clockwise or anticlockwise.

Questions

1 a Identify the three amino acids shown on the chromatogram in **fig. 2.4.34** by calculating the R_f values and using the data in **table 2.4.1**.

b What does the chromatogram suggest about the relative amount of each amino acid?

c Why would it be difficult to distinguish between glycine and serine using this solvent?

2 When carrying out a paper chromatography experiment of the products of hydrolysis of amino acids, it is important to touch the paper only on the edges and in places where the solvent will not reach. Why is this?

3 a Refer to **table 2.4.1** and identify any amino acids with two chiral centres.

b How would you expect this 'dual chirality' to affect the role of these amino acids in biological systems?

4 The ability to identify amino acids is important in forensic science. Why is this?

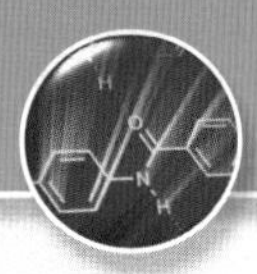

Proteins [5.4.2i (iv)]

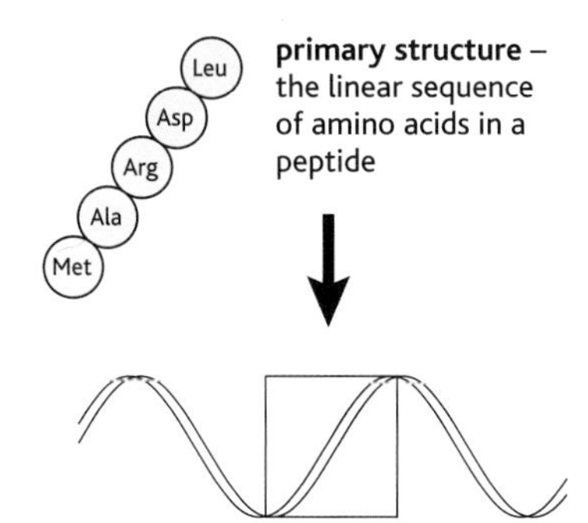

primary structure – the linear sequence of amino acids in a peptide

secondary structure – the repeating pattern in the structure of the peptide chain, for example an α-helix

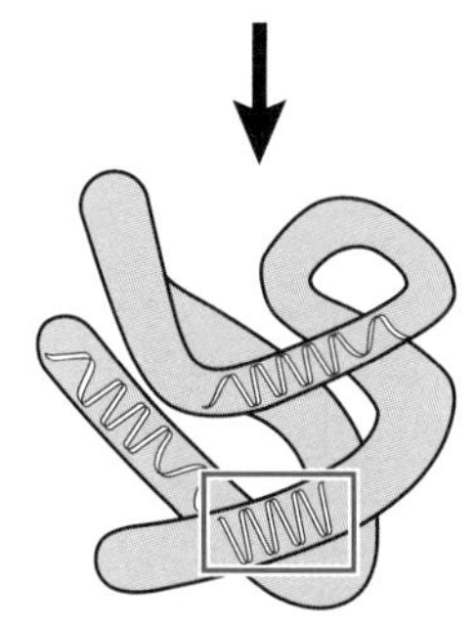

tertiary structure – the three-dimensional folding of the secondary structure

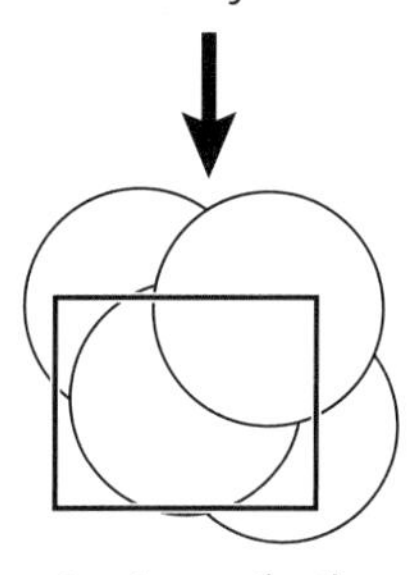

quaternary structure – the three-dimensional arrangement of more than one tertiary polypeptide

fig. 2.4.39 It isn't just the sequence of amino acids but also the arrangement of the polypeptide chains that determines the structure and characteristics of a protein.

Proteins are complex molecules formed by individual amino acids joining by condensation reactions. Every protein is a chain of amino acids in a particular sequence. Determining the nature of a particular protein is a problem of identifying not just the constituent amino acids but also their sequence in the chain.

The amino acid sequence of a protein is known as its **primary structure**. The polypeptide chains themselves form a 3-D arrangement known as the **secondary structure** – these include helices and folded sheets held together by hydrogen bonds between the amino acids. The **tertiary structure** is the way these chains are folded into more complex shapes – globular proteins are folded into spherical shapes; fibrous proteins are long and tough and are important in tissues such as ligaments, tendons and hair. These tertiary structures are held together both by hydrogen bonds and by sulfur bridges and ionic bonds. Finally, some proteins have a **quaternary structure** in which several polypeptide chains are fitted together – haemoglobin, the blood pigment, has four of these chains in its quaternary structure with iron-containing haem molecules incorporated in each.

Within protein structures there are hydrogen bonds and sulfur bridges. A simple demonstration of the difference in strength of sulfur bridges and hydrogen bonds is shown by blow-dried hair and hair that has been treated with a 'perm'. When you wash your hair, you disrupt hydrogen bonds in the protein and then these reform in different places when your hair is blown dry – this gives the shape you want. Next time you wash your hair it returns to its natural style because the original hydrogen bonds reform. If you have a perm, the chemicals in the solutions break the sulfur bridges between the polypeptide coils and reform them in different places. When you wash your hair it will still return to the desired shape because the sulfur bridges remain intact.

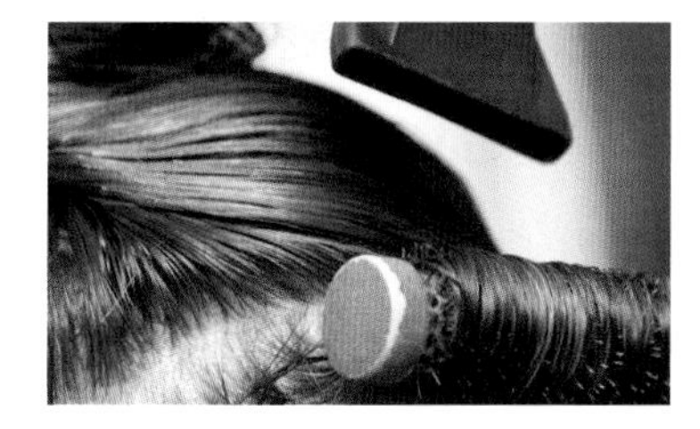

fig. 2.4.40 Hair is a complex protein molecule and the way you style it depends on some complex biochemistry.

Analysing protein structure

Protein	Where found	Molar mass/g	Approx. number of amino acid units
Insulin	Pancreas	5700	51
Haemoglobin	Blood	66000	574
Urease	Soya beans	480000	4500

Table 2.4.2 Information about three proteins – insulin, haemoglobin and urease.

Table 2.4.2 shows information about three proteins. As you can see, although insulin is a complex molecule it is much smaller and easier to analyse than proteins such as urease! Insulin was the first protein for which the amino acid sequence was determined in some groundbreaking work by Frederick Sanger.

HSW Composition and structure of insulin

phe		gly
val		ile
asp		val
gin		glu
his		gin
leu		cys
cys	—S—S—	cys
gly		thr
ser		ser
his		val
leu		cys
val		ser
glu		leu
ala		tyr
leu		gin
tyr		leu
leu		glu
val		asp
cys	—S	tyr
gly	S—	cys
gly		asp
arg		
gly		
phe		
phe		
tyr		
thr		
pro		
lys		
thr		

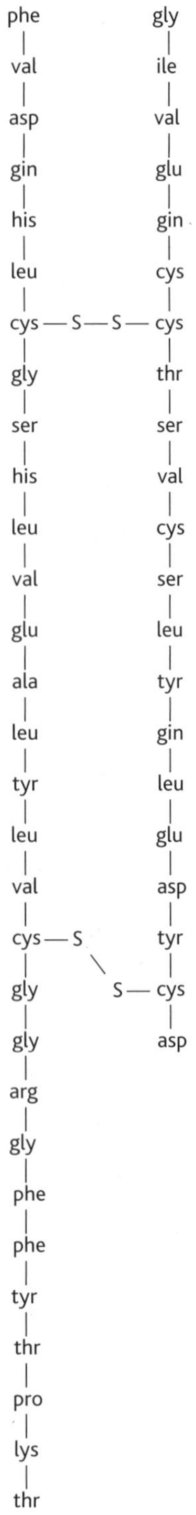

fig. 2.4.42 The sequence of amino acids in insulin from a sheep. The three-letter codes used for amino acids in table 2.4.1 have been used to simplify the diagram.

It is possible to work out from the molar mass and other experimental evidence that the molecular formula of insulin is $C_{256}H_{381}N_{65}O_{79}S_6$. Frederick Sanger (see fig. 2.4.41) first determined which amino acids are present in the compound by hydrolysis with hydrochloric acid and paper chromatography. These studies showed that the insulin molecule was made up of 17 different amino acids in different quantities, ranging from 6 molecules of cysteine to 1 molecule of lysine, totalling 51 amino acid molecules – and, to his satisfaction, his calculated molecular mass matched the accepted value.

fig. 2.4.41 Frederick Sanger first worked out the composition and structure of insulin in 1955.

Using other techniques, Sanger was then able to establish the sequence of amino acids in insulin. One of these involved the partial hydrolysis of the insulin molecule into shorter polypeptides, and determining the sequence in these.

Today we have a range of modern techniques to do the painstaking work of Sanger. For example, **mass spectrometry** provides a rapid way of finding the molecular masses of the fragment ions, and this can quickly lead to determining the different amino-acid linkages – and only a very small sample is needed (5–30 × 10^{-9} mol). Sanger had to use laborious manual methods to work out the sequences of proteins in polypeptides. Now this can be done much more easily, analysing in minutes what took early scientists months of careful work.

SC You have found out how paper chromatography can be used to separate and identify amino acids. Another way is to use electrophoresis. You had a brief introduction to this process in the previous section. Research and find out more about how this can be done and why it happens.

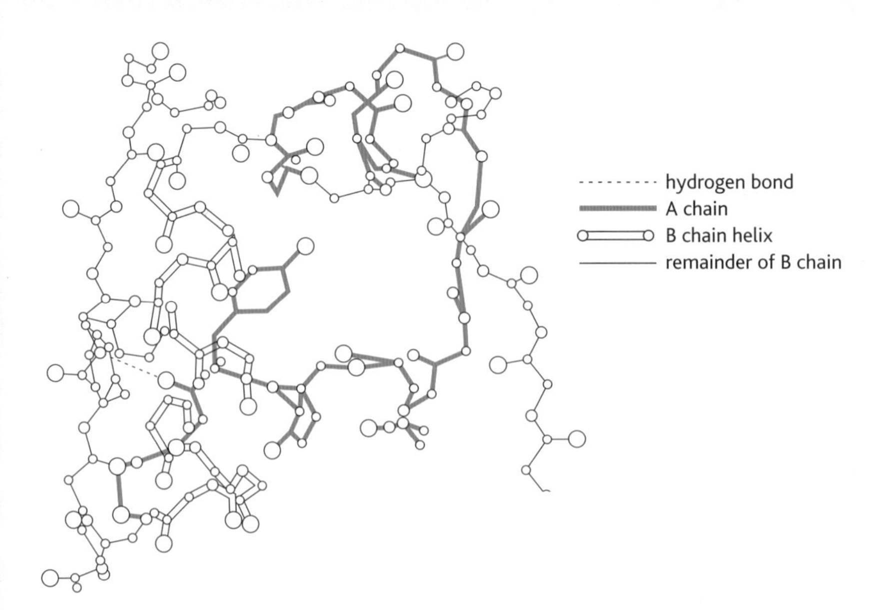

fig. 2.4.43 The three-dimensional structure of insulin. There are hydrogen bonds between the peptide groups that are important in maintaining the structure.

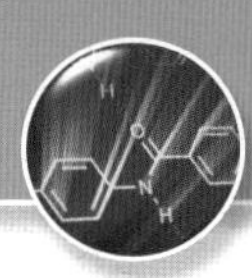

Protein synthesis

Proteins are synthesised continually in the human body. The instructions for a required protein sequence are carried on the DNA which makes up the chromosomes of cells. The instructions are carried to the cell organelles that carry out the synthesis by a closely related compound called RNA. The processes depend on the presence of many different enzymes.

The different amino acids are joined (see **fig. 2.4.44**) by peptide bonds that are formed when the –COOH group on one amino acid reacts with the $-NH_2$ group of another in a condensation reaction, producing one molecule of water and a peptide group.

$$H_2N-CH(R)-COOH + H_2N-CH(R')-COOH \rightarrow H_2N-CH(R)-CO-NH-CH(R')-COOH + H_2O$$

fig. 2.4.44 **The joining of amino acids produces a peptide group and a molecule of water in a condensation reaction.**

When chemists attempt to synthesise even a small peptide they find it a difficult thing to do. Suppose a chemist wants to synthesise a polypeptide consisting of four amino acids in the order gly-ala-gly-ala. In the first step, a glycine molecule and an alanine molecule react to form the dipeptide gly-ala. Steps have to be taken to ensure that the combination happens the right way round. This is done by blocking one end of each molecule by attaching an unreactive group.

Then another glycine molecule is added to get gly-ala-gly, and finally a further alanine molecule to give the desired gly-ala-gly-ala polypeptide.

This would seem to be a straightforward synthesis. However, the conditions needed to carry out the condensation reactions are such that you are at risk of rupturing one of the peptide bonds produced earlier. Obviously, the longer the chain becomes, the greater is the risk of this happening.

SC Investigate a globular protein and a fibrous protein found in the human body. Describe their structure and how this is related to their function in the body.

Questions

1 Suggest why it is easier to identify the composition and structure of insulin compared to, say, haemoglobin.

2 a Draw the dipeptide produced when two serine molecules are joined.

 b Draw a ring around the peptide linkage.

3 a Why was Frederick Sanger's work such a breakthrough?

 b Analysis of proteins is now a much simpler process. Compare what had to be done in Sanger's day with modern techniques.

2.5 Organic synthesis

The importance of synthetic organic chemistry [5.4.3a]

HSW Artificial sweeteners

Scientists are constantly working on the development of useful products that can then be manufactured – from new medicines to new building materials, from artificial joints to new fuels, chemists are looking for new ideas. Many of the processes used in the chemical industry today rely on **organic synthesis**. The starting materials are often derived from the products of petroleum refining, from coal or from natural fats and oils. The ideal reaction is as straightforward as possible – when reactants are converted to products in a simple, one-step process it is relatively easy to get a very high yield and also to control the reacting conditions relatively cheaply. However, organic synthesis reactions are rarely simple – in fact, they are notoriously difficult. They usually involve a series of steps and this, in turn, affects the overall yield adversely. Each individual step has a limited yield to pass on to the next reaction in the chain. The individual reactions are also often difficult to carry out because many organic molecules are much more sensitive to temperature, pH and so on than inorganic molecules. This means that it is technically difficult to control the reaction conditions for each stage of the reaction – as a result the yield is reduced even further at each stage!

In 1965 Robert Burns Woodward received the Nobel prize for his work on developing ways of synthesising a number of difficult and valuable molecules. His lifetime successes included the synthesis of quinine (to treat malaria), chlorophyll, strychnine, the steroid cortisone and the antibiotic cephalosporin. The discovery of a synthetic process is often a mixture of accidental discovery, good fortune and outstanding chemistry!

We are going to consider the processes for making different **sweetening agents** using organic synthesis. The major traditional sweetening agent in our diet has been sucrose, $C_{12}H_{22}O_{11}$, from sugar cane or sugar beet. Many people now use artificial sweeteners such as saccharin, cyclamates or aspartame – fig. 2.5.1 shows a range of such sweeteners. These provide sweetness with fewer calories, so can help people control their weight.

fig. 2.5.1 An assortment of artificial sweeteners.

The first – saccharin

Saccharin is much sweeter than sucrose but it has a bitter aftertaste, especially in high concentrations. It is an **arene** compound called benzoic sulfinide, $C_7H_5NO_3S$.

fig. 2.5.2 Saccharin – the first artificial sweetener.

It is unstable when heated but it does not react chemically with other food ingredients, and it stores well. Saccharin is important for diabetics because it goes through the digestive system unchanged – it has even been shown to trigger the formation of additional insulin in rats. Unlike sucrose, it has no energy value.

Saccharin was first produced by Constantin Fahlberg in 1879. Fahlberg worked on derivatives of coal tar with Ira Remsen. Remsen actually discovered the sweetness of the substance but it was Fahlberg who took out patents in many countries and became rich.

There are different ways of manufacturing saccharin. One method uses anthranilic acid, or 2-aminobenzoic acid (fig. 2.5.3), in successive reactions with nitrous acid, sulfur dioxide, chlorine, and finally ammonia.

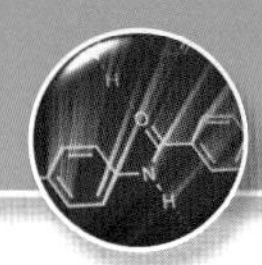

fig. 2.5.3 **Anthranilic acid – the starting point for making saccharin.**

It was not until sugar was rationed in World War I that saccharin became widely used. Its popularity increased during the 1960s and 1970s when people wanting to lose weight realised that it was a calorie-free sweetener.

Throughout the 1960s, various laboratory studies suggested that saccharin may cause cancers. One study in 1977 indicated an increased rate of bladder cancer in rats that had been fed on large doses of saccharin and a ban on its use was proposed. Saccharin was the only artificial sweetener available at the time and there was strong public opposition to the proposed ban, especially from diabetics. To date, no study has shown a clear causal relationship between saccharin consumption at normal concentrations and health risks in humans.

Cyclamate

The sweetness of cyclamate, like that of many artificial sweeteners, was discovered by accident. Sveda was working in the laboratory on the synthesis of anti-fever medicines using cyclamates. He put his cigarette down on the bench and when he put it back in his mouth he noticed that it had a sweet taste.

fig. 2.5.4 **Cyclamate is an artificial sweetener that was discovered in 1937 by graduate student Michael Sveda.**

In 1958 it was marketed to diabetics as an alternative to sucrose and saccharin. Cyclamate is sweeter than sucrose, but not as sweet as saccharin. It has less of an aftertaste than saccharin.

Cyclamate is the sodium or calcium salt of cyclamic acid (cyclohexanesulfamic acid). It is prepared by the sulfonation of cyclohexylamine using sulfur trioxide (fig. 2.5.5). Cyclohexamine is produced by the hydrogenation of phenylamine.

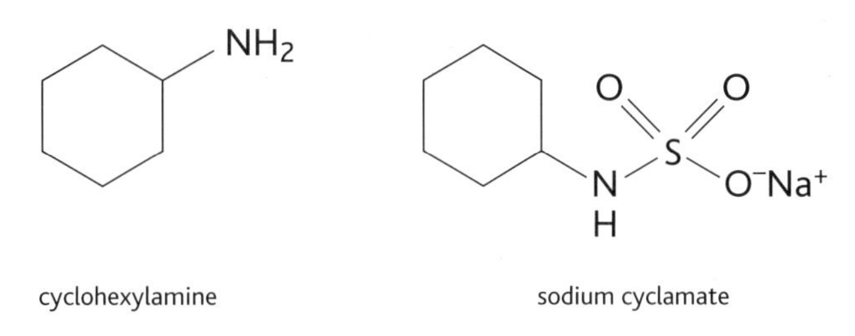

fig. 2.5.5 **The skeletal formulae for cyclohexylamine and sodium cyclamate.**

In 1966, a study reported that some intestinal bacteria could desulfonate cyclamate to produce cyclohexylamine – a chemical that is believed to be toxic in animals. Cyclamate is approved as a sweetener in more than 55 countries, but is banned in food products in the US.

Aspartame

Aspartame is an artificial sweetener used today in many reduced-calorie foods. Strictly speaking it is a dipeptide produced from the combination of two chemically combined, naturally occurring amino acids – phenylalanine and aspartic acid. Again, it was accidentally discovered by James Schlatter, an American drug researcher, in 1965. He was working on an anti-ulcer drug and spilled some aspartame on his hand. He did not wash it off and later detected a sweet taste when licking his fingers to separate sheets of paper.

Aspartame is a white, odourless, crystalline powder – it is about 200 times sweeter than sucrose and dissolves in water readily. It has a sweet taste without leaving a bitter aftertaste like other artificial sweeteners. These properties make it good to use in place of sucrose in many food recipes. However, it tends to interact with other chemicals in food – it cannot, for example, be used in food intended to be cooked in a microwave or baked because it decomposes when heated strongly. Even after extensive tests, there are no reasons to believe that aspartame is unsafe.

phenylalanine + aspartic acid + CH_3OH (methanol) → aspartame + $2H_2O$

The three naturally occurring raw materials used to make aspartame are aspartic acid, phenylalanine and methanol. The process involves three stages – fermentation, synthesis and purification.

Fermentation

Two of the starting materials for making aspartame are L-aspartic acid and L-phenylalanine. These are amino acids that can be produced by certain bacteria. Over the course of approximately three days, the amino acids are harvested and the bacteria that produced them are destroyed. The conditions in which the bacteria grow and develop include warmth and carbohydrate foods such as cane molasses, glucose or sucrose. The fermenting mixture also contains substances such as urea to provide the nitrogen needed to make amino acids.

Both amino acids are chiral, so there are two optical isomers of each. The bacterial fermentation produces a racemic mixture. The enantiomers have to be separated before the synthesis begins because only the L enantiomers are useful in producing a sweetener. If the D enantiomers are used, the resulting molecule does not fit into the sweetness receptors on the tongue, and so it does not taste sweet – making it useless as an artificial sweetener!

Synthesis

Aspartame can be made by a number of different pathways. The first stages of the most common synthesis path prepare the two amino acids. L-phenylalanine is treated with methanol and hydrochloric acid to produce L-phenylalanine methyl ester. This must react with the acid group of L-aspartic acid to give aspartame. So the aspartic acid is modified by adding benzyl groups to protect the amine group, preventing it from getting involved in the main reaction.

The modified amino acids are then pumped into a reactor tank and mixed for 24 hours at room temperature before the temperature is increased to approximately 65°C for a further 24 hours. After this, the reaction mixture is cooled to room temperature and diluted with a suitable solvent before cooling to about −18°C.

An aspartame intermediate crystallises out at this temperature, and the crystals are collected by filtration and dried before they are modified further to produce aspartame. The intermediate compound reacts with aqueous ethanoic acid in the presence of a palladium metal catalyst and hydrogen to produce aspartame. The process takes place in a large tank over about 12 hours.

To get the pure end product, the reaction mixture is filtered to remove the palladium metal catalyst. This leaves a solid residue which is purified by dissolving it in a hot, aqueous ethanol solution. The mixture is then allowed to cool and recrystallise. The crystals of aspartame are filtered and dried to provide the finished product.

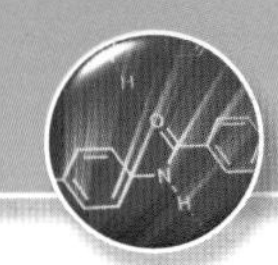

Aspartame must be produced from the **L-forms** of aspartic acid and phenylalanine, not from the **D-forms**. Investigate the process by which L-phenylalanine is separated from the racemic mixture produced by the bacteria in the initial fermenter.

Branded sweeteners also contain acesulfame-K. Find out what you can about this substance.

This brief look at artificial sweeteners suggests that organic synthesis is important both in the production of useful materials and in research. However, it is not as 'planned' as might be commonly believed. Many of the substances in the examples mentioned were found by accident rather than designed by chemists.

Organic synthesis often involves many stages and is both slow and expensive. This is typical of the production of many organic chemicals for food and pharmaceuticals.

The development of an organic synthesis is often a complex process. Investigate the work of Robert Burns Woodward and describe in detail how he developed the synthesis of quinine (including the context in which he was working, and the other scientists with whom he interacted) and one other organic synthesis of your choice.

Questions

1 Draw a flow diagram to summarise the process of aspartame production, showing the conditions required, the reactants and the products at each stage.

2 Why is the synthesis of saccharin from anthranilic acid unlikely to be attractive to a chemical manufacturer?

3 **a** Compare the synthesis of cyclamate sweeteners with the synthesis of aspartame.

b Which process would you expect to have the higher yield? Explain your answer.

c The choice of sweetener to use is not just a function of the economics of the industrial process. Summarise the benefits and disadvantages of saccharine, cyclamate and aspartame as artificial sweeteners.

4 Why is it necessary to block groups on each molecule before reacting them together in the synthesis of aspartame?

Identifying organic molecules for synthesis [5.4.3b, c]

As you read earlier, organic synthesis is difficult to carry out. To develop a successful synthesis – and to convert it into an equally successful industrial process – it is very important to select the best possible route at each stage of the reaction. Knowing the exact chemical composition and structure of the compound to be made will help in developing a suitable method for its synthesis. It is also important to try a number of different synthetic routes through all the intermediate stages.

Sensitive methods of chemical analysis

The most commonly used strategy in developing an organic synthesis is to work backwards from the desired end product. Precise analysis is needed to identify each functional group in the molecule to be synthesised, as well as the basic carbon skeleton you are working towards. You then work backwards, deciding at each step what type of reaction is needed to build up the required carbon skeleton or to add the right functional groups. As each stage is tried, sensitive analysis must be carried out to check that the correct products are formed in the best possible proportions. Several different synthetic routes are usually tried for each stage, and the results are used to help to determine which is eventually used.

All the methods of analysis you have met during your A-level chemistry course – from titration through to spectroscopy – are brought into play when new synthetic routes are being devised. It is very important that any potentially toxic or hazardous products of the reactions are identified at this stage. Another reason for sensitive analysis is the need for **stereo-selectivity** – it can be very important to select the correct optical isomer of a compound, particularly if it is to be active in the body.

Quantitative methods to find the percentage of elements in an organic compound

There are methods available for finding the percentage of carbon, hydrogen, nitrogen, halogens and sulfur in an organic compound. Having found the percentage of all other elements, the percentage of oxygen can then be found by calculating the difference from 100%. Some of this was covered in your AS course.

Combustion analysis

The percentages of carbon and hydrogen in an organic compound can be found by weighing the products after complete combustion of the organic compound. A known mass of the organic compound is mixed with dry copper(II) oxide (to supply the oxygen) and the mixture is heated. The apparatus is designed to absorb the water produced in magnesium chlorate(V) tubes, which are weighed before and after. The carbon dioxide produced is weighed in soda-asbestos tubes. From the masses of these two products, the percentages of carbon and hydrogen can be found.

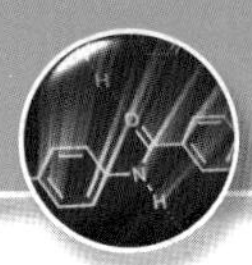

Worked example

Percentage composition

0.0234 g of an organic compound produced 0.0792 g of carbon dioxide and 0.0162 g of water. Given that the relative atomic masses of carbon, hydrogen and oxygen are 12, 1 and 16 respectively, calculate the percentage of carbon and hydrogen in the compound.

44 g of carbon dioxide (CO_2) contains 12 g of carbon.

0.0792 g of carbon dioxide contains $\frac{12}{44} \times 0.0792$ g

= 0.0216 g carbon.

Percentage of carbon in compound

$= \frac{0.0216\text{ g}}{0.0234\text{ g}} \times 100 = \mathbf{92.3\%}$

18 g of water (H_2O) contains 2 g of hydrogen.

0.0162 g of water contains $\frac{2}{18} \times 0.0162$ g

= 0.0018 g hydrogen.

Percentage of hydrogen in compound

$= \frac{0.0018\text{ g}}{0.0234\text{ g}} \times 100 = \mathbf{7.7\%}$

Note that the percentages add up to 100% so there is no oxygen in the compound – it is a hydrocarbon.

Calculating the empirical formula from the percentage composition

When you have the percentage composition of all the elements in an organic compound you can calculate the **empirical formula**. This gives the correct ratio of the combining atoms of the different elements in the compound.

Dividing the percentages by the appropriate relative atomic masses gives the combining ratio of the atoms. Using the data above:

Element	C	H
%	92.3	7.7
A_r	12	1
Combining ratio	$\frac{92.3}{12}$ = 7.7	$\frac{7.7}{1}$ = 7.7
Simplest whole number	1	1

So the empirical formula of the compound is CH.

Worked example

Empirical formula

An organic compound contains 40.1% carbon, 6.6% hydrogen and 53.3% oxygen. Calculate the empirical formula.

Element	C	H	O
%	40.1	6.6	53.3
A_r	12	1	16
Combining ratio	$\frac{40.1}{12}$ = 3.34	$\frac{6.6}{1}$ = 6.6	$\frac{53.3}{16}$ = 3.33
Simplest whole number	1	2	1

The empirical formula is CH_2O.

Note that a reliable way of working out the simplest whole numbers from the actual combining ratios is to divide by the smallest of the combining ratios.

Calculating the molecular formula from the empirical formula

The **molecular formula** of a compound is the same as its empirical formula or some exact multiple of it. In the first example above, the molecular formula of the hydrocarbon is C_1H_1, C_2H_2 … C_8H_8 etc. In the second example, the molecular formula is either CH_2O or some multiple of that – for example $C_2H_4O_2$, $C_3H_6O_3$, $C_4H_8O_4$ etc.

A straightforward way of determining the molecular formula is to measure the relative molecular mass of the compound, either by traditional experiments or by using a **mass spectrometer**.

The relative molecular mass must be an exact multiple of the empirical formula mass. In the first example above, the molecular mass of the hydrocarbon must be a multiple of 13 – so if its actual relative molecular mass turns out to be 78 then its molecular formula must be C_6H_6. In the second example, if the relative molecular mass was found to be 60, then the molecular formula would be $C_2H_4O_2$.

Characteristic reactions of functional groups

Knowing the molecular formula, you can consider the possible **functional groups** that could be in the molecule and then carry out some tests to confirm your ideas.

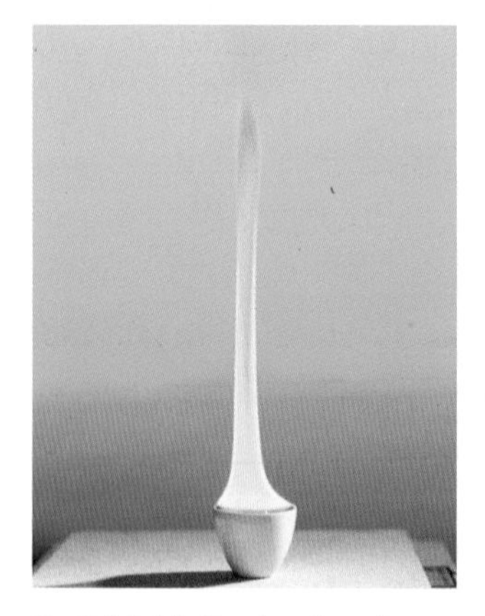

fig. 2.5.6 (a) Combustion of benzene produces a yellow smoky flame.

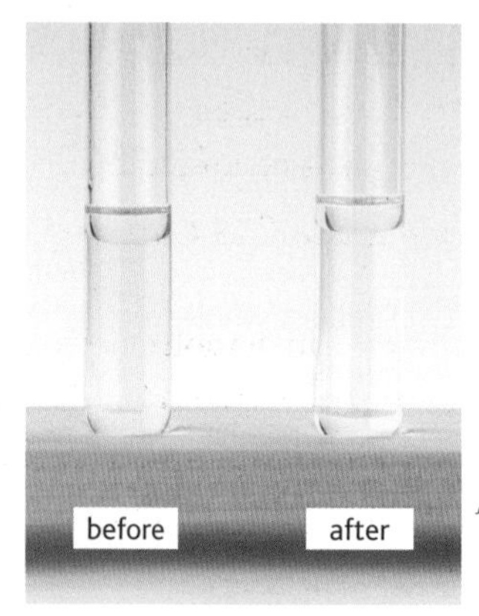

fig. 2.5.6 (b) Bromine is decolourised on mixing with alkenes.

fig. 2.5.6 (c) A carboxylic acid reacting with sodium carbonate.

fig. 2.5.6 (d) An alcohol reacting with phosphorus(V) chloride.

fig. 2.5.6 (e) A carbonyl compound reacting with 2,4-dinitrophenylhydrazine.

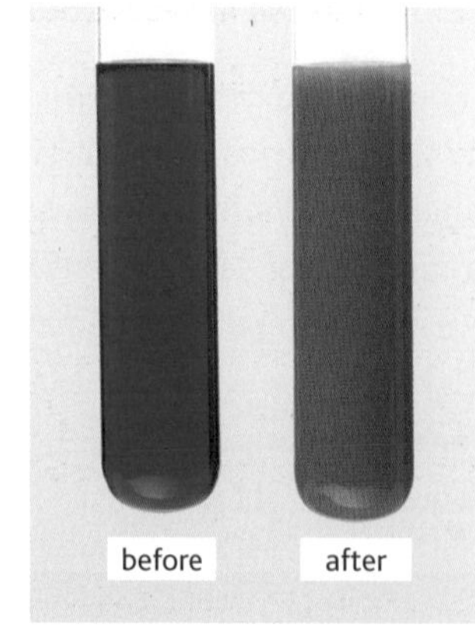

fig. 2.5.6 (f) Fehling's test for an aldehyde.

For example, if the compound is a hydrocarbon:

- if it burns with a smoky flame then it could be an aromatic compound
- if it is unsaturated and it decolourises bromine in solution then it could be an alkene.

If the compound contains carbon, hydrogen and oxygen:

- if carbon dioxide is produced with sodium carbonate then it could be a carboxylic acid
- if smoky white fumes of hydrogen chloride are produced with phosphorus(V) chloride then it could contain an –OH group
- if a yellow precipitate is produced when tested with 2,4-dinitrophenylhydrazine then it could be a carbonyl compound
- if it forms a red-brown precipitate with Fehling's solution then it could be an aldehyde.

Bear in mind that organic compounds may contain two or more functional groups, but carrying out simple and convenient tests such as these can enable you to make some predictions about the structure of an organic compound.

So in the worked examples on page 239, you would be surprised if the first compound did not burn with a smoky flame. And if the second compound was a colourless liquid that gives carbon dioxide gas with sodium hydrogencarbonate solution then it is probably an acid containing a carboxyl group – you could even be pretty sure that it was ethanoic acid, CH_3COOH.

When Frederick Sanger identified the elements in insulin (chapter 2.4) he would have used Lassaigne's tests – these break up an organic compound into units that can easily be tested to identify component elements. Research and find out about these tests.

Why is there no test for oxygen?

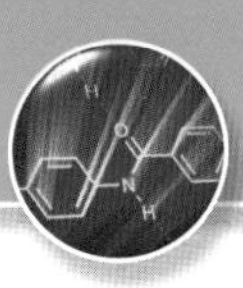

Use of modern techniques

The methods outlined above are very slow and laborious and great credit is due to past chemists who had the perseverance to use them. Today we have quicker and more reliable techniques:

- **Infrared spectroscopy** – *Edexcel AS Chemistry* pages 232–5 and chapter 1.7 in this book.
  Analysis of the IR spectrum of an organic compound will indicate particular functional groups and features in an organic compound. For example, a peak between 3300 and 3500 cm^{-1} suggests either an amine or an alcohol group – the breadth of the peak can be an indication of which ones.
- **Mass spectrometry** – *Edexcel AS Chemistry* pages 230–1 and chapter 1.7 in this book.
  This enables an accurate relative molecular mass to be determined, and then possible groups within the molecule from the masses of the ion fragments.
- **Nuclear magnetic resonance** – chapter 1.7 in this book.
  This can be used to identify groups of hydrogen atoms in an organic molecule – for example, if there are three hydrogen atoms in a similar environment then this indicates a $-CH_3$ group.

None of these methods gives the complete answer, but collectively they provide weighty evidence which, when combined with a chemist's knowledge and experience, allows a confident conclusion to be reached about molecular formula and structure.

Questions

1 A compound has the molecular formula $C_2H_4O_2$ but gives no carbon dioxide with sodium hydrogencarbonate. Suggest a suitable compound to fit this information and draw a displayed formula for it.

2 Complete combustion of 0.206 g of an organic compound yields 0.5666 g of carbon dioxide and 0.4635 g of water. Calculate the empirical formula of the organic compound.

3 A sample of an organic compound weighing 4.6 g contains 2.4 g of carbon and 0.6 g of hydrogen. Its relative molecular mass is 46.

   a Find the empirical and molecular formulae of this compound.

   b Draw the displayed formulae of two isomers.

   c Explain how phosphorus(V) chloride could be used to distinguish between these two isomers.

4 a An organic synthesis takes place in five steps. The first step gives an 80% yield, the second gives a 75% yield, the third 90%, the fourth and the fifth 80% each. What is the total overall percentage yield for the synthesis process?

   b Use your answer to help you explain why scientists look for the shortest possible route in an organic synthesis and pay attention to investigating the most efficient reaction at each stage of the process.

Predicting properties of organic compounds [5.4.3d (i)]

Modern techniques such as mass, infrared and/or nuclear magnetic resonance spectrometry are routinely used to build up a picture of the functional groups present in unknown organic compounds.

Needless to say, functional groups have an enormous influence on the behaviour of the compound of which they are a part. In this section we are going to work towards looking at compounds containing two or more functional groups and suggest how they are likely to interact to affect their properties and reactions.

Predicting the physical properties of compounds

Sometimes you may be presented with a number of compounds which all have the same molecular formula but have very different structural arrangements and functional groups – in other words, isomers of different types. As you know, some functional groups make the formation of hydrogen bonds between molecules much more likely, which increases the melting and boiling temperatures of the compounds. However, it is not always easy, or even possible, to predict how different functional groups will interact to affect the properties of an organic compound. For example, a compound with molecular formula $C_4H_8O_2$ could be, among others:

- butanoic acid, C_3H_7COOH
- methyl propanoate, $C_2H_5COOCH_3$
- propyl methanoate, $HCOOC_3H_7$.

The structural formulae of these isomers (**fig. 2.5.7**) reveal their different functional groups and this allows us to predict and compare their properties. Some of their physical properties are summarised in **table 2.5.1**.

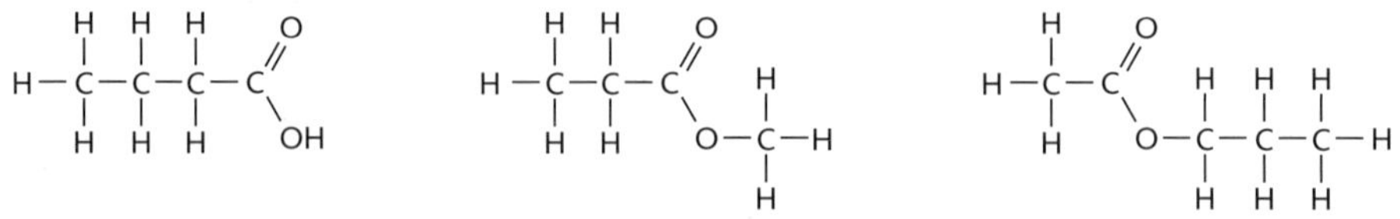

fig. 2.5.7 Structural formulae (left to right) of butanoic acid, methyl propanoate and propyl methanoate.

	Butanoic acid	Methyl propanoate	Propyl methanoate
Appearance	Colourless, flammable liquid	Colourless, flammable liquid	Colourless, flammable liquid
Melting temperature/°C	–4.4	–51.7	–92.9
Boiling temperature/°C	165.6	79.7	81.3
Density/g cm^{-3}	0.96	0.92	0.91

table 2.5.1

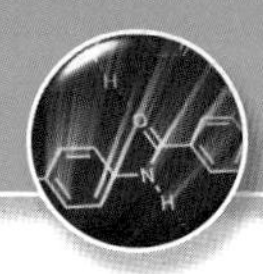

The melting and boiling temperatures of butanoic acid are clearly higher than those of the two esters. This can be explained by the ability of carboxylic acids to participate in hydrogen bonding. Esters cannot do this because the nature of their functional groups is different – they have no hydrogen atoms bonded to small, highly electronegative atoms with a lone pair of electrons.

SC Butanoic acid has one functional group (–COOH) but the two esters have two functional groups – one –COO group in common and an alkyl group, $-CH_3$ and $-C_3H_7$ respectively. Table 2.5.1 shows how this affects the physical properties of the compounds. How do the functional groups of the esters interact to affect the properties?

Predicting the chemical properties of compounds

Sometimes a chemical reagent can be used to show the presence of a particular functional group – for example, the presence of a carboxylic acid, aldehyde or ketone group is demonstrated quite easily.

Some functional groups impose a certain type of general behaviour on compounds that contain them. For example:

- some act as electrophiles, others as nucleophiles
- some are susceptible to addition reactions, others to substitution
- some are easily oxidised, others more easily reduced.

Figure 2.5.8 shows the two compounds having molecular formula C_7H_8O.

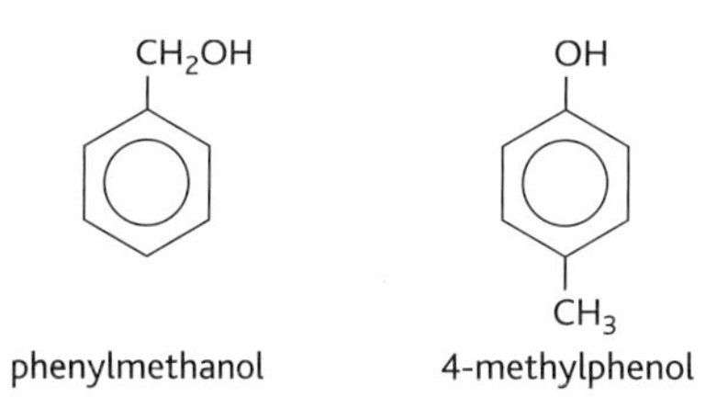

fig. 2.5.8 Phenylmethanol and 4-methylphenol.

Phenylmethanol is a liquid at room temperature and 4-methylphenol is a low melting temperature solid. Both are slightly soluble in water. The presence of the benzene ring disrupts the hydrogen bonding between the –OH groups and water molecules. They are both readily soluble in organic solvents.

Phenylmethanol is shown to be neutral when an aqueous solution of it is tested with a pH meter. However, aqueous 4-methylphenol is slightly acidic. This is because its –OH group is attached to the benzene ring and the lone pair on the oxygen atom are delocalised into the π ring system of the ring, making the O—H bond more polar. However, it is not acidic enough to release carbon dioxide from sodium hydrogencarbonate.

Phenylmethanol is a primary alcohol containing a $-CH_2OH$ group; 4-methylphenol is a phenol with an –OH attached to a benzene ring. Since both have an –OH group they will both react with phosphorus(V) chloride to produce a chlorine compound (see fig. 2.5.9) and hydrogen chloride.

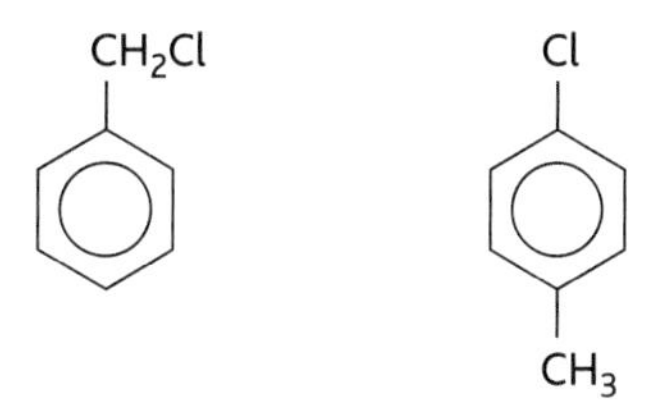

fig. 2.5.9 Products of the reaction of phosphorus(V) chloride with an alcohol and a phenol respectively.

Phenylmethanol will be oxidised to benzaldehyde (see fig. 2.5.10) and further oxidised to benzenecarboxylic acid using acidified potassium dichromate(VI) – accompanied by the colour changes you will be used to by now (see page 210 in *Edexcel AS Chemistry*). 4-methylphenol is not readily oxidised so there will be no colour changes with the same oxidising agent.

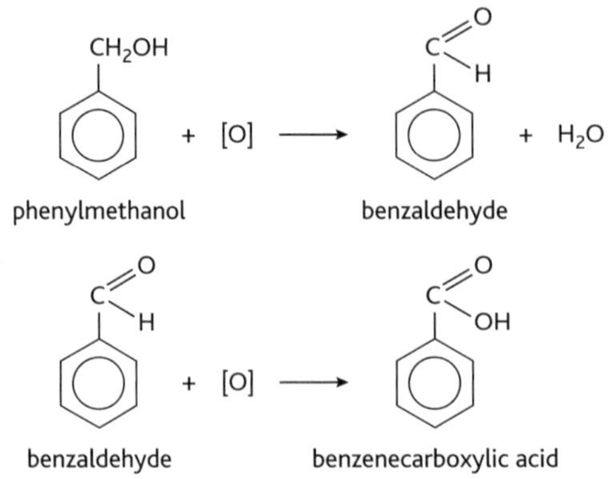

CH_2OH (phenylmethanol) + [O] → benzaldehyde + H_2O

benzaldehyde + [O] → benzenecarboxylic acid

fig. 2.5.10 Two stages in the oxidation of phenylmethanol.

Predicting the properties of natural molecules

All the examples you have looked at so far have used relatively small molecules. However, many molecules found in living organisms are much bigger and more complicated, often with several different functional groups which interact. Look at the structural formulae of the three compounds shown here, some of which you have met before and some of which you haven't.

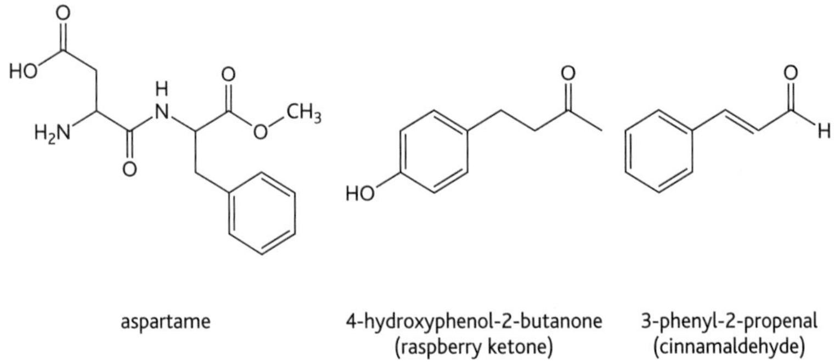

aspartame

4-hydroxyphenol-2-butanone (raspberry ketone)

3-phenyl-2-propenal (cinnamaldehyde)

For each molecule, list the functional groups and say how you would expect each to affect the physical and chemical properties of the compound. Then look up the data and compare their actual properties to your predictions.

Questions

1 Using your knowledge of the different functional groups in the molecules in fig. 2.5.7, predict what effect they will have on the chemical reactions of the molecules, then investigate and see how accurate your predictions are.

2 Make a table to summarise the influence of five different functional groups on the molecules to which they are attached.

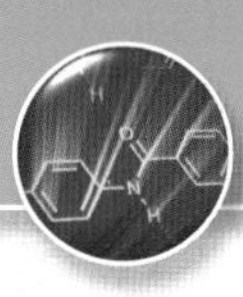

Planning synthetic routes [5.4.3d (ii), (iii)]

You probably realise by now that there is a vast range of possible organic compounds that can exist. Many new compounds are being made, and the existence of different stereoisomers only increases the range.

As you have seen in the previous sections, most organic compounds are made in a series of steps. A chemist wanting to make a new compound needs to plan the best route from the available starting materials. Until fairly recently, planning a route depended on the chemist's knowledge of different organic reactions. Now the chemist has access to computer programs which identify possible routes for making a compound and identifying possible starting materials. Here are some of the common steps which may be required in a complex synthesis, along with ways of carrying them out.

Adding or removing a carbon atom from an organic molecule

Many syntheses involve either adding or removing a carbon atom from molecules. There is a set way of doing each of these.

To add a carbon atom to a chain, start with the halogenoalkane and reflux it with potassium cyanide dissolved in ethanol. This will produce a **nitrile**. Hydrolysis of the nitrile by refluxing with dilute hydrochloric acid (or sodium hydroxide followed by acidification) produces the carboxylic acid:

$$R{-}X + KCN \longrightarrow RCN + KX$$

halogenoalkane → nitrile with extra C atom in chain

$$RCN + 2H_2O \longrightarrow RCOOH + NH_3$$

nitrile → carboxylic acid

Friedel-Crafts alkylation reactions (chapter 2.3) provide another means of adding one or more carbon atoms to an aromatic molecule.

Reducing the number of carbon atoms in a chain can be done by **Hofmann's degradation** in which an amide is heated with bromine and concentrated potassium hydroxide:

$$R{-}C(=O)NH_2 + Br_2 + 4KOH \longrightarrow R{-}NH_2 + K_2CO_3 + 2KBr + 2H_2O$$

amide → amine with one fewer C atom in chain

Another way of reducing the length of a carbon chain is by the **Hunsdiecker reaction**. If the silver salts of carboxylic acids are reacted with halogens such as bromine then an organic halide, carbon dioxide and an inorganic silver salt will result. The reaction is carried out under reflux in tetrachloromethane:

$$RCOOAg(aq) + Br_2(l) \xrightarrow{\text{reflux in } CCl_4} RBr(l) + CO_2(g) + AgBr(s)$$

The organic halide has a reduced number of carbon atoms in the skeleton, and can then be used in further synthetic steps.

SC The length of a carbon chain can be extended using Grignard reagents. Research and find out what Grignard reagents are, and how they are used to do this job. What other uses have these reagents?

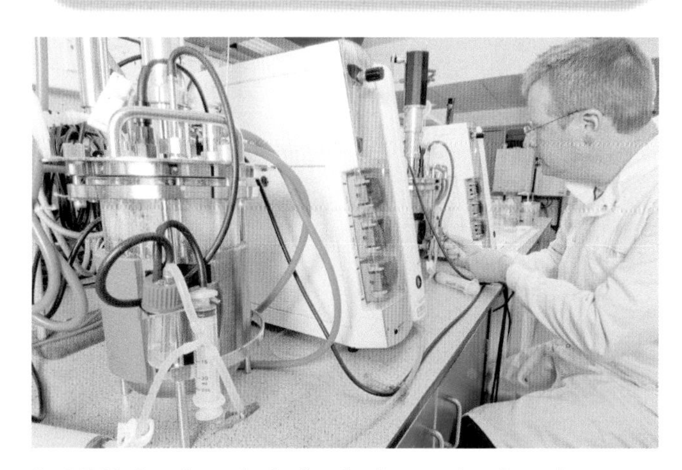

fig. 2.5.11 Organic synthesis often involves a series of complex reactions – and lots of apparatus!

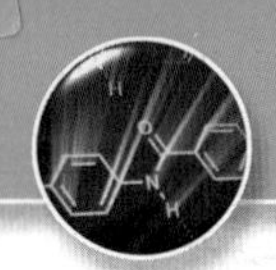

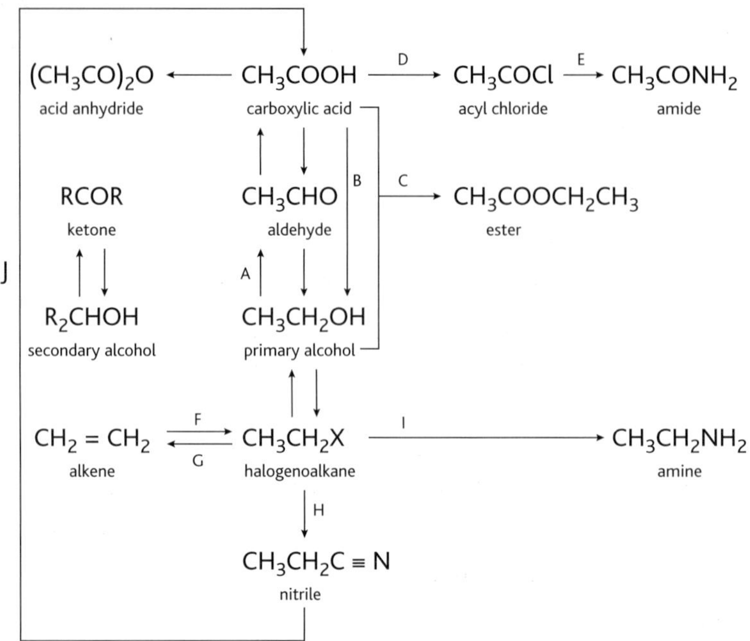

fig. 2.5.12 Some reaction schemes in aliphatic organic chemistry.

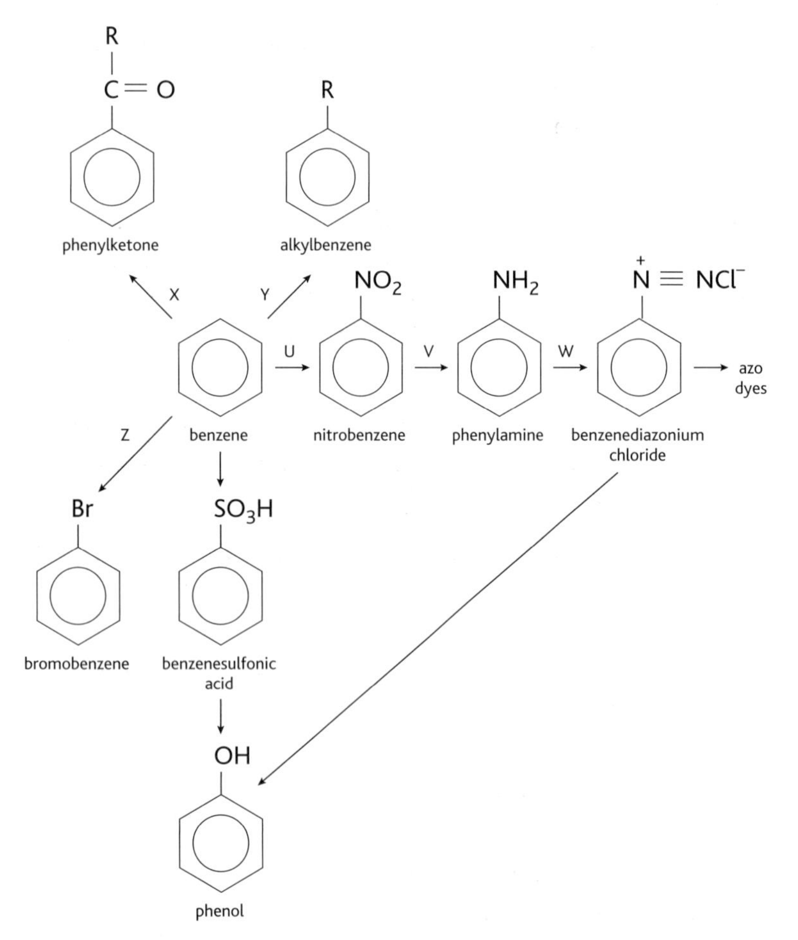

fig. 2.5.13 Some reaction schemes in aromatic organic chemistry.

Reaction pathways

Figures 2.5.12 and **2.5.13** show some of the common pathways in organic chemistry. The former uses ethanol as the starting material; the latter uses benzene to show corresponding aromatic reactions.

Choosing a synthesis route

Propanoic acid, CH_3CH_2COOH, can be added to animal feeds to inhibit the growth of moulds and bacteria. If there is a need for large-scale production of a compound such as this then there will probably be a variety of ways in which you could carry out the synthesis. For example, there may be different starting materials available, such as ethanoic acid, propan-1-ol or an ester of some sort.

Suppose it is apparent that ethanoic acid, CH_3COOH, would be a suitable raw material. We could use the following process (summarised in **fig. 2.5.14**).

- Reduce the acid to ethanol with lithium tetrahydridoaluminate in ether:

  $$CH_3COOH(l) + 4[H] \rightarrow CH_3CH_2OH(l) + H_2O(l)$$

- React ethanol with phosphorus(V) chloride to produce chloroethane:

  $$CH_3CH_2OH(l) + PCl_5(s) \rightarrow CH_3CH_2Cl(l) + HCl(g) + POCl_3(l)$$

- Reflux chloroethane with potassium cyanide in ethanol to produce propanenitrile – adding a carbon atom to the chain:

  $$CH_3CH_2Cl(l) + KCN(alc) \rightarrow CH_3CH_2CN(l) + KCl(alc)$$

- Hydrolyse propanenitrile by refluxing with dilute acid to produce propanoic acid:

  $$CH_3CH_2CN(l) + 2H_2O(l) \rightarrow CH_3CH_2COOH(l) + NH_3(g)$$

CH_3COOH ⟶ CH_3CH_2OH ⟶ CH_3CH_2Cl ⟶ CH_3CH_2CN ⟶ CH_3CH_2COOH

ethanoic acid | ethanol | chloroethane | propanenitrile | propanoic acid

fig. 2.5.14 Propanoic acid production using ethanoic acid as the starting point.

Alternatively, it may be that there is a reliable source of silver butanoate at a reasonable cost. In this case, the scheme might follow this plan.

- A carbon atom is removed from the chain in this reaction in a Hunsdiecker reaction by refluxing the ester with bromine in tetrachloromethane to form bromopropane:

$$CH_3CH_2CH_2COOAg(s) + Br_2(l) \xrightarrow{\text{reflux}} CH_3CH_2CH_2Br(l) + CO_2(g) + AgBr(s)$$

- Bromopropane is heated under reflux with aqueous sodium hydroxide solution to form propanol:

$$CH_3CH_2CH_2Br(l) + NaOH(aq) \xrightarrow{\text{reflux}} CH_3CH_2CH_2OH(l) + NaBr(aq)$$

- Propanol is oxidised by refluxing with acidified potassium dichromate(VI) solution to form propanoic acid:

$$CH_3CH_2CH_2OH(l) + [O] \xrightarrow{\text{reflux}} CH_3CH_2COOH(l) + H_2O(l)$$

The decision as to which route to follow for the synthesis of a compound for market will be influenced by many factors beyond the starting material. These cannot be covered in detail here but a general awareness of them is useful in appreciating the operation of the chemicals manufacturing industry.

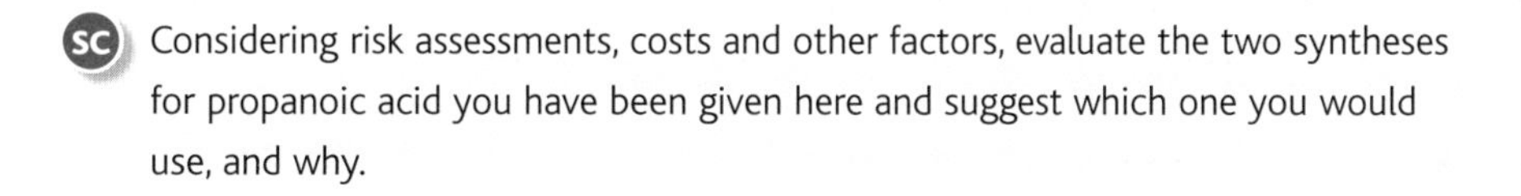

SC Considering risk assessments, costs and other factors, evaluate the two syntheses for propanoic acid you have been given here and suggest which one you would use, and why.

Questions

1 Give the reaction conditions for each of the reactions labelled A–J in fig. 2.5.12.

2 Give the reaction conditions for each of the reactions labelled U–Z in fig. 2.5.13.

3 Explain how you could convert benzene into phenylethanol (right), in two stages.

$C_6H_5CH(OH)CH_3$

4 Explain how you could convert propanoic acid into ethylamine in three stages.

5 Explain how you could convert ethanol into propylamine in three stages.

Synthesis of stereo-specific drugs [5.4.3d (v)]

We know that many chemicals have chiral centres and that because of this they can form enantiomers that have different effects on plane-polarised light. The shapes of molecules are very significant when dealing with drugs and medicines.

Enantiomeric drugs

We have seen, from the nucleophilic substitution reactions of halogenoalkanes, that the mechanism can be S_N1 or S_N2 (see *Edexcel AS Chemistry* page 226 and chapter 1.6 in this book).

If the reaction is S_N2 the molecule will effectively turn itself 'inside out' but the product should be a single isomer. However, if the mechanism is S_N1 then the intermediate formed has a planar structure. This means the attacking nucleophile has an equal chance of attacking either side of the intermediate. The result is a racemic mixture with equal amounts of both isomers.

Because of the specific ways in which drugs act on the body, usually only one of the enantiomers is effective – in fact, in some cases the one that is not effective may have serious side-effects if it is used. Thalidomide, prescribed to pregnant women in the 1950s with terrible effects, remains the most shocking example of the problems enantiomers can cause (see chapter 1.6). So, when producing drugs where only one enantiomer is effective, separating the isomers from the racemic mixture can be essential – despite reducing the yield by 50%.

Now that chemists are aware of the likelihood of stereo-specific molecules arising in drug-making processes, both new safeguards and new possibilities have arisen. All drugs are routinely tested for teratogenicity (effect on the foetus) during the development process. If different enantiomers are produced during the process, companies are required to separate and test them individually to identify any likely problems.

Making single enantiomers

Salbutamol (the drug in the ventolin inhalers often prescribed to people affected by asthma) is a racemic mixture containing one completely inactive enantiomer, while levalbuterol (sold as Xopenex) is a single enantiomer. Increasingly, as synthesis techniques improve, what is given to a patient is a single enantiomer rather than a racemic mixture. Up to 50% of new drug molecules are chiral – as a result the dose is generally smaller and any side-effects from the other enantiomers are eliminated.

HSW Should thalidomide still be used?

The historical problems have not stopped thalidomide being used as a drug because it has a number of therapeutic effects. For many years it was used in the treatment of leprosy in less economically developed countries (LEDCs). Although doctors are extremely careful not to use it on pregnant women, in LEDCs it is not always possible to monitor the use of the drug as closely as it is in the UK. More recent drugs control the symptoms of leprosy without the risk of foetal damage, although they are more expensive.

Thalidomide is also effective in the treatment of several forms of cancer. Its use is monitored carefully – currently in the UK it is made available only to named patients and others taking part in research trials. It is also being used as part of the battle against HIV/AIDS and research continues into a variety of ways in which it may be used.

Apart from having a devastating effect on the foetus, thalidomide is an extremely safe drug with very few side-effects and it may yet turn out to be extremely useful to society. However, there are some people – including some of the children affected by the drug – who feel that it should not be used because in their view the risks outweigh the benefits. Society has to decide whether this drug – with such potential for good and harm – should be allowed to make a comeback.

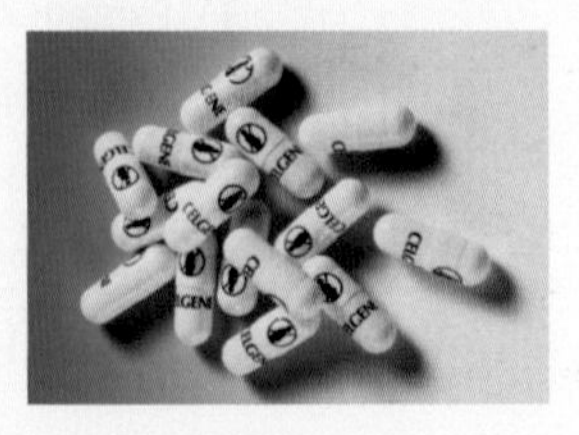

fig. 2.5.15 Look carefully at these thalidomide capsules and you will see the image of a pregnant woman crossed out – a clear warning that anyone who might be pregnant should not take this drug.

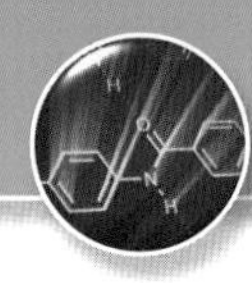

You might think that giving a reduced amount of a drug, because only the active enantiomer is present, would reduce costs but at the moment this is often not the case. It isn't easy to separate enantiomers – the methods most commonly used involve **thin-layer chromatography** using a very expensive solid phase, or a similarly difficult version of electrophoresis. The materials used need replacing regularly and the amount which can be treated at any one time is limited. All of this increases the cost of manufacturing the drug. **Asymmetric synthesis**, in which only one isomer is made during the reaction, is another alternative. Understanding the reaction mechanism may make it possible to synthesise the desired enantiomer by carefully controlling every step of the process – but even when possible this tends to be expensive, often involving extreme conditions, for example temperatures of –100°C.

Chemists are investigating new methods of synthesising single enantiomers. One of these, first published in 2008, looks at a synthesis that takes place in the much less severe temperature range of –40 to 0°C. Initial research reported that it was 98% effective in producing the desired enantiomers. The process involves adding 'chiral auxiliaries' – small molecules that attach to the molecules in the synthesis and help to direct the new functional groups to the desired positions on the molecule. The auxiliaries can be removed and recycled at the end of the synthesis, keeping costs down even more.

The initial research was carried out on ketones, but the research team hopes that their work will be relevant to the synthesis of many other optically active chiral molecules, including drugs, making them cheaper and safer for everyone.

Other new techniques involve enzymes that require temperatures of 30–40°C and give an almost 100% yield of a desired enantiomer. Yet more new processes involve the 'wrong' isomers being switched or recycled to be treated again, so improving the yield. As a result, single enantiomer drugs are becoming more achievable, more economically, all the time.

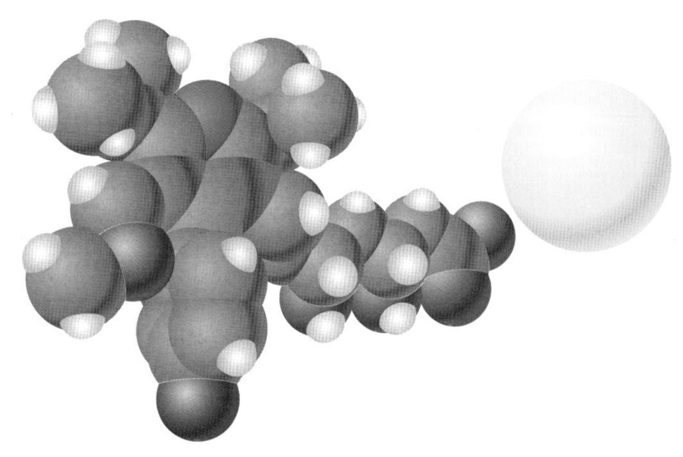

fig. 2.5.16 Single enantiomer drugs don't always mean safety. Cerivastatin, a cholesterol-lowering drug, was a single enantiomer. Used by thousands of people, it was discovered that for a tiny minority it destroyed their kidney function, and for some it led to death. It was withdrawn and is no longer prescribed.

SC Find out about the structure of thalidomide and explain how the thalidomide tragedy came about.

Investigate and write about the use of thalidomide in the treatment of HIV/AIDS.

Summarise the ethical arguments for and against the use of thalidomide as a treatment for HIV/AIDS or any other disease.

Questions

1 Why is the testing of the D enantiomer and of the L enantiomer for harmful side-effects not sufficient?

2 Explain carefully how understanding the reaction mechanism of the synthesis of a stereo-specific drug can help in the planning of a successful industrial process.

3 Some drugs are sold as racemic mixtures, others as single isomers. In some drugs, the dosage of a racemic mixture will be the same as the dosage of a single isomer version. For other drugs, the racemic mixture will be prescribed at double the dose. What does this tell you about the different isomers of the drug?

4 New drug syntheses, involving processes such as chiral auxiliaries and special catalysts, are developing all the time. Give one way in which these developments could reduce the likelihood of a thalidomide disaster from happening again, and one way in which they will have no effect on that risk.

Control measures for hazards in organic synthesis [5.4.3d (iv)]

During the process of an organic synthesis it is likely that you may need to use a number of hazardous chemicals, and that some reactions with a relatively high risk will be undertaken, particularly when done on an industrial scale. In *Edexcel AS Chemistry* (pages 96–8) you looked at the difference between **hazard** and **risk**, and at producing a risk assessment. In organic synthesis, with its many different steps and reacting conditions, you will find your risk assessment skills challenged to the full. When deciding which route to use to synthesise a particular chemical, you need to look at the risks posed in each stage of the process and then weigh up which route is the safest overall. However, one very risky step in an otherwise relatively risk-free process might be enough for you to choose another route!

The precautions taken in fig. 2.5.17 are similar to those you would take in the laboratory when carrying out preparations of organic compounds.

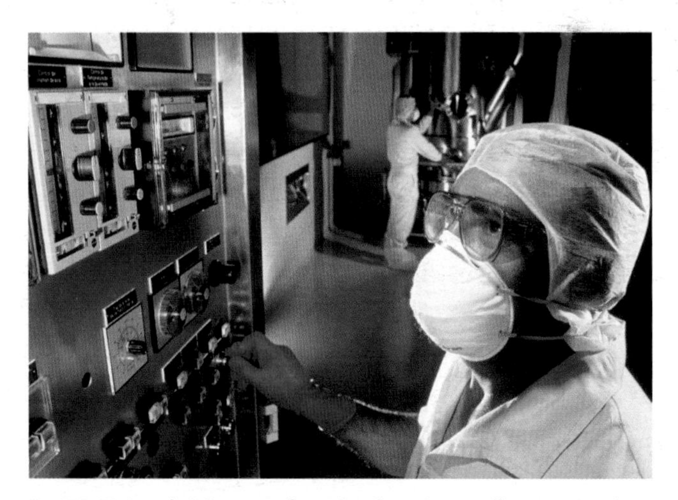

fig. 2.5.17 A technician manufacturing drugs in a sterile room wearing a laboratory coat, eye protection, gloves and a face mask.

In addition you might:

- *use a fume cupboard* – so any harmful gases are vented from the laboratory
- *reduce the scale of working* – working on a microscale makes the preparation faster, uses less raw material and, in case of accident, reduces the adverse effects
- *look for an alternative, and possibly safer route*, to the product
- *replace a reactant with a less toxic alternative* – for example, when carrying out ethanoylation (chapter 1.6) ethanoic anhydride is less toxic than ethanoyl chloride; the reaction is also more controllable.

Risks and hazards

It is important to distinguish between hazards and risks – a hazard is the potential to cause harm; risk is the likelihood of harm.

For example, potassium dichromate(VI) is frequently used as an oxidising agent in organic synthesis reactions but this useful solid is a highly toxic carcinogenic (cancer-causing) chemical. The extent of the hazard will be shown on a compound's **Hazcard** (see fig. 2.5.18).

However, potassium dichromate(VI) was used in early breathalysers (see chapter 2.1). It was sealed in a tube and so did not become airborne when air was drawn over it. Therefore, although potassium dichromate(VI) is a highly hazardous substance, its use in breathalysers did not present a risk to the subject.

Flour would not be commonly regarded as a hazard. However, a baker working in a bakery who is exposed to flour dust can develop all sorts of respiratory conditions. So, flour is a substance of low hazard but in a bakery it has high risk – and in the right conditions in a silo, flour can explode violently when mixed with air!

When preparing to carry out an organic synthesis – particularly a new one – it is extremely important to carry out a **risk assessment**. Even with care, organic compounds can produce some dangerous situations. For example, dichloromethane is a commonly used solvent in organic chemistry in universities and industry. However, a number of serious explosions have occurred when it is used in reactions involving azides, particularly when the temperature increased unexpectedly – this is partly because most people are unaware of this particular risk.

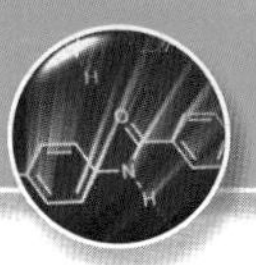

Chemical Safety Data: Potassium dichromate

Common synonyms	None
Formula	$K_2Cr_2O_7$
Principal hazards	** Potassium dichromate is toxic if swallowed, inhaled or absorbed through the skin. It is corrosive and may produce severe eye damage. ** Chromium (VI) compounds are carcinogens. ** Potassium dichromate may act as a sensitizer. ** This material is a strong oxidizing agent and reacts vigorously or explosively with a wide variety of reducing agents.

fig. 2.5.18 The Hazcard shows potassium dichromate(VI) to be a highly toxic carcinogen as a solid. This information must inform the way it is used.

Control measures used to reduce risk

As you saw earlier in this section, in planning any organic synthesis it is important to consider all the possible steps. One important aspect to consider is the nature of the reactants at each stage and the steps to take to reduce any risks – for example, refluxing to keep the reacting mixture cool or choosing a synthetic route with the least hazardous compounds. For example, in the alternative syntheses of propanoic acid you considered on pages 246–7, the hazard rating of the chemicals used in both routes is summarised in table 2.5.2.

When you look at this data you can see that the decision as to which synthesis to use is complicated. The use of reflux to keep temperatures low is understandable when the flammability of several of the reactants in taken into account, for example. The need to carry out reactions in a fume cupboard is also clarified.

Synthesis 1 – using ethanoic acid	Synthesis 2 – using silver butanoate
Ethanoic acid – harmful, corrosive	Silver butanoate – toxic
Lithium tetrahydridoaluminate – flammable, harmful, corrosive, strong reducing agent, reacts with water, reacts with air	Bromine – harmful
Ethoxyethane – harmful, highly flammable	Tetrachloromethane – toxic, dangerous for the environment
Ethanol – harmful, highly flammable	Bromopropane – harmful
Phosphorus(V) chloride – harmful, corrosive	Aqueous sodium hydroxide – harmful, corrosive
Chloroethane – toxic, highly flammable	Propanol – highly flammable
Potassium cyanide – toxic	Acidified potassium dichromate(VI) – very toxic, corrosive, oxidising
Propanenitrile – toxic, flammable	Potassium manganate(VII) – very toxic, corrosive, oxidising

table 2.5.2 Alternative syntheses of propanoic acid.

Manufacture of paracetamol

There are several different methods for preparing paracetamol in industry (see page 214). The method shown in fig. 2.5.19 is widely used, particularly in China which now supplies much of the world's paracetamol.

Investigate this synthesis further. Name all the chemicals used and research their hazard status. Find out more details about the reaction conditions and about the yields of the different intermediates. Explain why the reaction conditions are needed and how the method of carrying out each step controls the level of risk.

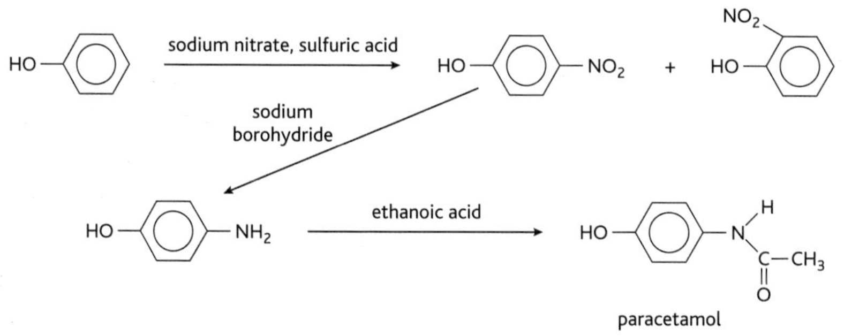

fig. 2.5.19 **A summary of the traditional industrial preparation of paracetamol.**

Questions

1 Look at the data in table 2.5.2 and use it, along with your knowledge of the two processes, to determine which synthetic route you would choose to use. Explain your decision carefully with regard to the hazards involved and the conditions needed to reduce the risk levels of the preparation.

2 In the scheme for the industrial preparation of paracetamol in fig. 2.5.19, assume that the nitration of phenol has a yield of 25% for 4-nitrophenol and of 36% for 2-nitrophenol. The reduction of 4-nitrophenol has a yield of 75% and the final stage a yield of 80%. Calculate the maximum mass of paracetamol that could be made from 1 tonne (1000 kg) of phenol.

3 In fig. 2.5.19 the conditions for the nitration of phenol are different from the conditions for the nitration of benzene (chapter 2.3, page 202). Explain why this is the case.

4 The flow diagram in fig. 2.5.20 summarises the steps in another method of producing paracetamol – this time from nitrobenzene:

  a Write down the reaction conditions for the steps labelled A–D.

  b Check up on the hazard levels of the compounds involved in this reaction and suggest ways in which the risks might be minimised by the choice of reaction conditions.

  c Suggest why the yield of this synthesis would not be very high.

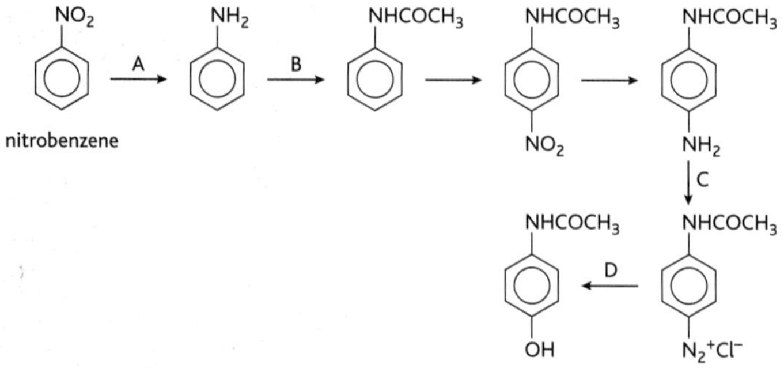

fig. 2.5.20

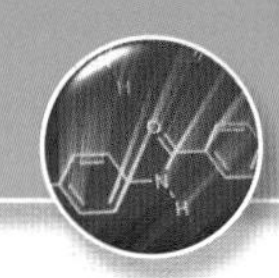

Combinatorial chemistry [5.4.3e]

Pharmaceutical companies spend millions of pounds on research into finding new or better drugs. Many of the compounds synthesised are never used. Research often shows that a drug made for one purpose can be used for other things – the humble aspirin is an analgesic (pain killer) but it can also be used to thin blood and prevent it clotting.

Any method that speeds up the synthesis of new compounds will save pharmaceutical companies a lot of money in getting new drugs to market faster. Over the last 20 years or so, **combinatorial chemistry** has done just that. It enables automated processes to carry out routine reactions very quickly and accurately. A 'lead' compound is identified (pronounced 'leed' – not one that contains Pb!) which will have some apparent therapeutic or biological activity. Combinatorial chemistry is then applied to make huge numbers of structurally similar but slightly different compounds all at the same time. Some of these may be much more effective than the original lead compound. They are then stored in huge data libraries as potential medicines, or moved forward into the early stages of drug development.

Before combinatorial chemistry was developed, each compound to be made was identified, a route for making it was planned, the synthesis was carried out, the product was purified and its structure confirmed – then it was ready for testing. A traditional chemist might synthesise 100–200 chemicals in a year compared with thousands, or even millions, of different combinations using a combinatorial robotic system. These can then be run through the sort of high-throughput screening systems you considered earlier in your chemistry course (see page 53 in *Edexcel AS Chemistry*) and all the molecules with therapeutic possibilities can be picked out.

Although combinatorial chemistry has actually had relatively little impact up to now on the development of new drugs, as techniques develop they are likely to become increasingly successful and important.

fig. 2.5.21 Long exposure of robotic drug-making equipment. The light trace in the upper centre shows the movement of the apparatus.

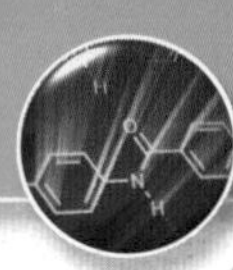

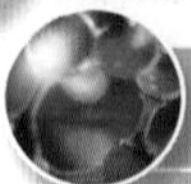

HSW The history of aspirin – making medicines the old-fashioned way

At least 7000 years ago, Hippocrates knew that willow bark and some other plants had pain-killing and fever-reducing properties. We now know that the active ingredient in willow bark (fig. 2.5.22 (a)) is related to 2-hydroxybenzenecarboxylic acid, commonly called salicylic acid (see *Edexcel AS Chemistry* page 95).

Salicylic acid (fig. 2.5.22 (b)) was isolated from the meadowsweet herb by German researchers in 1839. The problem with salicylic acid was that, although it helped to ease pain, it often caused bleeding in the stomach and intestines. Later it was realised that aspirin, being totally covalent, was not very soluble and so the sodium salt was made (fig. 2.5.22 (c)) – being soluble this was absorbed faster in the stomach, and so was quicker acting and more effective.

Around 1853 Charles Frederic Gerhardt first produced the chemical that we call aspirin, 2-ethanoylbenzenecarboxylic acid (fig. 2.5.22 (d)) but it was not until about 1890 that it was developed by the Bayer Company for pain relief. It is manufactured on an industrial scale by the reaction of 2-hydroxybenzenecarboxylic acid with ethanoic anhydride, which is both relatively cheap and not too reactive.

(a) CH_2OH, O-glucose

(b) O, C, OH, OH

(c) O, C, O^-Na^+, OH

(d) O, C, OH, O—C—CH_3, O

fig. 2.5.22 **Substances involved in the preparation of aspirin.**

Using combinatorial chemistry

Aspirin is an amazingly useful drug – but maybe there's an even more effective version, without so many side-effects on the gut wall? This is where combinatorial chemistry has such great potential. In this case salicylic acid is the 'lead' compound, which has a known and measured therapeutic effect. A pharmaceutical manufacturing company will have a directory of similar lead compounds. If they were seeking to produce a safer compound related to salicylic acid, or one with a stronger biological effect on pain, they might carry out a series of parallel reactions. They would react 2-hydroxybenzenecarboxylic acid (the precursor of aspirin) and many other related lead compounds with a series of different reactants – in this case perhaps acid anhydrides (see fig. 2.5.23).

All the chemicals are delivered with computer-controlled syringes and each of the reaction tubes contains a different product that can be identified. Preliminary tests on the effectiveness, or otherwise, of the new compounds can be carried out with enzymes, cell cultures or other techniques. Only products that are successful in the preliminary tests need be followed up and the rest can be discarded or saved in a chemical library.

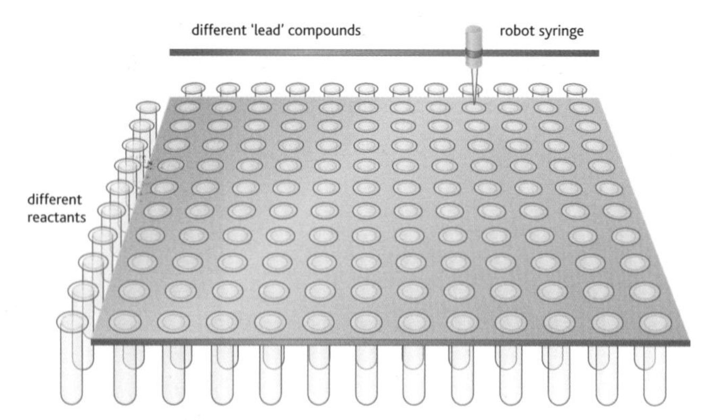

fig. 2.5.23 **Combinatorial robotic systems can permutate hundreds of chemicals in minutely different proportions under varying reaction conditions to form thousands of new compounds.**

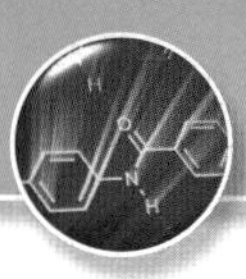

Solid phase reactions

The process can be further refined by getting small quantities of compounds to bond onto polymer beads and for the reaction to take place there. The beads are often polystyrene-based with the reactant molecules bonded to the beads by covalent bonds (see **fig. 2.5.24**).

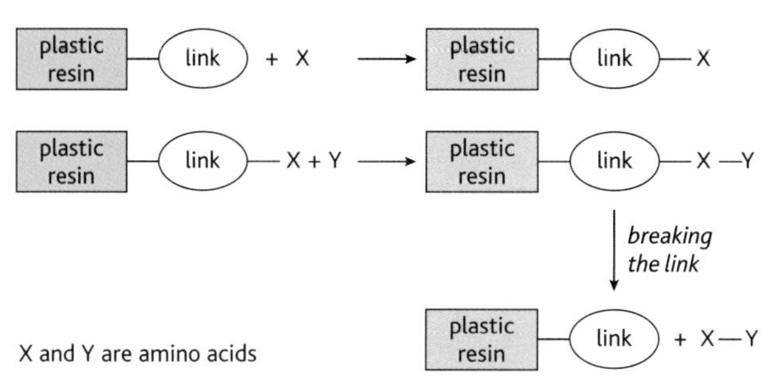

fig. 2.5.24 Synthesis of a dipeptide.

The product remains bonded to the beads. The excess reactants can simply be washed away, leaving only the desired product molecules bound to the solid phase. The product is released by breaking the bond with the beads.

One advantage of this method is that one end of the reacting molecule is fixed and all reactions happen at the other end of the molecule. This method was developed for the synthesis of peptides by an American chemist, Bruce Merrifield. Eventually he was able to synthesise simple polypeptides.

Merrifield also felt that the technique lent itself nicely to a mechanised and automated process. He devised the first prototype of an automated peptide synthesiser by 1965. Working with Bernd Gutte, Merrifield synthesised ribonuclease, a small stable protein with a known amino acid sequence and three-dimensional structure. Their synthetic ribonuclease was indistinguishable from the natural enzyme. Merrifield's work was recognised when he received the Nobel Prize for Chemistry in 1984 'for his development of methodology for chemical synthesis on a solid matrix'.

To demonstrate the power and potential of his method, Merrifield undertook the automatic synthesis of insulin in 1965. With 51 amino acid units and two peptide chains held together by two disulfide bridges, the molecule presented a formidable challenge. Although more than 5000 operations were involved in assembling the chains, most of these were carried out automatically over a period of a few days. Other researchers have been able to synthesise carbohydrates by similar methods. This technique has much potential for the future development of both biological molecules and new medicines. The future of chemistry is exciting!

Questions

1. Very tiny quantities of material are involved in the synthesis reactions described in this section – sometimes as little as 1×10^{-15} moles. Suggest a technique you have met in your chemistry studies that could be used to determine structure when such a small quantity of material is available.
2. Explain the advantages and disadvantages of combinatorial chemistry in the development of new medicines.
3. Most of the chemicals developed using combinatorial chemistry so far have been composed of molecules without chiral centres. How might this be linked to the fact that relatively few of them have shown therapeutic possibilities?

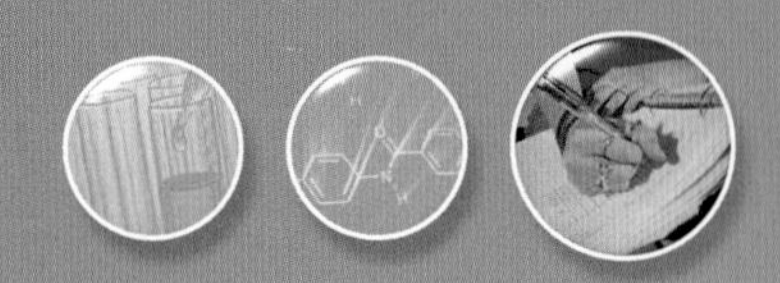

Examzone: Unit 4 Test 1 (chapters 1.1 to 1.5)

1 The kinetics of reactions can be followed in various ways. The techniques usually involve either sampling the reaction mixture at various times together with a means of slowing or stopping the reaction in the sample, or making use of some physical method for following the reaction continuously.

The iodination of propanone occurs in aqueous solution, and is catalysed by acid:

$$CH_3COCH_3(aq) + I_2(aq) \xrightarrow{H+} CH_3COCH_2I(aq) + HI(aq)$$

The rate is followed by finding the iodine concentration at various times.

a **i** What continuous method would you use to determine the iodine concentration? (2)

ii If you withdrew samples at various times and stopped the reaction with a suitable reagent, how might you then determine the iodine concentration? (2)

iii Suggest a reason why the continuous method might be preferable. (1)

b The variation of iodine concentration with time in such an experiment is shown below.

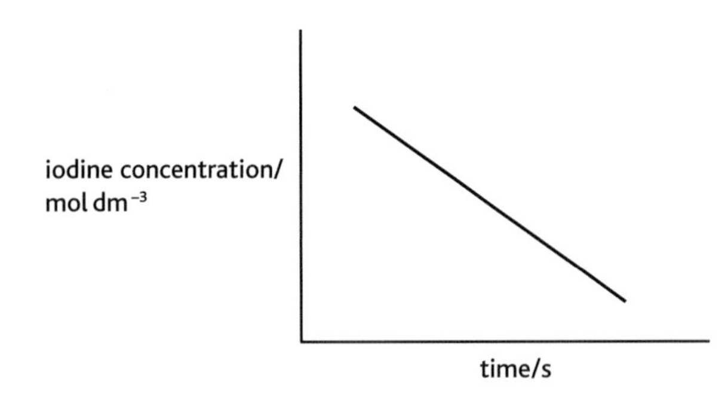

What does this tell you about:

i the rate of reaction over a period of time (1)

ii the order of reaction with respect to iodine (1)

iii the rate-determining step of the reaction (1)

iv the least number of steps in the reaction mechanism? (1)

c A suitable quenching solution for the method described in part **a(ii)** is sodium hydrogencarbonate (sodium bicarbonate) solution. It neutralises the acid catalyst without making the solution so alkaline that other reactions occur.

i If sodium hydroxide solution, a strong alkali, were used, propanone and iodine would give another product. Name it, and write an equation for the reaction occurring. (3)

ii This reaction is used to detect the presence of either of two possible structures in organic molecules. Give these structures. (2)

iii Describe briefly a test you could use to distinguish between these two structures. (1)

(Total 15 marks)

2 In an experiment designed to find the mechanism of the reaction between a halogenoalkane, RX, and hydroxide ions, the following data were obtained at constant temperature.

Initial concentration of RX/mol dm^{-3}	Initial concentration of OH^-/mol dm^{-3}	Initial rate/ mol dm^{-3} s^{-1}
0.01	0.04	8×10^{-6}
0.01	0.02	4×10^{-6}
0.005	0.04	4×10^{-6}

a Deduce the order with respect to RX and OH^-, and hence write the rate equation for the reaction. (3)

b Calculate a value for the rate constant for the reaction. (2)

(Total 5 marks)

3 **a** Propanoic acid is a *weak* acid. Explain the term 'weak'. (1)

b **i** Give an equation for the dissociation of propanoic acid, and hence an expression for its dissociation constant, K_a. (2)

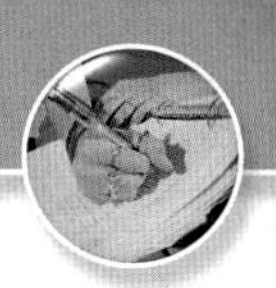

ii At 25°C, K_a for propanoic acid is $1.30 \times 10^{-5}\,mol\,dm^{-3}$. Find the pH of a solution of propanoic acid of concentration $0.0100\,mol\,dm^{-3}$. State any assumptions you make. **(3)**

iii Increasing the temperature of the propanoic acid solution causes the pH to decrease. What does this tell you about the enthalpy of dissociation? Justify your answer. **(3)**

(Total 9 marks)

4 a i Define pH. **(1)**

ii Define K_w, the ionic product of water. **(1)**

b Calculate the pH of the following solutions. (The ionic product of water, K_w, should be taken to be $1.00 \times 10^{-14}\,mol^2\,dm^{-6}$)

i A solution of sulfuric acid having a concentration of $0.100\,mol\,dm^{-3}$. **(1)**

ii A solution of sodium hydroxide having a concentration of $0.0500\,mol\,dm^{-3}$. **(2)**

c i What is the principal property of a buffer solution? **(1)**

ii The dissociation constant for ethanoic acid is $1.80 \times 10^{-5}\,mol\,dm^{-3}$. Calculate the pH of a buffer solution which has a concentration of $0.0150\,mol\,dm^{-3}$ with respect to ethanoic acid and $0.0550\,mol\,dm^{-3}$ with respect to sodium ethanoate. **(3)**

(Total 9 marks)

5 Ammonia is manufactured from hydrogen and nitrogen in the Haber Process:

$$N_2(g) + 3H_2(g) \rightleftharpoons 2NH_3(g) \qquad \Delta H = -92\,kJ\,mol^{-1}$$

a What is meant by the term 'dynamic equilibrium'? **(2)**

b State the conditions employed industrially in the manufacture of ammonia, and justify them on physico-chemical grounds. **(5)**

c What effect does a catalyst have on the rate of achievement of the equilibrium and the composition of the equilibrium mixture? **(2)**

(Total 9 marks)

6 Hydrogen and iodine react together to give an equilibrium:

$$H_2(g) + I_2(g) \rightleftharpoons 2HI(g)$$

a Write an expression for K_p for this equilibrium, giving consideration to its units. **(2)**

b When 0.50 mol of I_2 and 0.50 mol of H_2 were mixed in a closed container at 723 K and 2 atm pressure, 0.11 mol of I_2 were found to be present when equilibrium was established.

i Calculate the partial pressures of I_2, H_2 and HI in the equilibrium mixture. **(3)**

ii Hence calculate the value of K_p at 723 K. **(2)**

c In an experiment to establish the equilibrium concentrations in part **b**, the reaction was allowed to reach equilibrium at 723 K and was then quenched by addition to a known, large volume of water. The concentration of iodine in this solution was then determined by titration with standard sodium thiosulfate solution.

i Write an equation for the reaction between sodium thiosulfate and iodine. **(2)**

ii What indicator would you use? Give the colour change at the end point. **(2)**

iii In this titration and in titrations involving potassium manganate(VII), a colour change occurs during the reaction. Why is an indicator usually added in iodine/thiosulfate titrations but not in titrations involving potassium manganate(VII)? **(2)**

(Total 13 marks)

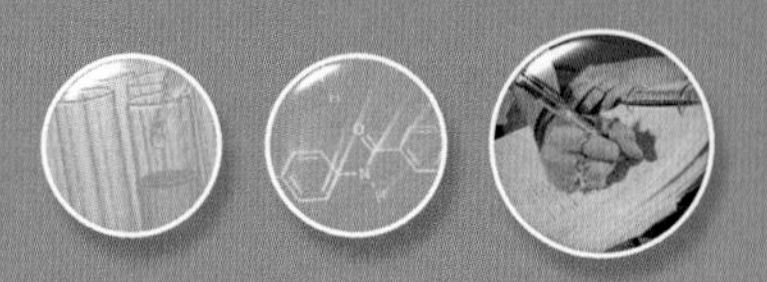

Examzone: Unit 4 Test 2 (chapters 1.6 and 1.7)

1 a What is meant by a 'chiral' compound? (1)

Consider the following series of reactions and answer the questions which follow.

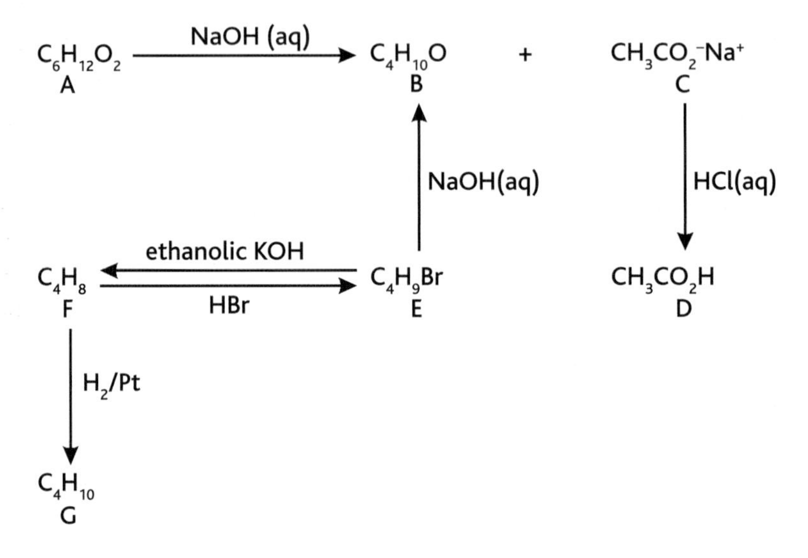

b Compound **E** displays optical isomerism.

i State what this means. (1)

ii Sketch the optical isomers of **E**. (2)

c The reaction of **E** to give **B** is a nucleophilic substitution.

i What is meant by the term 'nucleophile'? (1)

ii Give the structural formula for **B**. (1)

iii Give a simple chemical test for the functional group in **B**, and say what you would see. (2)

d The type of reaction exemplified by **A** → **B** + **C** is important in the manufacture of soap.

i What type of reaction is this? (1)

ii **A** could be reacted with aqueous acid to give **B** and **D**. If the same quantity of **A** was treated with aqueous acid, instead of aqueous alkali, how would the yield of **B** differ? Explain your answer. (1)

e Consider the reaction **C** → **D**.

i Name **D**. (1)

ii Identify the acid–base conjugate pair in this reaction. (1)

iii Explain why the reaction occurs. (2)

(Total 14 marks)

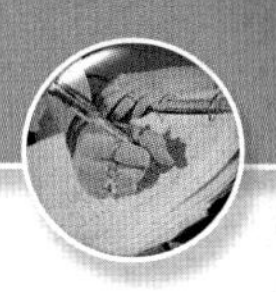

2 The steps in the process of recrystallisation are detailed below:

A Add a **little** solvent to the sample.
B Dry the product.
C Allow the solution to cool.
D Filter cold.
E Wash the residue with a **small** amount of cold solvent.
F Warm until the sample dissolves.
G Filter while the solution is still hot.

a Put these steps for recrystallisation in the correct order. **(2)**

b Give the reasons for the words in bold type in step **A** and in step **E**. **(2)**

(Total 4 marks)

3 An alcohol, C_3H_8O, was analysed by low-resolution NMR spectroscopy. The spectrum is given below.

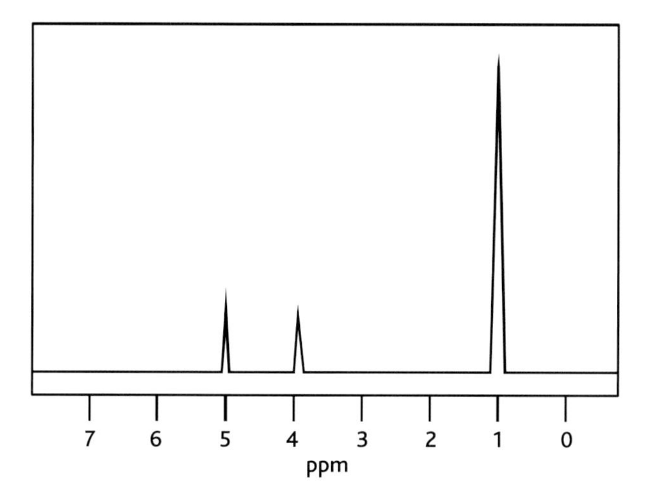

The chemical shifts of some hydrogen nuclei are given in the following table (where R represents an alkyl group).

Group	δ/ppm
CH_3–R	0.8–1.2
R–CH_2–R	1.1–1.5
CH–R_3	1.5

Group	δ/ppm
R–OH	1.0–6.0
R–CH_2–OH	3.3–4.0
R_2–CH–OH	3.2–4.1

a Show that the alcohol must be propan-2-ol. **(2)**

b What would the observation be if propan-2-ol is reacted with aqueous iodine in the presence of alkali? **(2)**

c Draw the structure of the ester formed from the reaction of propan-2-ol with propanoic acid in the presence of concentrated sulfuric acid. **(1)**

d State the name of the organic product if propan-2-ol is heated with concentrated sulfuric acid and classify the type of reaction occurring. **(2)**

(Total 7 marks)

4 Propanone can be prepared in the laboratory by oxidising propan-2-ol. A student was given the following instructions and data.

- Dissolve 10 g of potassium dichromate(VI) in 30 cm^3 of water.
- Add 12 cm^3 of propan-2-ol followed by 6 cm^3 of concentrated sulfuric acid.
- Boil the mixture for 15 minutes.
- The boiling temperatures of the volatile substances are:

Substance	Boiling temperature/°C
propan-2-ol	82
water	100
propanone	56

a What safety precaution, in addition to wearing eye protection, must be taken when adding the concentrated sulfuric acid? **(1)**

b **i** Draw a diagram of the apparatus that you would use to boil the mixture for 15 minutes. **(2)**

ii Outline how you would obtain propanone from this aqueous mixture after it had been boiled for 15 minutes. **(2)**

c The sample of propanone was dried. Describe a test that you would do to show that your final sample of propanone did not contain any propan-2-ol. **(2)**

(Total 7 marks)

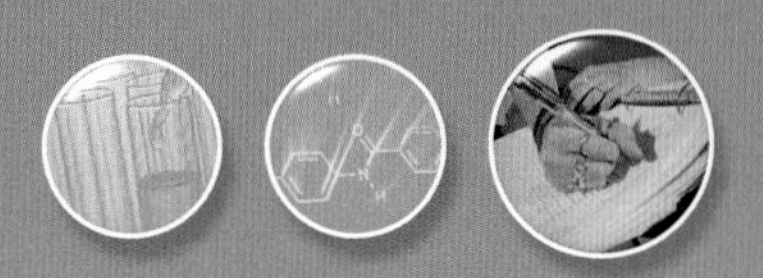

Examzone: Unit 5 Test 1 (chapters 2.1 and 2.2)

1 The following data concern the redox chemistry of halogen elements and sodium thiosulfate.

	$E^\ominus$/V
$Cl_2(g) + 2e^- \rightleftharpoons 2Cl^-(aq)$	+1.36
$Br_2(l) + 2e^- \rightleftharpoons 2Br^-(aq)$	+1.07
$2SO_4^{2-}(aq) + 10H^+(aq) + 8e^- \rightleftharpoons S_2O_3^{2-}(aq) + 5H_2O(l)$	+0.57
$I_2(s) + 2e^- \rightleftharpoons 2I^-(aq)$	+0.54
$S_4O_6^{2-}(aq) + 2e^- \rightleftharpoons 2S_2O_3^{2-}(aq)$	+0.09

a What is the maximum change in oxidation number of sulfur which can be brought about by the action of sodium thiosulfate with:

i iodine **(2)**

ii chlorine? **(3)**

Write full ionic equations for each reaction.

b Suggest a reason why bromine cannot be estimated by direct titration with sodium thiosulfate in the same way that iodine can. **(2)**

(Total 7 marks)

2 a What do you understand by the term 'standard electrode potential'? **(2)**

b If a metal is immersed in a solution of its ions, the potential set up between the metal and the solution cannot be measured without using a reference electrode. Explain why this is so. **(2)**

c i What is meant by 'disproportionation'? **(1)**

ii Use the following data to deduce whether MnO_2 will disproportionate in acidic solution.

$MnO_4^-(aq) + 4H^+(aq) + 3e^- \rightleftharpoons MnO_2 + 2H_2O(l)$ $E^\ominus = +1.70\,V$

$MnO_2(s) + 4H^+(aq) + 2e^- \rightleftharpoons Mn^{2+}(aq) + 2H_2O(l)$ $E^\ominus = +1.23\,V$ **(2)**

(Total 7 marks)

3 Some modern British coins are made from an alloy, nickel–brass, which consists essentially of the metals copper, nickel and zinc. A one pound coin weighing 9.50 g was completely dissolved in concentrated nitric acid, in which all three metals dissolve, to give solution **A**.

Dilute sodium hydroxide solution was then added carefully with stirring, until present in excess. Zinc hydroxide is amphoteric. The precipitate formed, **B**, was filtered off from the supernatant liquid, **C**. The precipitate, **B**, was quantitatively transferred to a graduated flask of 500 cm^3 capacity. Dilute sulfuric acid was then added to dissolve the whole of the precipitate, **B**, and the solution made up to 500 cm^3 with distilled water.

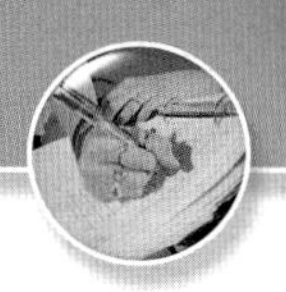

25.0 cm^3 portions of this solution were pipetted into a titration flask and an excess of potassium iodide solution added. The liberated iodine was then titrated against sodium thiosulfate solution of concentration 0.100 mol dm^{-3}. 18.7 cm^3 of the sodium thiosulfate solution was required for complete reaction.

	$E^\ominus$/V
$Ni^{2+}(aq) + 2e^- \rightleftharpoons Ni(s)$	−2.71
$Zn^{2+}(aq) + 2e^- \rightleftharpoons Zn(s)$	−0.76
$\frac{1}{2}S_4O_6^{2-}(aq) + e^- \rightleftharpoons S_2O_3^{2-}$	+0.09
$Cu^{2+}(aq) + e^- \rightleftharpoons Cu^+(aq)$	+0.15
$Cu^{2+}(aq) + 2e^- \rightleftharpoons Cu(s)$	+0.34
$\frac{1}{2}I_2(aq) + e^- \rightleftharpoons I^-(aq)$	+0.54
$NO_3^-(aq) + 2H^+(aq) + e^- \rightleftharpoons NO_2(g) + H_2O(l)$	+0.81

a **i** Using the appropriate half-equations, write an equation for the reaction of any one of the metals in nickel–brass with concentrated nitric acid. (2)

ii What type of reaction is taking place? (1)

b Identify by giving full formulae:

i the complex cations present in **A** (2)

ii the precipitates in **B** (2)

iii any metal-containing anion in **C**. (1)

c **i** Write an equation for the precipitation of any one of the metal ions in **A** with sodium hydroxide. (2)

ii What type of reaction is occurring in **c(i)**? (1)

d Suggest an explanation why it is necessary to add sodium hydroxide, followed by dilute sulfuric acid, before performing the titration. (3)

e On addition of the potassium iodide solution, the only reaction which occurs is:

$2Cu^{2+}(aq) + 4I^-(aq) \rightarrow 2CuI(s) + I_2(aq)$

i Write an equation for the reaction between sodium thiosulfate and the liberated iodine. What indicator would you use in this titration? At what stage would you add it? Give a reason for your answer. (3)

ii Calculate the percentage of copper in the alloy. (5)

iii Suggest why this reaction occurs in the light of the $E^\ominus$ values given. (3)

(Total 25 marks)

4 State three properties which distinguish transition metals from main group metals. (3)

(Total 3 marks)

5 Look at this reaction scheme:

$[Cu(H_2O)_6]^{2+}(aq) \rightarrow$ blue precipitate $\rightarrow$ deep blue solution

A **B** **C**

a Name the types of reaction involved in the conversion of:

i **A** to **B** (1)

ii **B** to **C** (1)

b Give the formula of compound **B**. (1)

c Draw the structure of the ion responsible for the colour in solution **C** and show its shape. (2)

(Total 5 marks)

6 **a** Suggest why the hydrated ion $[V(H_2O)_6]^{2+}$ is coloured. (1)

b Name the types of bonding within ions of this type. (2)

(Total 3 marks)

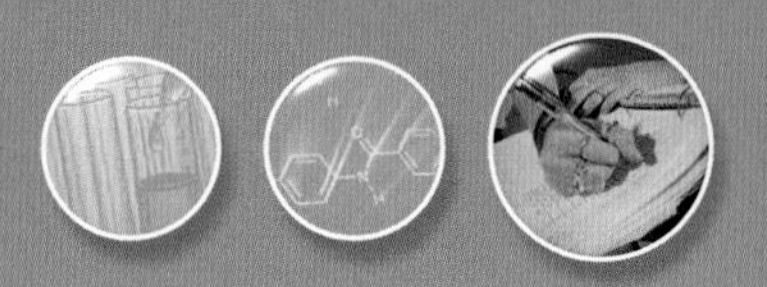

Examzone: Unit 5 Test 2 (chapters 2.3 to 2.5)

1 a The enthalpy change for the reaction

$CH_2{=}CH_2 + H_2 \rightarrow CH_3CH_3$

is –120 kJ mol^{-1}, whereas that for the reduction of benzene, C_6H_6, to cyclohexane, C_6H_{12}, is –208 kJ mol^{-1}. What may be deduced from the fact that this value is not three times the first one? **(2)**

b i State the conditions under which benzene may be nitrated to form mononitrobenzene. **(2)**

ii Both of the reagents that are used to nitrate benzene are usually regarded as acids. However, in this instance, one of them behaves as a base. Show how this is so. **(2)**

iii Give the mechanism for the nitration of benzene. **(2)**

iv Explain why benzene tends to undergo substitution rather than addition reactions. **(1)**

(Total 9 marks)

2 The following is adapted from a textbook of practical chemistry which gives details for the preparation of the azo dye phenylazo-2-naphthol.

Dissolve 2.5 g of phenylamine in a mixture of 8 cm^3 of concentrated hydrochloric acid and 8 cm^3 of water in a small beaker. Place in an ice bath; ignore any crystals that may appear. When the temperature is between 0 °C and 5 °C, add drop by drop a solution of 2 g of sodium nitrite dissolved in 10 cm^3 of water, not allowing the temperature to rise above 5 °C. Addition of the sodium nitrite solution should continue until, after a wait of 3-4 minutes, a drop of the reaction mixture gives an immediate blue coloration with starch–iodide paper.

Prepare a solution of 3.9 g of 2-naphthol in 10% aqueous sodium hydroxide solution in a 250 cm^3 beaker, and cool in ice bath to below 5 °C; add 10–15 g of crushed ice to this solution. Stir the mixture, and add the diazonium salt prepared as above very slowly; red crystals of the azo compound will separate. When addition is complete, allow the mixture to stand in ice for 10 minutes and then filter the product, using gentle suction on a Buchner funnel. Wash with water. The product may be recrystallised from glacial ethanoic acid. It has a melting temperature of 131 °C.

a Phenylamine is toxic by inhalation and skin absorption; concentrated hydrochloric acid is corrosive and gives harmful fumes. What specific precautions would you therefore take when doing this experiment? **(2)**

b Write the equation for the reaction between phenylamine and hydrochloric acid. **(1)**

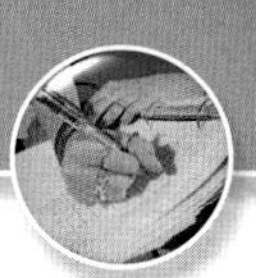

c **i** Why should the reaction mixture be kept between 0 °C and 5 °C? **(2)**

ii Is the diazotisation reaction exothermic or endothermic? How do you know? **(2)**

d Give the equation, using structural formulae, for the diazo coupling reaction between benzene diazonium chloride and 2-naphthol. **(2)**

(Total 9 marks)

3 Outline how the following conversions could be carried out in the laboratory, giving reagents, conditions and equations.

a **i**

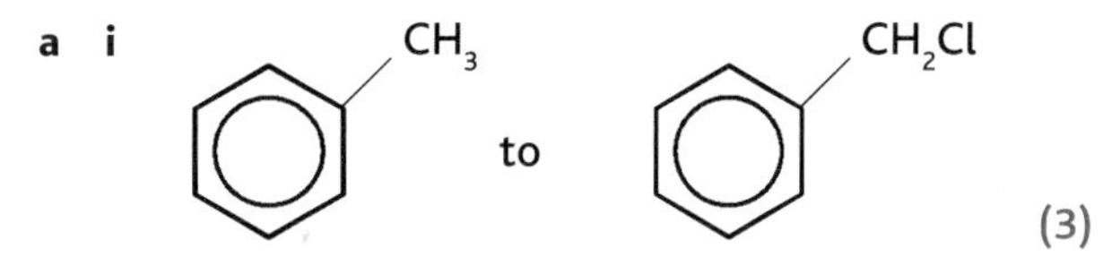

(3)

ii Given that the boiling temperatures of these two compounds are 111°C and 179°C, respectively, how would you isolate a pure sample of the chlorohydrocarbon? **(2)**

b $CH_3CH_2CH_2NH_2$ to $CH_3CH_2CH_2NH_3{}^+Cl^-$ **(3)**

(Total 8 marks)

4 Consider the compound **A**.

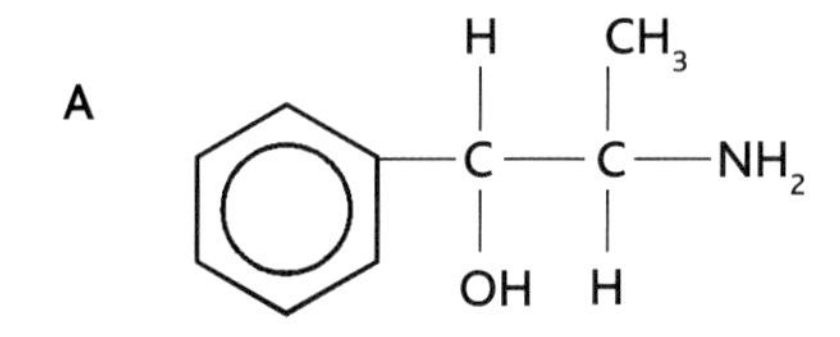

It is related to the hormone adrenaline.

a Draw the structures of the organic product(s) which you expect from the reaction of **A** with:

i phosphorus pentachloride **(1)**

ii dilute hydrochloric acid **(1)**

iii ethanoyl chloride **(2)**

iv hot alkaline potassium manganate(VII) **(1)**

v hot concentrated sulfuric acid **(1)**

b Suppose that you have to purify a sample of **A** by recrystallisation from trichloromethane. This solvent is toxic by inhalation and skin absorption but is not flammable.

i What safety precautions would you take using this solvent? **(2)**

ii Describe in detail how you would recrystallise a sample of about 5 g of **A**. **(5)**

iii What simple test would you use to determine the purity of your recrystallised material? **(2)**

(Total 15 marks)

5 This question relates to the following reaction scheme:

$$C_2H_5Br \xrightarrow{\text{step 1}} C_2H_5CN \xrightarrow{\text{step 2}} C_2H_5CH_2NH_2 \xrightarrow{\text{step 3}} C_2H_5CH_2NHCOCH_3$$

a Give the reagents, the conditions required and the equation for:

i Step 2 **(3)**

ii Step 3 **(3)**

b What drug is based on the compound formed in step 3? **(1)**

(Total 7 marks)

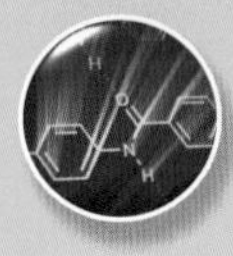
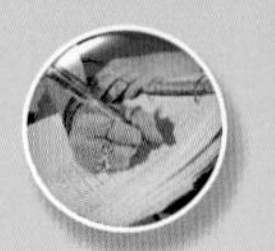

Index

The Periodic Table of Elements

Key

relative atomic mass **atomic symbol** name atomic (proton) number

1 (1)	2 (2)	(3)	(4)	(5)	(6)	(7)	(8)	(9)	(10)	(11)	(12)	3 (13)	4 (14)	5 (15)	6 (16)	7 (17)	0 (8) (18)
							1.0 **H** hydrogen 1										4.0 **He** helium 2
6.9 **Li** lithium 3	9.0 **Be** beryllium 4											10.8 **B** boron 5	12.0 **C** carbon 6	14.0 **N** nitrogen 7	16.0 **O** oxygen 8	19.0 **F** fluorine 9	20.2 **Ne** neon 10
23.0 **Na** sodium 11	24.3 **Mg** magnesium 12											27.0 **Al** aluminium 13	28.1 **Si** silicon 14	31.0 **P** phosphorus 15	32.1 **S** sulfur 16	35.5 **Cl** chlorine 17	39.9 **Ar** argon 18
39.1 **K** potassium 19	40.1 **Ca** calcium 20	45.0 **Sc** scandium 21	47.9 **Ti** titanium 22	50.9 **V** vanadium 23	52.0 **Cr** chromium 24	54.9 **Mn** manganese 25	55.8 **Fe** iron 26	58.9 **Co** cobalt 27	58.7 **Ni** nickel 28	63.5 **Cu** copper 29	65.4 **Zn** zinc 30	69.7 **Ga** gallium 31	72.6 **Ge** germanium 32	74.9 **As** arsenic 33	79.0 **Se** selenium 34	79.9 **Br** bromine 35	83.8 **Kr** krypton 36
85.5 **Rb** rubidium 37	87.6 **Sr** strontium 38	88.9 **Y** yttrium 39	91.2 **Zr** zirconium 40	92.9 **Nb** niobium 41	95.9 **Mo** molybdenum 42	[98] **Tc** technetium 43	101.1 **Ru** ruthenium 44	102.9 **Rh** rhodium 45	106.4 **Pd** palladium 46	107.9 **Ag** silver 47	112.4 **Cd** cadmium 48	114.8 **In** indium 49	118.7 **Sn** tin 50	121.8 **Sb** antimony 51	127.6 **Te** tellurium 52	126.9 **I** iodine 53	131.3 **Xe** xenon 54
132.9 **Cs** caesium 55	137.3 **Ba** barium 56	138.9 **La*** lanthanum 57	178.5 **Hf** hafnium 72	180.9 **Ta** tantalum 73	183.8 **W** tungsten 74	186.2 **Re** rhenium 75	190.2 **Os** osmium 76	192.2 **Ir** iridium 77	195.1 **Pt** platinum 78	197.0 **Au** gold 79	200.6 **Hg** mercury 80	204.4 **Tl** thallium 81	207.2 **Pb** lead 82	209.0 **Bi** bismuth 83	[209] **Po** polonium 84	[210] **At** astatine 85	[222] **Rn** radon 86
[223] **Fr** francium 87	[226] **Ra** radium 88	[227] **Ac*** actinium 89	[261] **Rf** rutherfordium 104	[262] **Db** dubnium 105	[266] **Sg** seaborgium 106	[264] **Bh** bohrium 107	[277] **Hs** hassium 108	[268] **Mt** meitnerium 109	[271] **Ds** darmstadtium 110	[272] **Rg** roentgenium 111							

Elements with atomic numbers 112-116 have been reported but not fully authenticated

* Lanthanide series	140 **Ce** cerium 58	141 **Pr** praseodymium 59	144 **Nd** neodymium 60	[147] **Pm** promethium 61	150 **Sm** samarium 62	152 **Eu** europium 63	157 **Gd** gadolinium 64	159 **Tb** terbium 65	163 **Dy** dysprosium 66	165 **Ho** holmium 67	167 **Er** erbium 68	169 **Tm** thulium 69	173 **Yb** ytterbium 70	175 **Lu** lutetium 71
* Actinide series	232 **Th** thorium 90	[231] **Pa** protactinium 91	238 **U** uranium 92	[237] **Np** neptunium 93	[242] **Pu** plutonium 94	[243] **Am** americium 95	[247] **Cm** curium 96	[245] **Bk** berkelium 97	[251] **Cf** californium 98	[254] **Es** einsteinium 99	[253] **Fm** fermium 100	[256] **Md** mendelevium 101	[254] **No** nobelium 102	[257] **Lr** lawrencium 103

Published by Pearson Education Limited, a company incorporated in England and Wales, having its registered office at Edinburgh Gate, Harlow, Essex, CM20 2JE. Registered company number: 872828

Edexcel is a registered trade mark of Edexcel Limited

Text © Pearson Education Limited 2009

First published 2009

12 11 10

10 9 8 7 6 5 4

British Library Cataloguing in Publication Data

A catalogue record for this book is available from the British Library.

ISBN 978 1 408206 05 8

Edited by Tony Clappison

Designed and typeset by 320 Design Limited

Original illustrations © Pearson Education Limited 2009

Illustrated by Oxford Designers and Illustrators

Picture research by Louise Edgeworth

Printed and bound by Scotprint, UK

Acknowledgements

The authors and publisher would like to thank John Apsey for his help in reviewing this book.

The authors and publisher would like to thank the following individuals and organisations for permission to reproduce copyright materials including photographs:

(Key: b–bottom; c–centre; l–left; r–right; t–top)

Alamy Images: Dave Bevan 78t; Bon Appetit 189; Peter Bowater 20; Bubbles Photolibrary 225; Diana Bier Museum 99; Jon Faulknor 208; George Impey 9bl, 78b; Georgios Kollidas 126tr; Letterbox Digital 188; Ian Miles-Flashpoint Pictures 220; David J. Green - nature 9c, 59t; Lourens Smak 231; The Print Collector 219r; Richard Lawrence Wade 193; Janine Wiedel Photolibrary 206; **Art Directors and TRIP photo Library:** Helene Rogers 192, 222t; **Trevor Clifford:** 12, 100, 107, 118, 120l, 181t, 194br, 222b, 227, 240cl, 240tr; **Corbis:** Tim Farrell/Star Ledger 221; Brownie Harris 215; Image Source 124; Herbert Kehrer/zefa 120r; Randy Lincks 62; **Mary Evans Picture Library:** 72b; **Jonny Fleetwood:** 9tl, 11b; **Getty Images:** Ben Edwards/The Image Bank 126b; SMC Images/The Image Bank 181b; **iStockphoto:** 203; Wilson Valentin 180l; **Jupiter Unlimited:** Stockxpert 180r; **No Trace:** 39; **Ordnance Survey:** Reproduced by permission of Ordnance Survey 2009. All rights reserved. 155; **PA Photos:** AP Photo/John Duricka 235; **Photolibrary.com:** Digital Vision/Rob Melnychuk 95; Focus Database 207l; Index Stock Imagery 59b; **POD – Pearson Online Database:** Peter Gould 34b, 65, 90, 175 (1), 175 (2), 175 (3), 175 (4); **Reuters:** Celgene Corp 248; **Rex Features:** 10; Design Pics Inc 54; Mujo Korach/IBL 34t; **Science Photo Library Ltd:** 76, 194bl, 194t, 232; Alex Bartel 145tl, 165; George Bernard 145c, 200; Martin Bond 123; Leslie J Borg 146; Dr Jeremy Burgess 117; J-L Charmet 153; Martyn F Chillmaid 168, 179; Carlos Dominguez 151; Michael Donne 166; A. Dowsett, Health Protection Agency 91; E.R.Degginger 152; Dr Tim Evans 214, 224t; Kenneth Eward/Biografx 224b; Eye of Science 190; Mauro Fermariello 145b, 229; Simon Fraser 143, 245; Adrienne Hart-Davis 218; Laguna Design 207r (1), 207r (2), 207r (3), 212; Andrew Lambert Photography 11t, 41b, 72t, 108, 110l, 111, 127, 145tr, 150, 159, 172, 186t, 199, 202, 223, 240bl, 240br, 240tl; Bill Longcore 161; Damien Lovegrove 92; Jerry Mason 147, 163, 186b; Andrew McClenaghan 234; Peter Menzel 253; Hank Morgan 126tl; PASIEKA 103; Phantatomix 191; Alexis Rosenfeld 97; Saturn Stills 104r; Science, Industry & Business Library/New York Public Library 47; SOVEREIGN, ISM 9br, 135; Volker Steger 104l; TH FOTO-WERBUNG 149; Geoff Tompkinson 250; Vanessa Vick 67; Charles D Winters 9tr, 41t, 74, 77b, 77t, 129, 219l, 240cr; **shutterstock:** Olivier Le Queinec 6 (icon); Alexis Puentes 18 (icon); **Xiuyan Yang:** 140

Cover images: *Front:* **Science Photo Library Ltd:** Colin Cuthbert

All other images © Pearson Education

Every effort has been made to contact copyright holders of material reproduced in this book. Any omissions will be rectified in subsequent printings if notice is given to the publishers.

Disclaimer

This Edexcel publication offers high-quality support for the delivery of Edexcel qualifications.

Edexcel endorsement does not mean that this material is essential to achieve any Edexcel qualification, nor does it mean that this is the only suitable material available to support any Edexcel qualification. No endorsed material will be used verbatim in setting any Edexcel examination/assessment and any resource lists produced by Edexcel shall include this and other appropriate texts.

Copies of official specifications for all Edexcel qualifications may be found on the Edexcel website – www.edexcel.org.uk.

Single User Licence Agreement:
Edexcel A2 Chemistry Students' Book with FREE ActiveBook CD-ROM

Warning:
This is a legally binding agreement between You (the user or purchasing institution) and Pearson Education Limited of Edinburgh Gate, Harlow, Essex, CM20 2JE, United Kingdom ('PEL').

By retaining this Licence, any software media or accompanying written materials or carrying out any of the permitted activities You are agreeing to be bound by the terms and conditions of this Licence. If You do not agree to the terms and conditions of this Licence, do not continue to use the Edexcel A2 Chemistry Students' Book with FREE ActiveBook CD-ROM and promptly return the entire publication (this Licence and all software, written materials, packaging and any other component received with it) with Your sales receipt to Your supplier for a full refund.

Intellectual Property Rights:
This Edexcel A2 Chemistry Students' Book with FREE ActiveBook CD-ROM consists of copyright software and data. All intellectual property rights, including the copyright is owned by PEL or its licensors and shall remain vested in them at all times. You only own the disk on which the software is supplied. If You do not continue to do only what You are allowed to do as contained in this Licence you will be in breach of the Licence and PEL shall have the right to terminate this Licence by written notice and take action to recover from you any damages suffered by PEL as a result of your breach.

The PEL name, PEL logo, Edexcel name, Edexcel logo and all other trademarks appearing on the software and Edexcel A2 Chemistry Students' Book with FREE ActiveBook CD-ROM are trademarks of PEL. You shall not utilise any such trademarks for any purpose whatsoever other than as they appear on the software and Edexcel A2 Chemistry Students' Book with FREE ActiveBook CD-ROM.

Yes, You can:
1. use this Edexcel A2 Chemistry Students' Book with FREE ActiveBook CD-ROM on Your own personal computer as a single individual user. You may make a copy of the Edexcel A2 Chemistry Students' Book with FREE ActiveBook CD-ROM in machine readable form for backup purposes only. The backup copy must include all copyright information contained in the original.

No, You cannot:
1. copy this Edexcel A2 Chemistry Students' Book with FREE ActiveBook CD-ROM (other than making one copy for back-up purposes as set out in the Yes, You can table above);

2. alter, disassemble, or modify this Edexcel A2 Chemistry Students' Book with FREE ActiveBook CD- ROM, or in any way reverse engineer, decompile or create a derivative product from the contents of the database or any software included in it:

3. include any materials or software data from the Edexcel A2 Chemistry Students' Book with FREE ActiveBook CD-ROM in any other product or software materials;

4. rent, hire, lend, sub-licence or sell the Edexcel A2 Chemistry Students' Book with FREE ActiveBook CD-ROM;

5. copy any part of the documentation except where specifically indicated otherwise;

6. use the software in any way not specified above without the prior written consent of PEL;

7. Subject the software, Edexcel A2 Chemistry Students' Book with FREE ActiveBook CD-ROM or any PEL content to any derogatory treatment or use them in such a way that would bring PEL into disrepute or cause PEL to incur liability to any third party.

Grant of Licence:
PEL grants You, provided You only do what is allowed under the 'Yes, You can' table above, and do nothing under the 'No, You cannot' table above, a non-exclusive, non-transferable Licence to use this Edexcel A2 Chemistry Students' Book with FREE ActiveBook CD-ROM.

The terms and conditions of this Licence become operative when using this Edexcel A2 Chemistry Students' Book with FREE ActiveBook CD-ROM.

Limited Warranty:
PEL warrants that the disk or CD-ROM on which the software is supplied is free from defects in material and workmanship in normal use for ninety (90) days from the date You receive it. This warranty is limited to You and is not transferable.

This limited warranty is void if any damage has resulted from accident, abuse, misapplication, service or modification by someone other than PEL. In no event shall PEL be liable for any damages whatsoever arising out of installation of the software, even if advised of the possibility of such damages. PEL will not be liable for any loss or damage of any nature suffered by any party as a result of reliance upon or reproduction of any errors in the content of the publication.

PEL does not warrant that the functions of the software meet Your requirements or that the media is compatible with any computer system on which it is used or that the operation of the software will be unlimited or error free. You assume responsibility for selecting the software to achieve Your intended results and for the installation of, the use of and the results obtained from the software.

PEL shall not be liable for any loss or damage of any kind (except for personal injury or death) arising from the use of this Edexcel A2 Chemistry Students' Book with FREE ActiveBook CD-ROM or from errors, deficiencies or faults therein, whether such loss or damage is caused by negligence or otherwise.

The entire liability of PEL and your only remedy shall be replacement free of charge of the components that do not meet this warranty.

No information or advice (oral, written or otherwise) given by PEL or PEL's agents shall create a warranty or in any way increase the scope of this warranty.

To the extent the law permits, PEL disclaims all other warranties, either express or implied, including by way of example and not limitation, warranties of merchantability and fitness for a particular purpose in respect of this Edexcel A2 Chemistry Students' Book with FREE ActiveBook CD-ROM.

Termination:
This Licence shall automatically terminate without notice from PEL if You fail to comply with any of its provisions or the purchasing institution becomes insolvent or subject to receivership, liquidation or similar external administration. PEL may also terminate this Licence by notice in writing. Upon termination for whatever reason You agree to destroy the Edexcel A2 Chemistry Students' Book with FREE ActiveBook CD-ROM and any back-up copies and delete any part of the Edexcel A2 Chemistry Students' Book with FREE ActiveBook CD-ROM stored on your computer.

Governing Law:
This Licence will be governed by and construed in accordance with English law.